Tight and taut submanifolds form an important class of manifolds with special curvature properties, one that has been studied intensively by differential geometers since the 1950's. They are in some ways the simplest figures after convex bodies: for example, tight manifolds in Euclidean space are characterized by the fact that their intersection with any half-space is connected. Examples include many well-known manifolds such as spheres, Veronese surfaces, isoparametric hypersurfaces and the cyclides of Dupin.

This book contains six in-depth articles by leading experts in the field and an extensive bibliography. It is dedicated to the memory of Nicolaas H. Kuiper, and the first paper is an unfinished but insightful exposition of the subject of tight immersions and maps, written by Kuiper. Other papers survey topics such as the smooth and polyhedral portions of the theory of tight immersions; taut, Dupin, and isoparametric submanifolds of Euclidean space; taut submanifolds of arbitrary complete Riemannian manifolds; and real hypersurfaces in complex space forms with constant principal curvatures. Taken together these articles provide a comprehensive survey of the field and point toward several directions for future research.

T0210883

Mathematical Sciences Research Institute
Publications

32

Tight and Taut Submanifolds

Mathematical Sciences Research Institute
Publications

Volumes 1 through 27 are available from Springer-Verlag

Tight and Taut Submanifolds

Papers in memory of Nicolaas H. Kuiper

Edited by

Thomas E. Cecil

College of the Holy Cross

Shiing-shen Chern

Emeritus, University of California, Berkeley

Thomas E. Cecil
College of the Holy Cross

Shiing-shen Chern
University of California

Mathematical Sciences Research
  Institute
1000 Centennial Drive
Berkeley, CA 94720

MSRI Editorial Committee
T. Y. Lam (chair)
Alexandre Chorin
Robert Osserman
Peter Sarnak

MSRI Book Series Editor
Silvio Levy

The Mathematical Sciences Research Institute wishes to acknowledge
support by the National Science Foundation.

CAMBRIDGE UNIVERSITY PRESS
Cambridge, New York, Melbourne, Madrid, Cape Town,
Singapore, São Paulo, Delhi, Tokyo, Mexico City

Cambridge University Press
The Edinburgh Building, Cambridge CB2 8RU, UK

Published in the United States of America by Cambridge University Press, New York

www.cambridge.org
Information on this title: www.cambridge.org/9780521175159

© Mathematical Sciences Research Institute 1997

This publication is in copyright. Subject to statutory exception
and to the provisions of relevant collective licensing agreements,
no reproduction of any part may take place without the written
permission of the copyright holder.

First published 1997
First paperback edition 2011

A catalogue record for this publication is available from the British Library

ISBN 978-0-521-62047-5 Hardback
ISBN 978-0-521-17515-9 Paperback

Cambridge University Press has no responsibility for the persistence or
accuracy of URLs for external or third-party internet websites referred to in
this publication, and does not guarantee that any content on such websites is,
or will remain, accurate or appropriate.

Tight and Taut Submanifolds
MSRI Publications
Volume **32**, 1997

Contents

Nicolaas Hendrik Kuiper, 1920–1994

Preface

This book grew out of a series of talks and closely related papers on tight and
taut submanifolds given at the Workshop on Differential Systems, Submanifolds,
and Control Theory held at MSRI on March 1–4, 1994. The workshop was
organized by Robert Bryant and Shiing-shen Chern.

The book is dedicated to the memory of Professor Nicolaas H. Kuiper, who
died on December 12, 1994. Kuiper made major contributions to the field of
tight and taut submanifolds over an extended period of time. In particular,
his technique of the analysis of topsets became an essential tool in almost all
work in the area of tight immersions and maps. The book begins with a short
description of Kuiper's life and work, written by Thomas Banchoff. Six of the
seven subsequent articles cover various parts of the broad field of tight and taut
submanifolds. The book concludes with an extensive bibliography of the field
compiled by Wolfgang Kühnel and Thomas Cecil, a list of Kuiper's publications,
and one of his doctoral students.

The first paper in the collection is an unfinished manuscript written by Kuiper
himself. The paper was intended to be a survey of the field, and it is based on
the Roever Lectures in Geometry that Kuiper gave at Washington University in
St. Louis during the period January 20–24, 1986. In its current state the article
is a masterly introduction to the subject and a good exposition of some more
advanced topics, concentrating on topological aspects, in particular the analysis
of topsets. It also contains a detailed proof of Kuiper's remarkable result that a
tight two-dimensional surface substantially immersed in $\mathbb{R}^5$ must be a Veronese
surface. We have made a few editorial notes in the text to aid the reader at
appropriate points.

The second paper in the book, by Thomas Banchoff and Wolfgang Kühnel, is
a comprehensive survey of the smooth and polyhedral portions of the theory of
tight immersions, including many open questions. The article is self-contained,
and there is some overlap with Kuiper's, since both begin with the basic definitions and examples. However, the two works are written from different points
of view, and the polyhedral case is given far more emphasis in the second. Although many aspects of the smooth and polyhedral theories are similar, there
are also points of significant divergence between the two theories. Banchoff and
Kühnel pay particular attention to these points of contrast.

An important special case of the difference between the smooth and polyhedral theories is the subject of the third paper, by Davide Cervone. In a paper published in 1992, François Haab resolved a problem posed by Kuiper in the early 1960's by proving that there does not exist a tight smooth immersion of the real projective plane with one handle ($\chi = -1$) into $\mathbb{R}^3$. Surprisingly, Cervone produced a tight polyhedral immersion of the same surface into $\mathbb{R}^3$. Here Cervone describes his example in detail and provides a careful analysis of the difference between the smooth and polyhedral theories in this important case.

The fourth paper in the collection, by Thomas Cecil, is a survey of the closely related notions of taut and Dupin submanifolds in Euclidean space. This is a rich theory with many beautiful examples from the theory of isoparametric and homogeneous submanifolds. The relationship between the tautness and the Dupin condition is discussed thoroughly. There are both local and global aspects to the subject. Most local results have been obtained in the context of Lie sphere geometry, and this approach is described in the article in some detail.

In the next article, Chuu-Lian Terng and Gudlaugur Thorbergsson use the critical point theory of Raoul Bott and Hans Samelson to extend the notion of tautness to submanifolds of arbitrary complete Riemannian manifolds. They obtain several new classification results in this more general context. This far-reaching paper opens up many new avenues for research. It is followed by a short paper by Daniel Ruberman, where the author proves a topological result needed by Terng and Thorbergsson about null-homotopic embedded spheres of codimension one.

The final paper in the collection, by Ross Niebergall and Patrick Ryan, is a survey of results on real hypersurfaces in complex space forms with special curvature properties. This field has developed extensively over the past twenty years, and the authors provide a cohesive context for a wide range of results, leading to the frontiers of current research. Particular attention is given to Hopf hypersurfaces and hypersurfaces with constant principal curvatures. These are clearly related to isoparametric hypersurfaces in spheres, which play a prominent role in the theory of taut and Dupin submanifolds.

We wish to thank the authors for their contributions and for their help with various other aspects of the book. In particular, we appreciate the assistance of Thomas Banchoff and Wolfgang Kühnel in preparing Kuiper's article and the lists of his publications and doctoral students. We wish to thank Christine Heinitz, who prepared many of the figures for Kuiper's paper, and Davide Cervone, who did the same for Banchoff and Kühnel's. We also thank Silvio Levy, editor of the MSRI book series, for his assistance in preparing the book for publication, and Carol Oliveira for her help in typing Kuiper's manuscript.

Thomas E. Cecil
Shiing-shen Chern
Fall 1996

Remembering Nicolaas Kuiper

THOMAS F. BANCHOFF

Tight and taut immersions are a living and growing part of contemporary mathematics largely due to the legacy of Nicolaas Kuiper. He made central contributions to many different areas of mathematics during his long and productive career, but it is in tight and taut immersions that his geometric style showed forth in a special way. In that subject, his personal enthusiasm and extraordinary geometric insight combined to bring forth examples and theorems of great conceptual and visual appeal. He delighted in discovering new phenomena, and in presenting his examples using sketches and in cardboard or wire-frame models. He found surprising connections among apparently unrelated areas of mathematics, creating entirely new methods for handling a range of geometric structures: analytic, differentiable, once-differentiable, combinatorial, and topological. He was the first to appreciate the essentially geometric character of tightness, exploiting the relationship between the minimal total curvature condition for smooth submanifolds and critical point theory so that the notion could be extended to non-smooth objects. He guided generations of mathematicians who have followed his lead.

In several other subjects his contributions were necessarily abstract, for example in the embedding theorems he produced with John Nash, or the surprising result that the unit sphere in Hilbert space is contractible. Often he would listen to lectures on a new subject, read about it and study it, and come up with a crucial insight that no one else was close to realizing. He would then leave the field to other mathematicians, encouraging their efforts. He was particularly supportive of young colleagues from many different countries, especially while he was director of the Institut des Hautes Études Scientifiques.

What was special about the theory of tight and taut immersions that kept bringing him back to the subject over a period of more than thirty years? Certainly it had to do with the original examples that kept appearing, illuminating new parts of the subject. Many of these phenomena he discovered himself, but equally important were the examples found by others, which he then helped to bring into full flower. He had such a varied background that he could often see some potential relationships that just about anyone else would have missed.

Who else would have recognized the central significance of the Veronese surface as the unique smooth tight surface in five-dimensional space? The proof of that result was based on theorems in projective differential geometry that almost no one knew about, but he did, at just the right time.

Some mathematicians are renowned because they are the first to arrive at a goal that many others are seeking at the same time. Others are remarkable because they find things no one else even thought of looking for. Nico Kuiper will be remembered as one of the most original mathematicians of his time.

In 1946, Kuiper received his Ph.D. at the University of Leiden working under the direction of Willem van der Woude in the field of classical differential geometry. He then had the chance to come to the United States, first at the University of Michigan where he met Raoul Bott and his student Steve Smale, and then at the Institute for Advanced Study for a crucial interaction with Shiing-Shen Chern. It is easy to see that even at this early stage in his career, Kuiper had a characteristic way of working—he would pay attention to a result or an approach of another mathematician and find that he was rephrasing the concepts in his mind, asking new questions, and more often than not coming up with a fresh insight. It is not surprising that he became a coauthor of a great many papers over the years.

During the 1950's he taught mathematics and statistics at the Landbouw-hogeschool (Agricultural Institute) in Wageningen, contributing a number of geometric insights to the theory of design of experiments. In the 1960's, he was professor of pure mathematics at the University of Amsterdam, concentrating primarily in the fields of differential geometry, differential topology, and algebraic topology, and nurturing a number of doctoral students and post-doctoral visitors. From 1971 to 1985, he was Director of the Institute des Hautes Études Scientifiques at Bures-sur-Yvette near Paris, where he exercised leadership in the world community of mathematicians. He was made a Knight of the Order of the Golden Lion by the Dutch Government and he was a Chevalier in the French Legion of Honor. In 1984, he received an honorary degree from Brown University. After his retirement, he remained in France until 1991, when he returned to live in the Netherlands. He continued to participate in mathematical colloquia at the University of Utrecht.

Nico Kuiper was only 74 years old when he died on December 12, 1994 after a year-long illness that had taken away his strength but not his love for mathematics. On a personal note, I was privileged to be able to visit him and his wife Agnete at their home in the Netherlands after the International Congress of Mathematicians in Zürich in August of that year, and I could see how difficult it was for him to be confined by his illness. He had attended so many Congresses, and had taken a leadership role at an officer of the International Mathematical Union. Now he could only hear reports of the new developments in the many subjects in which he had made key contributions.

One result that pleased him especially was a breakthrough in the theory of tight immersions. Thirty years ago, he had singled out as a special challenge the question of finding a tight immersion of the real projective plane with one handle, the sole remaining case for surfaces in three-space. He and I had traded letters on this subject for years, and more than once one of us would present elaborate drawings of a purported solution, only to follow it the next day by a "disregard previous letter" after finding an unallowable local self-intersection. Just two years earlier, François Haab had shown that there was no tight smooth immersion of this surface into three-space, and Nico himself had been instrumental in working with Haab to extend these results. He was surprised and delighted to learn about the discovery by Davide Cervone of a tight polyhedral immersion of this surface into three-space. It is too bad that he did not have access to the Internet so he could work interactively with this beautiful example, but he did draw the diagrams based on the coordinate description. In previous years, he had only been able to look for examples with small numbers of vertices, and it impressed us both to see how computer graphics had increased the opportunities to find and manipulate complicated new examples. (This example and some crucial differences between the smooth and polyhedral cases are described in the article of Cervone in this volume.)

Nico Kuiper is a model for so many of us. He left us a legacy of inspiring mathematical results, and even more importantly, a lasting love of mathematics that we can only hope to pass on to those who come after us, remembering our great friend as we do so. With affectionate gratitude we dedicate this volume to his memory.

With wife Arletta and Tom Banchoff, at Brown University.

Tight and Taut Submanifolds
MSRI Publications
Volume **32**, 1997

Geometry in Curvature Theory

NICOLAAS H. KUIPER

ABSTRACT. This article is based on the Roever Lectures in Geometry given by Kuiper at Washington University, St. Louis, in January 1986. Although incomplete, it is an excellent exposition of the topics it does cover, starting with elementary versions of the notion of tightness and going through the analysis of topsets, the classification in low dimensions, the notions of total curvature for curves and surfaces in space, homological notions of tightness, the Morse inequalities, and Poincaré polynomials. It contains a detailed proof of Kuiper's remarkable result that a tight two-dimensional surface substantially immersed in $\mathbb{R}^5$ must be a Veronese surface.

EDITORS' NOTE. At the time of Kuiper's death in December, 1994, this paper existed in the form of an unfinished typescript. For inclusion in this volume, it was edited by Thomas Banchoff, Thomas Cecil, Wolfgang Kühnel, and Silvio Levy. A few Editors' Notes such as this one were included, mostly pointing to additional references. Several minor typos were corrected and the numbering was normalized for ease of reference; thus Sections 5 and 6 of the manuscript were renumbered 4 and 5, since there was no Section 4. The present illustrations were made by Christine Heinitz and by Levy, based on Kuiper's hand drawings.

1. Banchoff's two-piece property. Zero-tightness

Prerequisites and Notation. *Euclidean space* $E = E^N$ of dimension N is the real vector space $\mathbb{R}^N$ with norm $\|u\| = \sqrt{\Sigma_1^N (u^i)^2}$ for $u = (u^1, \ldots, u^N) \in \mathbb{R}^N$ and distance $\|v - u\|$ for $u, v \in \mathbb{R}^N$. The identification $\kappa : \mathbb{R}^N \to E^N$ can be replaced by any other preferred Euclidean coordinate system $\kappa \circ g : \mathbb{R}^N \to E^N$, where g is an isometry:

$$g(u) = u_0 + u \cdot g_0, \quad \text{for } u_0 \in \mathbb{R}^N \text{ and } g_0 \in O(N) \text{ an orthogonal matrix.}$$

We use E^N to emphasize Euclidean space aspects and $\mathbb{R}^N$ for vector space aspects. A set $X \subset E$ is called *convex* if

$$u + \lambda(v - u) \in X \quad \text{for all } u, v \in X \text{ and } 0 \leq \lambda \leq 1.$$

The smallest convex set containing $X \subset E$ is its *convex hull*, denoted $\mathcal{H}X$. The smallest affine subspace (also a Euclidean space) that contains X is its *span*, denoted $\mathrm{span}(X)$. If $\mathrm{span}(X) = E$, then X is called *substantial* in E. The boundary of $\mathcal{H}X$ in $\mathrm{span}(X)$ is called the *convex envelope* $\partial\mathcal{H}X$ of $X \subset E$. If X is one point then $\mathcal{H}X = X$ and $\partial\mathcal{H}X = \varnothing$.

The subspaces $\{u : \|u\| < r\}$ and $\{u : \|u\| = r\}$, for $r > 0$, are called the *N-ball* B^N and the $(N-1)$-*sphere* S^{N-1}, respectively. As metric spaces they are called the *round ball* and the *round sphere*. Let $z : \mathbb{R}^N \to \mathbb{R}$ be a *linear function*, $z(u) = \sum_i \zeta_i u^i$ for $\zeta_i \in \mathbb{R}$, with $\|z\|^2 = \sum \zeta_i^2 > 0$. The subspaces $\{u : z(u) \geq c\}$, $\{u : z(u) > c\}$ and $\{u : z(u) = c\}$ of E are called the *half-space* h, its interior the *open half-space* $\mathring{h}$, and its boundary the *hyperplane* ∂h, respectively. The function z is often called a *height function*.

A metrizable topological space X is called *separated* if it is the disjoint union of two nonempty open and closed subsets, say X_1 and X_2. If $U \subset X$ contains points $x_1 \in X_1$ and $x_2 \in X_2$, then $U \cap X_1$ and $U \cap X_2$ are disjoint open and closed in U, and so U is also separated. The space X is called *connected* if it is not separated. A connected nonempty open closed subset of a metrizable space X is called a *topological component* of X.

EXAMPLE. The plane set

$$\{(\xi,\eta) : \xi = 0 \text{ or } \eta = \sin(\xi^{-1})\}$$

is connected (but not pathwise connected).

CONSEQUENCE. If Y is a metrizable space and for any two points $y_1, y_2 \in Y$ there is a connected space $W(y_1, y_2) \subset Y$ containing y_1 and y_2 (in other words, if "any two points $y_1, y_2 \in Y$ can be connected in Y"), then Y is connected. Indeed Y separated would show an immediate contradiction.

Definitions and General Theorems. For given compact spaces X, either embedded in $E = E^N$ or given independently, we are interested in embeddings or other continuous maps in E with nice properties that generalize convexity. We introduce the important notion and tool called a *topset*:

DEFINITION. Suppose the half-space

$$h = h_z = \{u \in E : z(u) \geq c\}$$

for some linear function $z : E \to \mathbb{R}$ supports ("leans against") the compact set X, without containing it completely:

$$X \neq h \cap X = \partial h \cap X \neq \varnothing.$$

Then $X_z = h \cap X$ is called a (proper) *topset* of the set $X \subset E$. It is the set of points in X for which the function $z : X \to \mathbb{R}$ attains its maximal (top) value.

Note that $\mathring{h} \cap X = \varnothing$. More generally, if $f : X \to E$ is a continuous map of the compact space X into E, and $h \cap f(X)$ is a topset of $f(X) \subset E$, then

$$f^{-1}(h) = \{x \in X : f(x) \in h\} = f^{-1}(\partial h) \subset X$$

is called a *topset of the map* f. A topset of a topset is called a $top^2 set$. A top^j set for some $j \geq 1$ is called a $top^* set$. If the span of a top^* set X' of X has dimension k, then X' is called an E^k-$top^* set$.

REMARK. Let X be a substantial set of $N + 1$ points $e_0, \ldots, e_N$ in E^N. Then $\mathcal{H}X$, the convex hull, is an N-simplex σ_N. Any proper nonempty subset of X is a topset.

EXERCISE. Determine all topsets of a standard torus in E^3, obtained by rotating a circle around a disjoint line in its plane.

EXERCISE. Determine the topsets of the map $f : w \to w^3$ of the unit circle $\{w : |w| = 1\} \subset \mathbb{C}$ into $\mathbb{C} = \mathbb{R} \oplus \mathbb{R} = E^2$.

THEOREM 1.1. *The convex envelope of a compact set $X \subset E$ is the union of the convex hulls of its topsets:*

$$\partial \mathcal{H}X = \bigcup_z \mathcal{H}X_z.$$

The union may be taken only over all linear functions $z : E \to \mathbb{R}$ with norm $\|z\| = 1$. If X consists of one point, both sides are the empty set.

PROOF. We need to show the implications in both directions:

$$x \in \bigcup_z \mathcal{H}X_z \iff x \in \partial \mathcal{H}X.$$

Assume that the span of X is $\text{span}(X) = E = E^N$. For a topset $X_z = h \cap X$, let $x \in \mathcal{H}X_z$. Since h supports X, it supports also $\mathcal{H}X$. Then $x \in h \cap \mathcal{H}X \subset \partial \mathcal{H}X$. Conversely if $x \in \partial \mathcal{H}X$, then there is a $\mathcal{H}X$-supporting half-space h_z containing x, and

$$x \in h_z \cap \partial \mathcal{H}X = h_z \cap \mathcal{H}X = h_z \cap \mathcal{H}X_z = \mathcal{H}X_z. \qquad \square$$

Now we propose a preliminary generalization of convexity:

DEFINITION (for compact sets). A connected compact set $X \subset E$ is said to have the *two-piece property*, or TPP [Banchoff 1971b], and is called 0-*tight*, in case any of the following equivalent conditions hold:

(a) $h \cap X$ is connected for every half-space h.
(b) $\mathring{h} \cap X$ is connected for every h.
(c) The set difference $X \setminus \partial h$ has at most two components for every h (this is the two-piece property).
(d) In terms of Čech homology and any coefficient ring, the homomorphism $H_0(h \cap X) \to H_0(X)$ is injective for every h.

We mention this last condition now for the sake of completeness, but defer the relevant discussion till later (page 35).

PROOF OF EQUIVALENCE. (a) $\Rightarrow$ (b). If (a) holds then any two points in $\mathring{h} \cap X$ are contained for some half-space h_i in $h_i \cap X \subset \mathring{h} \cap X$, and they can be connected in $h_i \cap X$. Then $\mathring{h} \cap X$ is connected.

(b) $\Rightarrow$ (a). Suppose $h \cap X$ is not connected for some $h = \{u \in E : z(u) \geq c\}$, let and Y_1 and Y_2 be disjoint nonempty open closed subsets with union $Y_1 \cup Y_2 = h \cap X$. Let U_1 and $U_2 \subset X$ be disjoint nonempty open neighborhoods of the compact subsets Y_1 and Y_2. If $c - 2\varepsilon$ is the maximum of z on the compact set $X \setminus (U_1 \cup U_2)$ and $\mathring{h}_0 = \{u \in E : z(u) > c - \varepsilon\}$, then $\mathring{h}_0 \cap X$ is not connected. This contradicts (b).

The equivalence (b) $-$ (c) is tautological. $\square$

The same proof works for the equivalences in the following more general situation.

DEFINITION (for maps). A continuous map $f : X \to E$ of a connected compact space X in E has the *two-piece-property* (TPP) and is called 0-*tight* if any of the following equivalent conditions hold:

(a) $f^{-1}(h)$ is connected for any half-space h.
(b) $f^{-1}(\mathring{h})$ is connected for any h.
(c) $f^{-1}(E \setminus \partial h)$ has at most two components for any h (two-piece-property).
(d) $H_0(f^{-1}(h)) \to H_0(X)$ is injective for any h.

EXAMPLES. The following are 0-tight sets:

(1) a convex body $X = \mathcal{H}X \subset E^N$, for $N \geq 0$;
(2) a convex hypersurface $X = \partial \mathcal{H}X$ substantial in E^N, for $N \geq 2$ (convex curve for $N = 2$);
(3) a hemisphere, $\{u \in \mathbb{R}^3 : \|u\| = 1, z(u) \leq 0\}$;
(4) the standard round torus in E^3 (see the first exercise on page 3);
(5) the solid round ring (solid torus) bounded by the standard torus;
(6) the 1-*skeleton* $\mathrm{Sk}_1(\sigma_N)$ of the N-simplex $\sigma_N \subset E^N$; this is by definition the union of all edges of σ_N, and as a topological space it is a *complete graph* on $N + 1$ vertices.

REMARK. These are corollaries of the definition:

(1) 0-tightness is invariant under *linear embeddings* $i : \mathbb{R}^M \to \mathbb{R}^N$ and *projections* $p : \mathbb{R}^N \to \mathbb{R}^M$, where $M < N$. Indeed, if $f : X \to \mathbb{R}^M$ and $g : Y \to \mathbb{R}^N$ are 0-tight, then so are $i \circ f : X \to E^N$ and $p \circ g = Y \to E^M$.

An example of a 0-tight map (immersion) is the projection of the 1-skeleton $\mathrm{Sk}_1(\sigma_3)$ in E^3 onto the union of edges and diagonals of a convex 4-gon in a plane in E^3. Note that $f : X \to$ point $\in E^N$ is 0-tight for any connected compact X.

(2) 0-tightness is an *affine* and even a *projective property* in the following sense. Let P^N be a real projective N-space and P^{N-1} a hyperplane. Then $P^N \setminus P^{N-1}$

can be identified with E^N, and this identification is natural up to affine transformations $u \mapsto u_0 + gu$, where $u_0 \in \mathbb{R}^N$, $g \in \mathrm{GL}(n, \mathbb{R})$. Given $X \subset E^N$, let $\eta : P^N \to P^N$ be a projective transformation such that $\eta(X) \subset E^N \subset P^N$. Suppose $f : X \to E^N$ is 0-tight. Then also $\eta \circ f : X \to E^N$ is 0-tight by condition (c), which is expressed in terms of hyperplane sections.

THEOREM 1.2. *Any topset X_z of a 0-tight set $X \subset E$ or of a 0-tight map $f : X \to E$ is itself 0-tight. So is any top*set.*

PROOF. We deal with the case of a set; the proof for a map is the same. Suppose $X \subset E$ has a topset that is not 0-tight, say $X_z = h_z \cap X$, where $h_z = \{u \in E : z(u) \geq c_1\}$. See Figure 1. Then there exists a half-space $h_0 = \{u \in E : w(u) \geq c_2\}$ such that

$$\varnothing \neq h_0 \cap X_z \neq X_z,$$

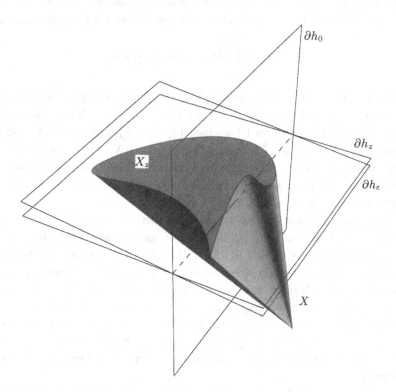

Figure 1. If a topset X_z of X is not 0-tight, some hyperplane ∂h_0 cuts X_z into more than two pieces. By tilting slightly the support hyperplane ∂h_z of X_z around the hinge $\partial h_z \cap \partial h_0$, one can obtain a hyperplane that cuts X into more than two pieces, so X is not 0-tight. (In the figure, the half-space h_0 is to the right of, and the half-spaces h_z and h_ε are above, their bounding planes.)

and $h_0 \cap X_z = h_0 \cap h_z \cap X = X_1 \cup X_2$ is separated with X_1 and X_2 open and closed in the relative topology, compact, and disjoint. There is a (small) open neighborhood U in X of $X_1 \cup X_2$ that is also separated:

$$U = U_1 \cup U_2, \quad U_1 \cap U_2 = \varnothing, \quad U_1 \supset X_1, \quad U_2 \supset X_2.$$

For small ε, the half-space

$$h_\varepsilon = \{u \in E : (z(u)-c_1) + \varepsilon(w(u)-c_2) \geq 0\}$$

meets X in $h_\varepsilon \cap X \supset h_0 \cap h_z \cap X$ and

$$h_\varepsilon \cap X \subset U = U_1 \cup U_2.$$

Therefore $h_\varepsilon \cap X$ is not connected, so X is not 0-tight. $\qquad\square$

THEOREM 1.3 (CLASSIFICATION OF PLANE 0-TIGHT SETS). *Any plane compact 0-tight substantial set $X \subset E^2$ can be obtained from its convex hull $\mathcal{H}X$ by deleting a countable family of disjoint open convex subsets from the interior:*

$$X = \mathcal{H}X \setminus \bigcup_{i=0}^{r} U_i, \quad \text{where } 0 \leq r \leq \infty.$$

PROOF. Let X be substantial and 0-tight in E^2. Every topset X_z of X is 0-tight and lies in a supporting line ∂h_z, so it is either a point or a line segment, and in any case convex. By Theorem 1.1 we have $\bigcup_z X_z = \bigcup_z \mathcal{H}X_z = \partial \mathcal{H}X$, and $\partial \mathcal{H}X$ is contained in X. The set X is then obtained from $\mathcal{H}X$ by deleting disjoint open connected sets (holes) U. Any embedded circle in a hole U does not separate X, and can be contracted inside U to a point. So each hole U is contractible and homeomorphic to an open disc.

Now suppose U is not convex, so there are distinct $u_1, u_2 \in U$ and $0 < \lambda < 1$ such that $u_1 + \lambda(u_2 - u_1) \notin U$. The smallest such value λ for given u_1 and u_2 yields a point $x = u_1 + \lambda(u_2 - u_1)$ in X. Let h and h^- be the two half-planes having as common boundary the line $u_1 u_2 = \partial h = \partial h^-$. Connect u_1 and u_2 by an embedded polygonal arc $\beta \subset U$ that meets the line ∂h transversally in every intersection point (see Figure 2). Since X is 0-tight, $h \cap X$ and $h^- \cap X$ are connected. Then x and $h \cap \partial \mathcal{H}X$ can be connected in $h \cap X$ and (even better) in the component of $(h \cap \mathcal{H}X) \setminus \beta$, which contains $h \cap X$. In $h \cap X$, the points x and $h \cap \partial \mathcal{H}X$ can be connected by a polygonal arc α, which meets ∂h only in the point x.

There is another such polygonal arc α^- in $(h^- \cap \mathcal{H}X) \setminus \beta$ connecting x with $h^- \cap \partial \mathcal{H}X$. The union $\alpha \cup \alpha^-$ lies in $\mathcal{H}X \setminus \beta$ and divides the segment $\partial h \cap \mathcal{H}X$, as well as $\mathcal{H}X$, into two parts, one containing u_1 and the other containing u_2. This contradicts the existence of the arc β from u_1 to u_2. So all holes U are open convex discs.

Any collection of disjoint open sets in the plane is countable. $\qquad\square$

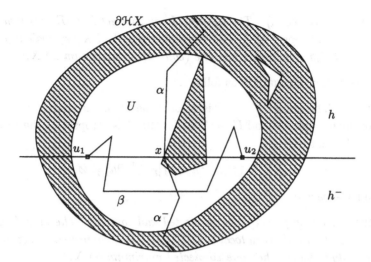

Figure 2. The holes in a 0-tight plane set must be convex (see the proof of Theorem 1.3).

EXAMPLE. For later reference we mention a curious example of 0-tight plane set, the *limit Swiss cheese*. A Swiss cheese is a round 2-ball (disc) in E^2 from which a union of disjoint open round discs is deleted; see Figure 3. Touching of discs in their boundaries is permitted. If the union of the open discs is everywhere dense, the resulting 0-tight set is called a *limit Swiss cheese*.

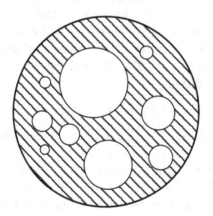

Figure 3. A Swiss cheese has the TPP.

DEFINITION. The subspace $Y \subset X \subset E^N$ is a *local topset* of X if Y has an open neighborhood U in X such that Y is a topset of $\bar{U} \subset E$ (here as elsewhere the bar indicates closure). That means

$$X \neq Y = h \cap \bar{U} = \partial h \cap \bar{U} \neq \emptyset$$

for some half-space h. Note that Y is then open and closed in $\partial h \cap X$.

THEOREM 1.4 (TOPSETS). *The connected compact set $X \subset E$ is 0-tight if and only if every local topset of X is a connected topset of X; equivalently, if and only if every height function z has one (connected) maximum on X.*

The generalization for maps is as follows:

DEFINITION. The subset $Y \subset X$ is a *local topset* of the map $f : X \to E$ if Y has an open neighborhood U in X such that Y is a topset of the restriction $f|_{\bar{U}} : \bar{U} \to E$, that is

$$X \neq Y = (f|_{\bar{U}})^{-1}h = (f|_{\bar{U}})^{-1}(\partial h) \neq \varnothing,$$

for some half-space h.

THEOREM 1.4 (MAPS). *Let X be connected and compact. The map $f : X \to E$ is 0-tight if and only if every local topset Y of f is a connected topset of f; that is, if and only if every z has one connected maximum on X.*

REMARKS. If the continuous function $f : X \to \mathbb{R}$ has exactly one local topset, and so does the function $-f$, then f is 0-tight. The map $z \mapsto z^3$ for $|z| = 1$ (see the second exercise on page 3) is not 0-tight.

PROOF OF THEOREM 1.4 FOR SETS. If X is not 0-tight, there is a half-space $h' = \{u \in E : z(u) \geq c'\}$ for which $h' \cap X$ is separated and is the disjoint union of two open closed subsets X_1 and X_2. Let $c \geq c'$ be the maximal value for which $h \cap X_1$ and $h \cap X_2$ are both nonempty, where $h = \{u \in E : z(u) \geq c\}$. One at least of $h \cap X_1$ and $h \cap X_2$ is then a local topset and not a connected topset.

Conversely, if $Y \subset X$ is a local topset in $\partial h \cap X$ and $\partial h \cap X$ is not a connected topset, then $\partial h \cap X = Y \cup Z$ is the disjoint union of Y and Z and $h \cap X$ is the disjoint union of Y and $h \cap X \setminus Y \supset Z$, both open and closed. So $h \cap X$ is separated and X is not 0-tight. $\qquad\square$

EXERCISE. Prove Theorem 1.4 for maps.

EXAMPLE. Let $\sigma_4 = \mathcal{H}(\{e_1, \ldots, e_5\}) \subset E^4$ be a four-simplex, and let M be the union of five triangles $\mathcal{H}(\{e_i, e_{i+1}, e_{i+2}\})$, for $i = 1, \ldots, 5$ (indices being taken modulo 5). Then M is a 0-tight Möbius band, substantial in E^4. Observe that M contains the 1-skeleton $\mathrm{Sk}_1(\sigma_4)$. Figure 4 shows a projection in E^3 (a 0-tight embedding), as well as a projection in E^2 (a 0-tight map) with folds along the edges e_1e_2, e_2e_3, e_3e_4, e_4e_5, and e_5e_1. The boundary of M is the polygon $e_1e_3e_5e_4e_2e_1$. To prove that $M \subset E^4$ is 0-tight, we observe that any local topset Y of M contains at least one vertex and cannot cut any opposite edge in σ_N transversally. Then Y lies in a supporting half-space h and $Y = h \cap M = \partial h \cap M$.

REMARK. For the same reason any union X of $\mathrm{Sk}_1(\sigma_N) \subset E^N$ with some of the simplices of σ_N of various dimensions is 0-tight. In particular, $\mathrm{Sk}_i(\sigma_N)$ is 0-tight, for $1 \leq i \leq N$.

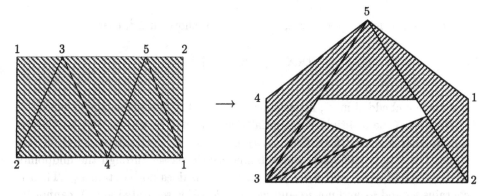

Figure 4. A Möbius band made from five of 2-faces of a tetrahedron. The embedded in E^3 is 0-tight, as is the projection onto the plane of the paper.

Zero-Tight Balls and Spheres of Dimension 1 and 2. We now prove some classification theorems with the help of our tool, the topsets.

A 0-tight embedded arc (1-ball) $X \subset E^N$ in E^N is necessarily a straight line segment. If not, there is a point $y \in X$ not on the line connecting the endpoints x_0 and x_1, and a half-space h containing x_0 and x_1 but not y. So $h \cap X$ is separated and X is not 0-tight.

THEOREM 1.5 (0-TIGHT CIRCLES, SPHERES, BALLS).

(i) *A 0-tight embedded closed curve in E^N is a plane convex curve.*

(ii) *A 0-tight substantial 2-sphere in E^N is a convex surface $X = \partial \mathcal{H} X$ in 3-space E^3.*

(iii) [Lastufka 1981] *A 0-tight substantial 2-ball (disc) in E^N is either*

   (a) *a convex plane disc in E^2, or*

   (b) $X = \partial \mathcal{H} X \setminus U$ *in E^3, where the deleted set U is an open disc of a plane convex topset $(\partial \mathcal{H} X)_z = \mathcal{H} X_z$ (see Figure 5).*

Figure 5. Zero-tight disc in E^3.

PROOF. If $X \subset E$ is 0-tight, then by Theorem 1.2 so are any of its top*sets. An E_0-top*set of X is a point; an E_1-top*set of X is a line segment; possible E_2-top*sets of X are described in Theorem 1.3.

(i) If X is a 0-tight closed curve, or "circle" for short, in E^2, then

$$\partial \mathcal{H} X = \bigcup_z \mathcal{H} X_z = \bigcup_z X_z$$

is a circle embedded in X, hence equal to X, and X is a plane convex curve.

Now let X be a 0-tight substantial circle X in E^N, where $N \geq 3$. Choose a nonplanar 4-gon with vertices u_1, u_2, u_3, u_4 on X, cyclically ordered on this circle. Let h be a half-space whose boundary ∂h passes through the midpoints of $u_1 u_2$, $u_2 u_3$, $u_3 u_4$, and $u_4 u_1$, and such that h does not contain u_1. Then h contains u_2 and u_4 and not u_1 and u_3; thus $h \cap X$ is separated and X cannot be 0-tight.

(ii) Let X be a 0-tight 2-sphere in E^N. First let $N = 3$. Any topset of X is a point, a line segment, or an E_2-topset $Y = X_z$. We show that in the latter case Y has to be convex. Indeed, X contains the convex envelope $\partial \mathcal{H} Y \subset Y \subset X$. If Y is not convex then $\mathcal{H} Y \setminus Y$ contains at least one hole U as component and the open half-space $\mathring{h} = E^3 \setminus h_z$ intersects X in $X \setminus Y$, which has nonempty open pieces in X, separated by Y, also separated by the circle $\partial U \subset X = S^2$. This contradicts 0-tightness. So all topsets X_z of X are convex and the "convex" surface $\partial \mathcal{H} X$ is contained in the 2-sphere X. Then X is equal to this convex surface $\partial \mathcal{H} X$.

Next suppose X is a 0-tight substantial 2-sphere in E^N for $N > 3$. Let Y be as before a nonconvex E_2-top*set. Then $\partial \mathcal{H} Y \subset Y \subset E^2 \subset E^N$. Choose x_1 and x_2 in $X \setminus Y \subset X \setminus \partial \mathcal{H} Y$ in different components of $X \setminus Y$. There exists a half-space h such that $x_1, x_2 \in h$, but $h \cap Y = \varnothing$. Then $h \cap X = h \cap (X \setminus Y)$ is separated, contradicting 0-tightness of X. So all E^2-top*sets are convex. If Y is an E^3-top*set, then all topsets of Y are convex and the 2-sphere $\partial \mathcal{H} Y \subset Y \subset X$ must be equal to X. It cannot be substantial in E^N, for $N > 3$. So there is no E_3-top*set. Let k be either N or the smallest number $k < N$ with $k > 3$ for which there is an E_k-top*set Y. Then all topsets of X and of Y, are convex. So $\partial \mathcal{H} X \subset X$ and $\partial \mathcal{H} Y \subset Y \subset X$. But the dimension of $\partial \mathcal{H} Y$ is $k - 1 > 2$. This is absurd for dimension reasons. The desired result follows.

(iii) A 0-tight 2-disc in E^2 is a convex disc by Theorem 1.2. We can therefore assume the 0-tight disc X embedded in E^N, where $N \geq 3$. First let $N = 3$. A 0-tight nonconvex top*set X_z is necessarily an E_2-topset X_z, obtained from the convex disc $\mathcal{H} X_z$ by deleting one or more open sets U. If U is one of them and ∂U is not the boundary ∂X of the disc X, then $X \setminus X_z = X \setminus h_z$ contains at least two points x_1 and x_2, which are separated by ∂U. As $X \setminus h_z = (E \setminus h_z) \cap X$ is then separated and $E \setminus h_z$ is an open half-space, this contradicts 0-tightness. There is only *one* boundary ∂X for X, so that only one topset of X is nonconvex and it has only one convex hole U. This is the conclusion of the theorem for $N = 3$.

Finally let X be a 0-tight substantial 2-disc in E^N, where $N > 3$. Let Y be a nonconvex E^2-top*set. As before, $\mathcal{H}Y \setminus Y$ contains at least one hole U as component. If ∂U is not the boundary ∂X of X, then $X \setminus Y$ has at least two points x_1 and x_2 that are separated in $X \setminus Y \subset X \setminus \partial U$. As in (ii) we find a contradiction. Also there can be at most one hole U. Since ∂U must be the boundary of X, we can fill in U to obtain $U \cup X$, a 0-tight embedded S^2 in E^N, for $N > 3$. This led to a contradiction in (ii), and part (iii) of Theorem 1.5 is proved. $\qquad\qquad\Box$

PROBLEM. Let Y be a plane 2-disc from which a number of disjoint open 2-discs are deleted. Determine all continuous 0-tight embeddings of Y in E^N. Lastufka [1981] found that all 0-tight immersions are embeddings.

Background: Manifolds and the Topological Classification of Surfaces. Although we are mainly interested in surfaces, we make some general remarks on higher-dimensional manifolds.

A compact n-*dimensional topological manifold* (or simply n-*manifold*) is a metrizable topological space M such that each point $x \in M$ can be assigned a homeomorphism or *chart*

$$\kappa_x : U_x \to \kappa_x(U_x) \subset h^n = \{u \in \mathbb{R}^n : u^n \geq 0\} \subset \mathbb{R}^n$$

of some open neighborhood U_x onto $\kappa_x(U_x)$, an open set in h^n. The maps

$$\kappa_{yx} = \kappa_y \kappa_x^{-1} : \kappa_x(U_x \cap U_y) \to \kappa_y(U_x \cap U_y),$$

for $x, y \in M$, are local homeomorphisms in h^n. The pieces $\kappa_x^{-1}(\partial h^n)$ constitute the *boundary* ∂M of M. If $\partial M = \varnothing$, we say that M is a *closed manifold*, that is, compact without boundary. If $\partial M \neq \varnothing$, it is easy to see that ∂M is a closed $(n-1)$-manifold.

If choices for κ_x are given such that every change-of-coordinate map κ_{yx} is smooth (by which we mean C^∞, that is, such that all derivatives exist), we can consider the set of *all* functions locally generated from functions $u^i \circ \kappa_x : U_x \to \mathbb{R}$, $i = 1, \ldots, n$ by smooth compositions ψ (smooth functions of n variables $u^1, \ldots, u^n$): $\psi(u^1 \circ \kappa_x, \ldots, u^n \circ \kappa_x)$. This set of functions is called a *smoothing* of M, and M with this smoothing is called a *smooth manifold*. Any smooth closed manifold determines a topological manifold by forgetting the smoothness. It is known that for $n \leq 3$ every closed topological n-manifold can so be obtained from some smooth n-manifold, which smooth n-manifold is moreover unique but for differentiable equivalence. For $n \geq 4$ there are topological closed manifolds for which existence fails, and others for which existence holds but not uniqueness. (The result for $n = 4$ is due to deep work of M. Freedman [1982] and S. Donaldson [1983; 1986].)

A *finite simplicial complex* W can be defined as a union of affine subsimplices $\mathcal{H}(e_{i_0}, \ldots, e_{i_r})$ of various dimensions $r \geq 0$ of an N-simplex $\sigma_N = \mathcal{H}(e_0, \ldots, e_N)$ in E^N for some N. A homeomorphism $\lambda : W \to X$ of W onto a given topological space X is called a *triangulation* of X. If $\nu : W_1 \to W$ is a triangulation of W that is affine on each simplex of W_1 and restricts to a triangulation onto each simplex of W, then the triangulation $\lambda \circ \nu : W_1 \to X$ is called a *subdivision* of $\lambda : W \to X$.

A triangulation of a closed n-manifold is called *Brouwer* if the union of all n-simplices with a common vertex, admits a homeomorphic embedding into E^n for which the image of each n-simplex is an affine n-simplex in E^n. One can prove that every smooth n-manifold M has a Brouwer triangulation with smooth embeddings for each simplex, which is unique modulo subdivision and modulo diffeomorphisms of M.

A topological n-manifold has, for $n \leq 3$, a unique Brouwer triangulation [Moise 1952]. This is not so for $n \geq 4$. Some subdivision of a Brouwer triangulation of a closed n-manifold has in any case a compatible smoothing for $n \leq 7$, which is unique for $n \leq 6$, but not so for larger dimensions. The n-sphere has a few-vertex triangulation as the boundary of the n-simplex $\sigma_n \subset E^n$. This is a Brouwer triangulation. A few-vertex triangulation of the Möbius band was seen in Figure 4. In these examples the number of vertices is minimal.

An *orientation* of a manifold M at a point $x \in M$ is a choice of one of the two generators of the homology group $H_0(M, M \backslash x; \mathbb{Z}) \approx \mathbb{Z}$. For an embedded arc $I \subset M$ with end points x_0 and x_1, the axioms of homology [Spanier 1966, p. 294 ff.] give natural isomorphisms by inclusions of spaces

$$\mathbb{Z} \approx H_0(M, M \backslash x_0; \mathbb{Z}) \leftarrow H_0(M, M \backslash I; \mathbb{Z}) \to H_0(M, M \backslash x_1; \mathbb{Z}),$$

which permit unique transport of an orientation from x_0 to x_1. Transport along homotopic paths gives the same result. If the result is the same for any paths connecting two points x_0 and x_1 then it defines a global orientation on M, now assumed connected, and M is called *orientable* and *oriented* by this choice. The local orientations determine a two-point bundle (double covering of M) and M is orientable if and only if this bundle is a product bundle. The Möbius band is not orientable.

Let B_x be an embedded n-ball around $x \in M$, with boundary the embedded $(n-1)$-sphere $\partial B_x = S_x$. Then there are natural isomorphisms (by excision and exactness)

$$H_0(M, M \backslash x; \mathbb{Z}) = H_0(B_x, S_x; \mathbb{Z}) \to H_0(S_x; \mathbb{Z}) \approx \mathbb{Z}.$$

Thus an orientation of M at x determines a generator of $H_0(S_x; \mathbb{Z}) = \mathbb{Z}$.

Given connected closed topological n-manifolds M_1 and M_2, one constructs a new n-manifold $M_1 \# M_2$ called the *connected sum* as follows (Figure 6). For $i = 1, 2$, delete from M_i, a small open ball U_i interior of an n-ball $B_i \subset M_i$ with boundary an $(n-1)$-sphere $S_i = \partial B_i$, and glue $M_1 \backslash U_1$ to $M_2 \backslash U_2$ by a homeomorphism $\lambda : \partial U_1 \to \partial U_2$. If M_1 and M_2 are orientable and oriented, one chooses λ moreover in such a way that for $x_1 \in (M_1 \backslash B_1)$ and $x_2 \in (M_2 \backslash B_2)$, there is

Figure 6. The connected sum of two closed surfaces.

an orientation on $M_1 \# M_2$ such that local generators that define orientations correspond in the sequence of isomorphisms:

$$\mathbb{Z} \longleftrightarrow H_0(M_1, M_1 \backslash x_1; \mathbb{Z}) \longleftrightarrow H_0(M_1 \# M_2, M_1 \# M_2 \backslash x_1; \mathbb{Z})$$
$$\longleftrightarrow H_0(M_1 \# M_2, M_1 \# M_2 \backslash x_2; \mathbb{Z}) \longleftrightarrow H_0(M_2, M_2 \backslash x_2; \mathbb{Z}_2).$$

It is known (only recently for $n = 4$) that the connected sum for topological closed manifolds so defined is unique up to equivalence. If at least one of M_1 and M_2 is nonorientable, then we get always the same (unique) connected sum. For closed oriented n-manifolds M_1 and M_2 and $n \geq 3$, the orientations may give two nonhomeomorphic results. For any two closed surfaces ($n = 2$) the connected sum is however a unique closed surface.

REMARK. A continuous map $f : X \to Y$ is called an *embedding* if $f : X \to f(X)$ is a homeomorphism. It is called an *immersion* if $f|_{U_x} : U_x \to Y$ is an embedding for some neighborhood U_x of any $x \in X$. If X and Y are manifolds of dimension n and $N \geq n$, the immersion f is called *tame* in case for any $x \in X$ there is a commutative diagram concerning neighborhoods $U_x \subset X$ and $U_y \subset Y$, $y = f(x)$, charts κ_x and κ_y, and a linear embedding i :

$$
\begin{array}{ccc}
U_x & \xrightarrow{\ f\ } & U_y \\
\downarrow{\scriptstyle \kappa_x} & & \downarrow{\scriptstyle \kappa_y} \\
\mathbb{R}^n & \xrightarrow{\ i\ } & \mathbb{R}^N.
\end{array}
$$

These definitions for topological manifolds have natural counterparts in the smooth context. Most of these facts were known for dimension $n \neq 4$ around 1971. For $n = 4$ the breakthrough came since 1982 with the work of Casson [1986], Freedman [1982] and Donaldson [Donaldson 1983; 1986].

Every orientable closed surface is the (repeated) connected sum of a two-sphere S^2 and some number $g \geq 0$ of tori T:

$$M_g = T \# T \cdots \# T, \qquad M_0 = S^2.$$

We say M_g has *genus* g and has *Euler characteristic* $\chi = 2 - 2g$. Every nonorientable closed surface is a connected sum of the form

$$P \# M_g, \qquad \text{for some } g \geq 0,$$

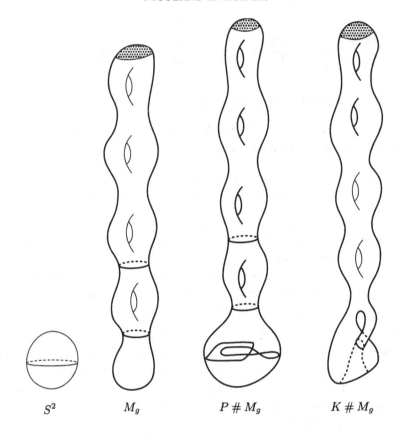

$$S^2 \qquad M_g \qquad P \# M_g \qquad K \# M_g$$

Figure 7. The classification of closed surfaces.

or of the form

$$P \# P \# M_g = K \# M_g, \qquad \text{for some } g \geq 0,$$

where P is the real projective plane and $K = P \# P$ is the Klein bottle. Surfaces of these types have Euler characteristic $\chi = 1 - 2g$ and $\chi = -2g$, respectively. See Figure 7.

Any compact surface with boundary is obtained from a unique closed surface by deleting the interiors of a finite number $r \geq 1$ of disjoint embedded 2-balls (discs). The same classifications for surfaces hold in the smooth as in the triangulated (modulo subdivision) context.

The Möbius band is a projective plane from which the interior of one open 2-ball is deleted, and the projective plane is obtained from the Möbius band by closing it with a disc.

Zero-Tight Immersions of Surfaces in the Plane. If a compact surface $(M, \partial M)$ admits an immersion $f : M \to E^2$ in the plane, it is orientable (because it gets, by virtue of the immersion, a unique orientation from an orientation of E^2) and has at least one boundary component (embedded circle), because any point on the boundary of $f(M)$ must be an image of boundary points of M by definition of immersion. For any such surface $M_g^{(r)}$ of genus g, with $r \geq 1$ interiors of disjoint discs deleted, there is an immersion in the plane, as seen in the examples in Figure 8.

Figure 8. Left: The surface with boundary obtained by removing an open disk from the torus $T^2 = M_1$ is denoted $M_1^{(1)}$. Middle: Immersion of $M_1^{(1)}$ in E^2. Right: Immersion of $M_2^{(1)}$ (genus-two surface minus a disk) in E^2.

Next let $f : X = M_g^{(r)} \to E^2$ be a 0-tight immersion. By Theorem 1.2 any topset $X_z = f^{-1}(\partial h_z)$ is connected and immersed in the line ∂h_z. So it is embedded as a convex set (a point or a segment). The union of these topsets, $f^{-1}\partial\mathcal{H}f(X)$, is then a topological circle in X embedded into the convex curve $\partial\mathcal{H}f(X)$, and this circle is one boundary component $\partial_1 X$ of $X = M_g^{(r)}$. If there are no other boundary components ($r = 1$), then $f : X \to E$ is an immersion onto the convex disc $\mathcal{H}f(X)$, and every image-point $Y \in \mathcal{H}f(X)$ is covered by the same finite number of points by the immersion f. This number is 1 because that's what it is on the boundary. Then f is an embedding onto $\mathcal{H}fX$ and X is a disc: $X = M_g^{(r)} = M_0^{(1)}$. In particular, $g = 0$. For $g \geq 1$, the 0-tight immersed surface must have at least two boundary circles, $\partial X = \partial_1 X \cup \cdots \cup \partial_r X$, and the first $f\partial_1 X = \partial\mathcal{H}fX$ is embedded and convex.

Next consider a boundary point x on another component $\partial_i X$, where $i \geq 2$. Let

$$\kappa_x : U_x \to h^2 = \{(u^1, u^2) \in \mathbb{R}^2 : u^2 \geq 0\}$$

be a chart for a connected neighborhood U_x of x in X, so small that the restriction $f|_{U_x}$ is an embedding into E. (See Figure 9.) We can assume that $f(\kappa_x^{-1}(\partial h^2))$ and $\kappa_x^{-1}(\partial h^2)$ are connected. The first of these sets is an open arc (1-manifold). By 0-tightness of f, no straight line segment $[y_1, y_2] \subset E$ with end-points y_1 and y_2 in $f(\kappa_x^{-1}(\partial h^2))$ can divide $f(U_x)$. That is, $(f(U_x)\backslash[y_1, y_2])$ must be connected. Components of $U_x \backslash [y_1, y_2]$ that touch $[y_1, y_2]$ in interior points must then be convex. Therefore $f(\kappa_x^{-1}(\partial h^2))$ is nonconvex (call it *concave*) with respect to the interior of $f(U_x)$. We conclude that f immerses each boundary circle $\partial_i(X)$ for $i \geq 2$ in a locally concave manner.

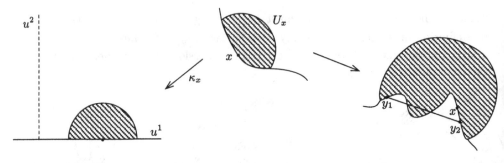

Figure 9. A 0-tight immersion of a surface with boundary in E^2 maps all but one boundary component in a concave manner.

Figure 10 shows 0-tight immersions of $M_1^{(2)}$ and $M_2^{(2)}$. Following the pattern of this figure we can conclude (and summarize):

THEOREM 1.6. *There are 0-tight immersions of the compact surfaces $X = M_g^{(r)}$ (genus g, $r \geq 1$ holes) in the plane if $g = 0$, $r \geq 1$ and if $g \geq 1$, $r \geq 2$, but not if $g \geq 1$, $r = 1$. The boundary components of $\partial X = \partial_1 X \cup \cdots \cup \partial_r X$ have one convex image $f(\partial_1 X)$ and further locally concave immersions $f : \partial_i X \to E$ for $i \geq 2$.*

EXERCISE. In Figure 10, left, the 0-tight immersion of $\mathrm{Sk}_1(\sigma_3)$ "embeds in" the 0-tight immersion of $M_1^{(2)}$.

Zero-Tight Immersions of Closed Surfaces in Space. In Figure 11 we show examples of 0-tight immersed surfaces in E^3. Every height function z is seen to have, and must have, exactly one local top set, a maximum, and one local minimum (maximum of $-z$). This characterizes 0-tightness by Theorem 1.4. We now prove for smooth immersions:

THEOREM 1.7 (smooth version). *If $f : M \to E^3$ is a 0-tight smooth immersion of a closed surface $M \neq S^2$, there is a decomposition of M as a union of surfaces with boundary $M = M^+ \cup M^-$, where M^+ is connected, $M^- = \bigcup_{j=1}^r M_j^-$, and*

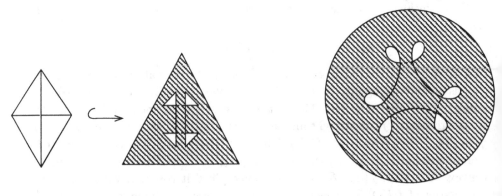

Figure 10. Zero-tight immersions of $M_1^{(2)}$ and $M_2^{(2)}$.

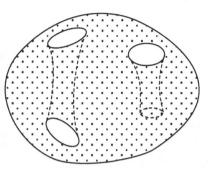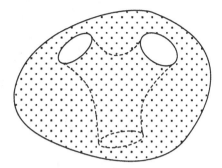

Figure 11. A 0-tightly smoothly immersed closed surface in E^3 splits into one component M^+ of positive curvature and one or more components M_j^- of negative curvature. These pieces join along plane curves X_i, called windows.

the boundary $\partial M^+ = M^+ \cap M^- = \bigcup_{i=1}^r X_i$ is a union of $r \geq 2$ closed curves with the following properties:

(i) *$M^+ \to fM^+ \subset E^3$ is embedded into the convex surface $\partial \mathcal{H} fM$ as the complement of the union of the interiors of r disjoint plane convex discs, called windows. The Gauss curvature $K(p)$ is nonnegative for $p \in M^+$ and nonpositive for $p \in M^-$.*

(ii) *No local topset $A \subset M$ is contained completely in $\mathring{M}^-$.*

(iii) *Each closed curve X_i carries an essential cycle of $H_1(M, \mathbb{Z}_2)$.*

(iv) *$\chi(M) = \chi(M^+) + \sum_{j=1}^r \chi(M_j^-)$, $\chi(M^+) \leq 0$, and $\chi(M_j^-) \leq 0$.*

(v) *$M_{K>0} = \partial \mathcal{H}(M)_{K>0}$.*

(For convenience we will write M^+ for fM^+ and X_i for fX_i.)

PROOF. The 0-tight restriction of f to an E^k-top*set X is a 0-tight immersion of a connected set onto one point for $k = 0$, and onto a straight segment for $k = 1$. So for $k = 0$ and $k = 1$, it is an embedding onto a convex set. An E^2-topset $X = M_z \subset M$ immerses 0-tightly into a subset of the plane convex set $\mathcal{H} fX$ and restricts to an embedding on (each of the points of) the circle $X' = f^{-1}\partial \mathcal{H} fX \subset X \subset M$. Suppose the circle X' bounds in M, that is $M \setminus X'$ has two components. If both components have image points under the topset level of z, then $z \circ f$ has two minima on M, and f is not 0-tight. So one of the components immerses into $\mathcal{H} fX$ and its boundary into $\partial \mathcal{H} fX$. Such an immersion is a topological covering of $\mathcal{H} fX$. As it covers each boundary point once, it is an embedding onto $\mathcal{H} fX$, and $X = f^{-1}\mathcal{H} fX$ is an E^2-topset disc $X \subset M$. The union of all convex topsets $X = fX \subset M$ so far discussed is, by definition, the set $M^+ = fM^+ \subset \partial \mathcal{H} fM$. It is bounded by plane convex curves $X' = fX' = \partial \mathcal{H} fX$, for which $X' \subset M$ is not bounding in M. The circle X' then carries a generator of $H_1(M, \mathbb{Z}_2)$ and is called an (essential) *top cycle*. This proves (iii). As (ii) is obvious, there remains the proof of the last part of (iv). Suppose M^+ or M_j^- has only one boundary component. Then this boundary component bounds in M and is a top cycle, a contradiction. $\qquad\square$

Figure 12. A 0-tight smooth immersion of the nonorientable surface of Euler characteristic -2 ($K \# T$). There are three windows (front, back, and top) and one curve of self-intersections, where the central column goes through the "ceiling".

COROLLARY. There is no smooth tight immersion of the projective plane P into E^3.

PROOF. By (iv) $\chi(M) \le 0$, but $\chi(P) = 1$. Another proof: Every closed embedded curve in P which is nonbounding has a Möbius band as neighborhood. Every top cycle for a smooth 0-tight surface in E^3 can have only a (trivial) band as neighborhood. "The M-normal bundle is trivial along the top cycle." □

For topological immersions, we mention without proof a more subtle result:

THEOREM 1.8 (topological version). *If $f : M \to E^3$ is a 0-tight C^0-immersion of a closed surface M, there is a decomposition of M as $M = M^+ \cup \mathring{M}^-$, $M^+ \cap \mathring{M}^- = \varnothing$, $\mathring{M}^- = \cup \mathring{M}_j^-$, M^- an open surface, satisfying furthermore the following properties:*

(i) *$M^+ = fM^+ \subset E^3$ is an embedded subset of M, embedded in the convex surface $\partial \mathcal{H} fM = \partial \mathcal{H} fM^+$, as the complement of the union of r disjoint interiors of plane convex sets $\mathcal{H} fX_j$. Here $X_j \subset M^+ \subset M$ is an embedded circle.*

(ii) *No local topset $A \subset M$ is completely contained in the open set $\mathring{M}^-$.*

(iii) *Each closed curve X_j compactifies one end of $\mathring{M}^-$ as a boundary.*

(iv) *$\chi(M) = \chi(M^+) + \sum_{j=1}^r \chi(M_j^-)$, $\chi(M^+) \le 0$, $\chi(M_j^-) \le 0$.*

See Figure 13 for an example, a 0-tight torus.

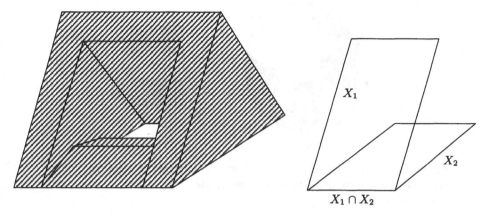

Figure 13. A 0-tightly embedded PL torus, showing that windows can intersect (Theorem 1.8).

2. Curvature and Knots

Definitions of Curvature. A *closed curve* $\gamma : S^1 \to \mathbb{R}^N$ in Euclidean space is a continuous immersion of the circle

$$S^1 = \{(\cos 2\pi t, \sin 2\pi t) : t \in \mathbb{R}\}.$$

It can be parametrized and oriented by t mod 1. For convenience, we only discuss the case $N = 3$.

A *polygon* P (*n-gon* γ_n) is a closed curve γ_n with vertices $u_i = \gamma_n(t_i) \in \mathbb{R}^3$, i mod n, for $0 \le t_1 < t_2 < \ldots < t_n < 1$, connected by straight line segments $\gamma_n(t)$, $t_{i-1} \le t \le t_i$.

The curvature on a curve is a measure. For a polygon the measure is concentrated in the vertices. Let $\alpha_i \in [0, \pi)$ be the angle between two successive edges meeting at u_i, that is the angle between the unit vectors v_{i-1} and v_i, where

$$v_j = \frac{u_j - u_{j-1}}{\|u_j - u_{j-1}\|} \in S^2, \quad j \text{ taken mod } n.$$

Choose an open interval U_i on γ_n between u_{i-1} and u_{i+1} that covers u_i. Then the *normalized curvature* on U_i is $\tau(\gamma_n, U_i) = \alpha_i/\pi$. The (normalized) *total curvature* of γ_n is $\tau(\gamma_n) = \sum_i (\alpha_i/\pi)$. We connect the points v_i and v_{i+1} in S^2 by a geodesic segment of length α_i in S^2 and obtain a "tangential image", a continuous image of an oriented circle. This leads to the following equivalent definition:

DEFINITION 2.1. The (normalized) *total absolute curvature* $\tau(\gamma_n)$ is $1/\pi$ times the length $L(\tan \gamma_n)$ of the tangential image $\tan \gamma_n : S^1 \to S^2$ in S^2.

Next we consider the normal unit vectors along the open edge from u_{i-1} to u_i. At each point they form a unit circle, which we carry over by parallel displacement to $0 \in \mathbb{R}^3$ to obtain a great circle $S^1(v_i)$ in S^2 orthogonal to $v_i \in S^2$. We rotate

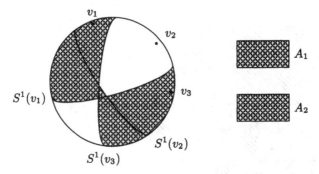

Figure 14. At a vertex of a polygonal line the direction changes from v_i to v_{i+1}. The smaller pair of segments bounded by the circles of normals $S^1(v_i)$ and $S^1(v_{i+1})$ is denoted A_i.

$S^1(v_i)$ into $S^1(v_{i+1})$ through circles $S^1(v)$, orthogonal to v, moving along the geodesic arc with ends v_i and v_{i+1} in S^2. The point set swept by the circles is denoted A_i. It consists of two congruent diametrical sectors (see Figure 14). Let $|A_i|$ be the area of A_i and $|S^2|$ the area of S^2. Clearly $|A_i|/|S^2| = \alpha_i/\pi$. So we have the following equivalent definition:

DEFINITION 2.2. Let $\tau(\gamma_n, U_i) = |A_i|/|S^2|$ be the swept-out area of the unit normal bundle on the interval U_i. The total absolute curvature is

$$\tau(\gamma_n) = \sum_i \tau(\gamma_n, U_i).$$

Critical Points. Let $z : \mathbb{R}^3 \to \mathbb{R}$ be a linear function satisfying $\|z\| = 1$, and denote the gradient of z by $z^* \in S^2$ (see right). The restriction $z|_{U_i}$ may or may not have a maximum or minimum on U_i. We assign to z (and to z^*) the number $\mu_z(\gamma_n, U_i)$ of such maxima and minima. Clearly $\mu_z(U_i) = 1$ if $z^* \in \mathring{A}_i$, the interior of A_i, and $\mu_z(U_i) = 0$ if $z^* \notin A_i$. Functions z that

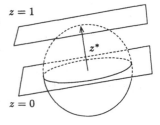

have a constant value on an edge of γ_n are called *degenerate*; the set of $z^* \in S^2$ for which this occurs has measure zero and can be neglected. Let $\mathcal{E}_z$ denote the expectation or mean value with respect to the standard SO(3)-invariant measure for z^* on S^2. If $|A_i|$ and $|S^2|$ denote the area of A_i and S^2, respectively, then clearly

$$\mathcal{E}_z(\mu_z(\gamma_n, U_i)) = |A_i|/|S^2| = \alpha_i/\pi.$$

Therefore we have another equivalent definition:

DEFINITION 2.3. The total absolute curvature is the mean number of critical points of z^* on γ, with respect to the standard measure on S^2.

$$\tau(\gamma_n, U_i) = \mathcal{E}_z \mu_z(\gamma_n, U_i), \qquad \tau(\gamma_n) = \mathcal{E}_z \mu_z(\gamma_n).$$

Total Curvature for General Closed Curves. Let $\gamma : S^1 \to \mathbb{R}^N$ be a closed continuous immersed curve. A polygon P with vertices $u_i = \gamma(t_i)$, i mod n, $0 \le t_1 \le \ldots \le t_n < 1$ is called an *inscribed polygon*; we denote this by $P \prec \gamma$. We propose:

DEFINITION 2.4. The *total curvature* of a continuous closed immersed curve $\gamma : S^1 \to \mathbb{R}^N$ is the least upper bound of $\tau(P)$ for inscribed polygons

$$\tau(\gamma) = \sup_{P \prec \gamma} \tau(P) \le \infty.$$

In this section we will recover the important Definition 2.3 for general closed curves thanks to the following result:

THEOREM 2.5. *If the total curvature $\tau(\gamma)$ of a continuous immersed curve $\gamma : S^1 \to \mathbb{R}^N$ is finite, it equals the mean number of maxima and minima:*

$$\tau(\gamma) = \mathcal{E}_z \mu_z(\gamma).$$

To begin the proof, we first prepare the definition of $\mu_z(\gamma)$ and again elaborate only the case $N = 3$. The function $z|_\gamma$ is called *degenerate* if it is constant on some maximal arc on γ. If this arc is on a straight line in $\mathbb{R}^3$, it "belongs to" at most a great circle of points $z^* \in S^2$. Such arcs can be counted by their nonincreasing lengths in the parameter t. The corresponding circles on S^2 are countable in number, and their union has measure zero in S^2. Any nonlinear planar arc "belongs to" two points z^* and $-z^*$ in S^2. Such arcs can be counted as well. It follows that the points z^* that belong to degenerate $z|_\gamma$ have a union of measure zero in S^2.

DEFINITION. *One maximum* of $z|_\gamma$ is a point or a maximal arc σ on γ in which z is constant, and such that some open neighborhood of σ gives no greater values for $z|_\gamma$. One maximum of $-z|_\gamma$ is called *one minimum* of $z|_\gamma$. If $z|_\gamma$ is constant we count one maximum and one minimum.

The total number of maxima and minima of $z|_\gamma$ is denoted $\mu_z(\gamma) \le \infty$. It is twice the number $b_z(\gamma)$ of maxima. Note that $b_z(\gamma)$ can be infinite for a nondegenerate function, when the isolated (maximal) points have one or more accumulation points.

For the proof of Theorem 2.5 we need some lemmas.

LEMMA 2.6. *If $P \prec \gamma$ is an inscribed polygon of γ, then $\mu_z(\gamma) \ge \mu_z(P)$ for all z.*

PROOF. Suppose $z|_P$ attains a maximum on P in one point, the vertex $u_i = \gamma(t_i)$. Then

$$z(\gamma(t_{i-1})) < z(\gamma(t_i)), \qquad z(\gamma(t_{i+1})) < z(\gamma(t_i)).$$

Therefore z has a maximum on the open interval $\{\gamma(t) : t_{i-1} < t < t_{i+1}\}$. If $z|_P$ attains a maximum at the maximal segment from u_i to u_{i+l} of P, then

$$z(\gamma(t_{i-1})) < z(\gamma(t_i)) = \ldots = z(\gamma(t_{i+l})) > z(\gamma(t_{i+l+1})),$$

and z has a maximum on the open interval $\{\gamma(t) : t_{i-1} < t < t_{i+l+1}\}$.

If z is constant on P the conclusion is obvious. The same applies to minima of z and the lemma follows. $\square$

COROLLARY 2.7. *If $P \prec P'$ is an inscribed polygon in the polygon P' (obtained, for example, by deleting one vertex of P'), then $\mathcal{E}_z \mu_z(P') \geq \mathcal{E}_z \mu_z(P)$, and so $\tau(P') \geq \tau(P)$.*

COROLLARY 2.8. $\mu_z(\gamma) \geq \sup_{P \prec \gamma} \mu_z(P)$.

Let $\{t_i : i = 1, 2, \ldots\}$, with $0 \leq t_i < 1$ and $t_j \neq t_h$ for $j \neq h$, be a countable dense subset of $[0, 1)$, and let $P_j \prec \gamma$ be the inscribed polygon with vertices $\gamma(t_1), \ldots, \gamma(t_j)$. Then P_j is said to *converge* to $\lim_{j \to \infty} P_j = \gamma$. From the definition of $\tau(\gamma)$ and Corollary 2.7, follows another equivalent definition of curvature:

DEFINITION 2.9. $\tau(\gamma) = \lim_{P \to \gamma} \tau(P) \leq \infty$.

COROLLARY 2.10. $\mu_z(\gamma) \geq \lim_{P \to \gamma} \mu_z(P)$.

Next suppose $z|_\gamma$ is nondegenerate. Then $z|_{P_j}$ has a maximum (minimum) as near as we please to any finite number of isolated maxima (minima) of $z|_\gamma$, for j sufficiently large. Therefore:

LEMMA 2.11. *If $z|_\gamma$ is nondegenerate, then $\mu_z(\gamma) \leq \lim_{P \to \gamma} \mu_z(P) \leq \infty$.*

Theorem 2.5 follows from Corollary 2.10 and Lemma 2.11.

EXERCISE. Define suitably the total curvature of an immersed closed curve in $\mathbb{R}^N$ and an immersed arc $\gamma : \{t : 0 \leq t \leq 1\} \to \mathbb{R}^N$, so that Theorem 2.5 applies also to arcs. *Hint:* Use $\mu_z(\text{interior of arc } \gamma)$ instead of $\mu_z(\gamma)$ to avoid curvature at the end points.

EXERCISE. Consider the arc $\gamma \subset \mathbb{R}^2$ defined by

$$u = (u^1, u^2) = \begin{cases} (t, 0) & \text{for } -1 \leq t \leq 0, \\ (t^\alpha, \sin 2\pi t^{-1}) & \text{for } 0 < t \leq 1. \end{cases}$$

Prove that:

(a) $\gamma(t)$ has a tangent for all t if and only if $\alpha > 1$.
(b) The tangent depends continuously on t if and only if $\alpha > 2$.
(c) $\tau(\gamma) < \infty$ if and only if $\alpha > 3$.
(d) γ has continuous second derivatives if and only if $\alpha > 4$.

Hint: Consider by way of comparison the inscribed polygons P_j, $j \to \infty$, with vertices for $t = 0$, $t = -1$, $t^{-1} = 1$, and for $t^{-1} = (1 + 2i)/4$ for $i = 4, 5, \ldots, j$.

As a corollary of Theorem 2.5 we have Fenchel's Theorem:

THEOREM 2.12 [Fenchel 1929]. *The total curvature of a continuous closed immersed curve $\gamma : S^1 \to \mathbb{R}^N$ is $\tau(\gamma) \geq 2$. Equality is attained if and only if γ is plane convex.*

Fenchel proved this for smooth curves in $\mathbb{R}^3$, Borsuk [1947] for $N \geq 3$.

PROOF. Any continuous function on the circle has at least one maximum and one minimum. So $\mu_z \geq 2$ for all $z^* \in S^2$, therefore

$$\tau(\gamma) = \mathcal{E}_z \mu_z(\gamma) \geq \mathcal{E}_z(2) = 2.$$

If $\tau(\gamma) = 2$, then $\tau(P) = 2$ for any inscribed polygon P as well. So $\mu_z(P) = 2$ and P has Banchoff's two-piece property, is 0-tight, hence plane convex by Theorem 1.5. Adding vertices yields polygons, say P_j, converging to γ. All P_j are convex in the same plane. So γ is plane convex, $N = 2$. □

Next we prepare the generalization of the curvature as π^{-1} times the length of a tangential image, a result due to van Rooij [1965]. If $u(t) \in \mathbb{R}^3$ is an immersed arc or a closed curve in $\mathbb{R}^3$ and

$$\frac{\big(u(t) - u(t_0)\big)(t - t_0)}{\|u(t) - u(t_0)\| \, |t - t_0|}$$

converges to a unit vector $v(t_0)$ for $t \to t_0+$, $t \to t_0-$ or $t \to t_0$, then $v(t_0)$ is called a *right, left* or *general tangent unit vector*, respectively. It is denoted by $\dot{u}_+(t_0), \dot{u}_-(t_0)$, and $\dot{u}(t_0)$, respectively. Here the dot does not (yet) mean differentiation! The parallel lines through $u(t_0)$ are called *right, left* and *general tangents*, respectively. A plane convex curve has a right and left tangent at every point.

LEMMA 2.13. *Let* $\gamma(t) = u(t)$, $0 \leq t < 1$, $t \bmod 1$, *be a continuous closed curve in* $\mathbb{R}^3$ *with* $\tau(\gamma) < \infty$. *Then:*

(i) γ *has a right and left tangent at every point* t.
(ii) *The set of points* t *for which* $\dot{u}_+(t) \neq \dot{u}_-(t)$ *is countable.*
(iii) *The right (left) tangent vector is continuous on the right (left).*

PROOF. (i) If $u(t)$ has no right tangent for $t = 0$, at say $u(0) = 0 \in \mathbb{R}^3$, there exists a sequence $t_1 > t_2 > \ldots > 0$ converging to 0 and unit vectors v and $w \neq v \in S^2$ such that

$$u(t_j)/\|u(t_j)\| \to \begin{cases} v & \text{for } j = 2i - 1 \to \infty, \\ w & \text{for } j = 2i \to \infty. \end{cases}$$

We can assume moreover that

$$\|u_{j+1}\| < \tfrac{1}{100}\|u_j\| \quad \text{for } j \geq 1.$$

Then the inscribed polygon P_k with vertices $\gamma(t_1), \ldots, \gamma(t_{k-1})$ and $\gamma(0) = 0$ has curvature $\tau(P_k)$ converging to ∞ for $j \to \infty$, a contradiction.

(ii) If $\dot{u}_-(t_0)$ and $\dot{u}_+(t_0)$ form an angle $\alpha(t_0) > 0$, then a polygon with vertices at $\gamma(t_0 - \delta), \gamma(t_0)$ and $\gamma(t_0 + \delta)$ will contribute, in the limit for $\delta \to 0$, an amount $\alpha(t_0)/\pi$ to $\tau(\gamma)$. The sum of such amounts is bounded by $\tau(\gamma)$. We can count such vertices by their nonincreasing contributions to $\tau(\gamma)$. The set of values t

for which $\dot{u}_+(t_0) \neq \dot{u}_-(t_0)$ is therefore countable. All other points have a unique tangent $\dot{u}(t) = \dot{u}_+(t) = \dot{u}_-(t)$.

(iii) If $\dot{u}_+(t)$ is not continuous on the right at $u(0) = 0 \in \mathbb{R}^3$ for $t = 0$, then there exist v and $w \neq v \in S^2$ and $t_j > t_{j+1} > \ldots > 0$ converging to 0, with limits

$$\dot{u}_+(t_{2i-1}) \to v, \quad \dot{u}_+(t_{2i}) \to w, \quad \text{for } i \to \infty,$$

and we can assume

$$\left\| \frac{u(t_{2i}) - u(t_{2i-1})}{\|u(t_{2i}) - u(t_{2i-1})\|} - \dot{u}_+(t_{2i-1}) \right\| < \tfrac{1}{10}\|v - w\| \quad \text{for all } i \geq 1.$$

The inscribed polygons P_j with vertices $u(t_i)$, $i = 1, \ldots, j$, and $u(0) = 0 \in \mathbb{R}^3$ have unbounded curvatures $\tau(P_j)$ for $j \to \infty$, a contradiction. $\qquad\square$

Let $\gamma(t) = u(t)$ be as before and assume $\tau(\gamma) < \infty$. Let $u(t_i)$, for $i = 1, \ldots, h$, $0 \leq h \leq \infty$, $0 < t_i < 1$, be the points where $\dot{u}_-(t_i)$ and $\dot{u}_+(t_i)$ form a positive angle. Insert an interval of length 2^{-i} at t_i in the parameter space of $t \in \mathbb{R}$ mod 1, so as to obtain a parameter space with a new parameter $t' \in \mathbb{R}$ modulo $(1 + \sum_{i=1}^h 2^{-i})$, $0 \leq h \leq \infty$. We define as follows a map $\tan \gamma$ that generalizes the tangential map for polygons. It has as embedded image the geodesic arc in S^2 between $\dot{u}_-(t)$ and $\dot{u}_+(t)$ for an inserted interval at $t = t_i$, where $\dot{u}_-(t) \neq \dot{u}_+(t)$. For other points, it has as image $\dot{u}(t) = \dot{u}_-(t) = \dot{u}_+(t)$. It is continuous in t' and is again called the *tangential map* $\tan \gamma : S^1 \to S^2$. We recover Definition 2.1 in this result:

THEOREM 2.14 [van Rooij 1965]. *If $\tau(\gamma) < \infty$, there is a continuous tangential map $\tan \gamma : S^1 \to S^2$. The length of the image $L(\tan \gamma)$ is bounded, and the total curvature $\tau(\gamma)$ is $\tau(\gamma) = L(\tan \gamma)/\pi$. (Here the length $L(\tan \gamma)$ is defined to be the least upper bound of the length of inscribed polygons.)*

PROOF. Let γ have the parameter t, $0 \leq t \leq 1$. Consider a sequence of j-gons $P_j \prec \gamma$ with vertices at $u(t_{2i-1,j})$ for $i = 1, \ldots, j$, with $0 < t_{2i-1,j} < t_{2i+1,j} < 1$. Suppose P_j converges to γ for $j \to \infty$. For each P_j add vertices and obtain a $2j$-gon $\bar{P}_{2j}$ with vertices at $u(t_{i,j})$ for $i = 1, \ldots, 2j$, $0 < t_{i,j} < t_{i+1,j} < 1$ and such that

$$\left\| \frac{u(t_{2i,j}) - u(t_{2i-1,j})}{\|u(t_{2i,j}) - u(t_{2i-1,j})\|} - \dot{u}_+(t_{2i-1,j}) \right\| < \varepsilon_j.$$

The points $\dot{u}_+(t_{2i-1,j})$ are image points of $\tan \gamma$. They determine an "inscribed polygon" T_j for $\tan \gamma$ with length $L(T_j)$. Note that some edges of T_j may be degenerated to a point. Clearly $L(T_j)$ has as limit

$$\lim_{j \to \infty} L(T_j) = L(\tan \gamma).$$

If ε_j is chosen very small, half of the points of $\tan P_{2j}$ are as near as we please to the vertices of T_j in S^2. Therefore we can assume

$$(1 - 2^{-j})L(T_j) \leq L(\tan \bar{P}_{2j}) = \tau(\bar{P}_{2j}),$$

and in the limit

$$L(\tan\gamma) \le \tau(\gamma) < \infty.$$

To prove the reverse inequality we start by observing that the properties (i)–(iii) in Lemma 2.11 imply that $\gamma(t)$ is rectifiable: it has length $L(\gamma) = \lim_{P \to \gamma} L(P)$ and can be parametrized by arclength $s(t)$ such that the right-hand derivative of $u(s)$ exists and equals $\left(du(s)/ds\right)_+ = \dot{u}_+(t)$, $s = s(t)$.

For convenience suppose $L(\gamma) = 1$, $s \equiv t$. Let $Q_j(t)$ be the unique polygonal arc in $\mathbb{R}^3$ for $0 \le t \le 1$, with vertices for $t = i2^{-j}$ for $i = 0, \dots, 2^j$ with edges of length 2^{-j} and such that the initial point is

$$Q(0) = u(0) = \gamma(0)$$

and

$$Q_j((i+1)2^{-j}) = Q_j(i2^{-j}) + 2^{-j}\dot{u}_+(i2^{-j}) \quad \text{for } i = 0, 1, \dots, 2^{j-1}.$$

If $z|_\gamma$ is nondegenerate, one can verify that

$$\lim_{j \to \infty} \mu_z(Q_j) \ge \mu_z(\gamma).$$

But $\mathcal{E}_z\mu_z(\gamma) = \tau(\gamma)$, and $\mathcal{E}_z\mu_z(Q_j) = \tau(Q_j) = L(Q_j)$ which for $j \to \infty$ converges to $L(\tan\gamma)$. Therefore

$$L(\tan\gamma) \ge \tau(\gamma).$$

This completes the proof of van Rooij's Theorem 2.14. $\qquad\square$

COROLLARY. Let $\gamma(t) = u(t)$ be a closed curve of class C^2. The usual definition of total curvature and our definition are equivalent:

$$\frac{1}{\pi}\int\left\|\frac{d^2u}{ds^2}\right\| ds = \frac{1}{\pi}L(\tan\gamma) = \tau(\gamma) = \mathcal{E}_z\mu_z(\gamma).$$

PROOF. If s is the arclength on γ, then $(\tan\gamma)(s) = du/ds \in S^2$, and the length of the curve $\tan\gamma$ is $\int \|d^2u/ds^2\| ds$. $\qquad\square$

EXERCISE. Use each of the definitions to calculate $\tau(\gamma)$ for the plane closed curve on the right, consisting of three semicircles (see Definition 2.4, Theorem 2.5 and Theorem 2.12).

EDITORS' NOTE. For results on the total curvature of knots, see [Fáry 1949; Milnor 1950; 1953]. For knotted surfaces, see [Kuiper and Meeks 1983; 1984; 1987].

3. Smooth Submanifolds

Definitions of Curvature. In Section 1 we defined topological n-manifolds M with charts $\kappa_x : U_x \to \kappa_x U_x \subset h^n \subset \mathbb{R}^n$. A two-manifold is called a *surface*. If the homeomorphisms

$$\kappa_{yx} = \kappa_y \kappa_x^{-1} : \kappa_x(U_x \cap U_y) \to \kappa_y(U_x \cap U_y)$$

all belong to the pseudogroup C^k (continuous k-th derivatives, for $k \leq \infty$), C^ω (analytic), or C^{Nash} (the graph of κ_{yx} in $\mathbb{R}^n \times \mathbb{R}^n$ is a nonsingular part of a real algebraic variety given by polynomial equations), then M has a C^k-*structure*, and is called a C^k-*manifold*. (Here k can be a nonnegative integer, ∞, ω or Nash.) Two C^k-manifolds M_1 and M_2 are C^k-*equivalent* if there is a homeomorphism $\varphi : M_1 \to M_2$ that is expressed in C^k-charts in a commutative diagram

$$
\begin{array}{ccc}
x \in U & \xrightarrow{\;\kappa_x\;} & \kappa_x(U) \subset \mathbb{R}^n \\
\Big\downarrow \varphi & & \Big\downarrow \psi \\
y \in V & \xrightarrow{\;\kappa_y\;} & \kappa_y(V) \subset \mathbb{R}^n
\end{array}
$$

where ψ is C^k. If $l < k$ then $M \in C^k$ determines a unique (up to equivalence) C^l-structure. Here $0 < 1 < \ldots < \infty < \omega < \text{Nash}$.

Another pseudogroup on $\mathbb{R}^n$, called PL, consists of homeomorphisms $\psi : U_1 \to U_2$, with U_1, U_2 open subsets in $\mathbb{R}^n$, that are affine on each affine simplex of a finite triangulation of some compact part of $\mathbb{R}^n$. Manifolds with such a structure are called *piecewise linear manifolds* or *PL-manifolds*. Obviously $0 < \text{PL}$.

For closed surfaces all C^k-equivalence classes for $k < \omega$ correspond one-to-one to all C^0-equivalence classes. Recall from Section 1 the classification of smooth closed surfaces, summarized in Figure 7.

The $\mathbb{Z}_2$-Betti numbers $\beta_i = \text{rank}\, H_i(M, \mathbb{Z}_2)$ of the surface are $\beta_0 = \beta_2 = 1$, $\beta_1 = 2 - \chi$. The alternating sum $\beta_0 - \beta_1 + \beta_2$ is χ, the Euler characteristic. The sum of the Betti numbers is $\beta = \beta_0 + \beta_1 + \beta_2 = 4 - \chi$. Note that the height function z on the smooth surfaces of Figure 7 has exactly β nondegenerate critical points, the minimum possible number. Note also that the surface obtained from M by deleting an open disc can be contracted over itself to a wedge of β_1 circles $S^1 \vee S^1 \vee \cdots \vee S^1$. These circles carry generators for $H_1(M, \mathbb{Z}_2)$. A C^k-map $f : M^n \to \mathbb{R}^N$ of an n-manifold is called a C^k-*immersion* for $k \geq 1$ if its derivative has maximal rank n. It is then clearly a tame immersion in the topological sense.

REMARK. There is no smooth (or even topological) embedding of a closed nonorientable surface in $\mathbb{R}^3$.

PROOF. Let $M \subset \mathbb{R}^3$ be such a smooth embedding. Take a small orthogonal vector v at a point $p \in M$. Move p along a closed embedded curve γ in M and drag v along, so that on returning to p the vector has the opposite direction. Move γ away from M in the direction of the transported vectors v. The end points at p of the curve γ' so obtained are connected by a straight segment orthogonal to M. Now you have a closed curve in $\mathbb{R}^3$ that meets M in one (odd number) point. Make it a smooth curve and move it far away into space $\mathbb{R}^3$, but in such a way that the number of intersection points changes each time by two ("transversal move"). The final curve γ'' meets M in an odd number of points, but also in no point, a contradiction. $\qquad\square$

If $f : M \to \mathbb{R}^3$ is a smooth immersion, there is a choice of charts near $p \in M$ and orthogonal Euclidean coordinates z, x, y for $E^3 = \mathbb{R}^3$ so that the surface has near P an equation

$$z = \tfrac{1}{2}(k_1 x^2 + k_2 y^2) + O(x^2 + y^2)^{3/2}.$$

The numbers k_1 and k_2 are the *principal curvatures* and their product is the *Gaussian curvature* (a density) $K = k_1 k_2$. It can be positive, zero or negative.

The *normalized absolute curvature* of an open set $U \subset M$ is defined by

$$\tau(U) = \int_U \frac{|K \, d\sigma|}{2\pi}$$

where $d\sigma$ is the volume element (a two-form) induced by the immersion in $\mathbb{R}^3$. It differs from the classical curvature by a normalization factor.

Consider all unit vectors (p, z^*) at points $p \in U$ orthogonal at $f(p)$ to $f(U) \subset \mathbb{R}^3$, for some small U for which $f : U \to f(U)$ is an embedding. These vectors, displaced parallel to $(p, 0)$, form a surface $\mathcal{N}(U) \subset M \times S^2 \subset M \times \mathbb{R}^3$ consisting of two parts, one for each side of $f(U)$ in $\mathbb{R}^3$. There is a natural projection

$$\gamma : \mathcal{N}(U) \to S^2.$$

The surface $\mathcal{N}(U)$ is called the *unit normal bundle* of M on U. If $K > 0$ on U (or $K < 0$ on U) then γ sweeps out a set $A \subset S^2$ consisting of two diametrically opposite parts. See Figure 15. Each is called a *Gauss-map image* of U. It is well

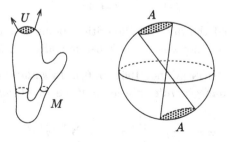

Figure 15. The Gauss map.

known that its total area $|A|$ determines the curvature:

$$\tau(U) = \int_U \frac{|K\, d\sigma|}{2\pi} = |A|/4\pi = |A|/|S^2|.$$

Another expression for the measure τ is

$$\tau(U) = c^{-1} \int_{\mathcal{N}(U)} |\nu^*(\omega)|,$$

where $c = |S^2| = 4\pi$, and $\nu^*(\omega)$ is the pull-back of the volume form ω on S^2. The unit normal bundle space $\mathcal{N}$ of f over M,

$$\mathcal{N} = \mathcal{N}(f) = \{(p, z^*) : z^* \text{ orthogonal to } f(M) \text{ at } f(p)\} \subset M \times \mathbb{R}^3$$

is a double covering of the surface M. It is connected if and only if M is nonorientable. We thus derive an equivalent definition of the total absolute curvature, in terms of the swept-out area of unit normal bundle:

DEFINITION 3.1. The *total absolute curvature* of a smooth immersion $f : M \to \mathbb{R}^3$ is

$$\tau(f) = c^{-1} \int_{\mathcal{N}} |\nu^*(\omega)| = \int_M |K\, d\sigma|/2\pi.$$

REMARK. A smooth C^∞-map ν such as our $\nu : \mathcal{N} \to S^2$ is called *critical* at a point $q \in \mathcal{N}$ if its derivative $d\nu$ does not have maximal rank. An image of such a point q is called a *critical value*. The theorem of Sard says that the set of critical values (here in S^2) has measure zero. A generalization by A. P. Morse [1939] says more, namely that the same conclusion holds if ν is a C^1-map between manifolds of the same dimension.

Morse's Lemma and Inequalities. A smooth function $\varphi : M \to \mathbb{R}$ is *non-degenerate* at a critical point $p \in M(d\varphi(p) = 0)$ if $d^2\varphi(p)$ has maximal rank. (Recall that $d^2\varphi(p)$ is expressed in coordinates by the matrix $\{\partial^2\varphi/\partial x^i \partial x^j\}$.) By linear algebra one has for some coordinate chart near p:

$$\varphi = \varphi(p) - \sum_{j=1}^{i}(x^j)^2 + \sum_{k=i+1}^{n}(x^k)^2 + O(\|x\|^2).$$

The number i is called the *index* of the critical point. We can do even better with the various results knows as *Morse's Lemma*, after M. Morse:

LEMMA 3.2 (C^∞ VERSION [Milnor 1963, p. 6]). *In suitable C^∞-coordinates, φ is expressed near a nondegenerate critical point $p \in M$ of index i by*

$$\varphi = \varphi(p) - \sum_{j=1}^{i}(x^j)^2 + \sum_{k=i+1}^{n}(x^k)^2.$$

LEMMA 3.2 (C^h VERSION [Kuiper 1966]). *A C^h-function $(2 \leq h \leq \omega)$ φ on M^n can with suitable C^{h-1} coordinates be expressed near a nondegenerate critical point $p \in M$ of index i by*

$$\varphi = \varphi(p) - \sum_{j=1}^{i} (x^j)^2 + \sum_{k=i+1}^{n} (x^k)^2.$$

So for C^h-functions $(2 \leq h < \infty)$ one loses one class of differentiability!

The function $\varphi : M \to \mathbb{R}$ is called *nondegenerate* if it is so at every critical point.

The C^h-Morse lemma for $2 < h < \omega$ also holds on infinite-dimensional manifolds modelled on Hilbert space [Palais 1963].

Let $f : M^2 \to \mathbb{R}^3$ be our smooth immersion. Consider as before linear functions $z : \mathbb{R}^3 \to \mathbb{R}$, $\|z\| = 1$. The real function $z \circ f : M \to \mathbb{R}$ is the *restriction* of z to M. It is clearly nondegenerate at points $p \in M$ where $K(p) \neq 0$. Take a small disc U on M where $K(p) > 0$ and count, for every point $z^* \in S^2$, the number of critical points $\mu_{iz}(U)$ of index i of the function z on U. See Figure 16. The sum $\mu_z(U) = \sum_i \mu_{iz}(U)$ is one or zero. It is one in the case where $z^* \in \mathring{A}$ and zero in the case where $z^* \notin A$. The value of the function $\mu_z(U)$ for $z^* \in S^2$ has therefore a mean (expectation) of

$$\tau(U) = \mathcal{E}_z \mu_z(U) = c^{-1} \int_{S^2} \mu_z(f)|\omega| = |A|/4\pi.$$

In this way we get positive contributions near points $p \in M$ where $K > 0$, or where $K < 0$. These points form an open set on M. We can now state the definition of total curvature in terms of the mean number of critical points:

DEFINITION 3.3. *The total curvature of the smooth (or C^2) immersion $f : M \to \mathbb{R}^3$ is*

$$\tau(f) = \mathcal{E}_z \mu_z(f) = c^{-1} \int_{S^2} \mu_z(f)|\omega|. \tag{3.1}$$

To show that this is equivalent to Definition 3.1, observe that the set $z^* \in S^2$ for which $z \circ f$ is a degenerate smooth (or C^2) function is just the set of critical values of the C^∞ (or C^1) map $\nu : \mathcal{N} \to S^2$. This set has measure zero and can

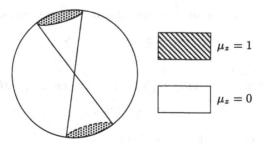

Figure 16. The number of critical points of z in a neighborhood $U \subset M$, as a function of z^*.

be neglected in the integral. The equality then follows from our discussion of the parts $K > 0$ and $K < 0$ of M.

We now formulate Definitions 3.1 and 3.3 for dimension n, but leave the proof of their equivalence to the reader.

Let $f : M \to \mathbb{R}^N$ be a smooth (or C^2) immersion of a closed manifold, $\mathcal{N} = (p, z^*) \subset M \times S^{N-1}$ the unit normal bundle, an S^{N-n-1}-bundle over M, and $\nu : \mathcal{N} \to S^{N-1}$ the projection into $S^{N-1} \subset \mathbb{R}^N$.

DEFINITION 3.4. The total (absolute) curvature of the smooth immersion $f :$ $M^n \to \mathbb{R}^N$ is

$$\tau(f) = c^{-1} \int_{\mathcal{N}} |\nu^*(\omega)| = \int_M |\tau(p)\,d\sigma|, \qquad c = |S^{N-1}|.$$

Here $\tau(p)$ is a continuous curvature density function, and $d\sigma$ is the volume element of M induced by the immersion f. $\tau(p)$ is obtained by integration over the fibres. For hypersurfaces it corresponds to $K = k_1 \cdots k_{N-1}$, with principal curvatures $k_1, \ldots, k_{N-1}$.

DEFINITION 3.5. The total curvature of the smooth immersion $f : M^n \to \mathbb{R}^N$ is

$$\tau(f) = \mathcal{E}_z \mu_z(f) = c^{-1} \int_{z^* \in S^{N-1}} \mu_z(f)|\omega|.$$

Clearly the definition implies

$$\tau(f) \geq \min_z \mu_z(f),$$

where the minimum is taken over all z such that $z \circ f$ is nondegenerate. If $\gamma^{(N)} = \gamma(M^n, \mathbb{R}^N)$ is the minimum of $\min_z \mu_z(f)$ for all immersions f and γ is the minimal number of critical points a nondegenerate function $\varphi : M^n \to \mathbb{R}$ can have, then $\min_z \mu_z(f) \geq \gamma^{(N)} \geq \gamma^{(N+1)} = \gamma \geq \beta$. Here $\gamma^{(N+1)} = \gamma$ because we can introduce the optimal φ as the $(N+1)$-st coordinate for $\mathbb{R}^{N+1}$. By the Morse inequalities (see Theorem 3.8 below) one has $\mu_z(f) \geq \beta$ for $\beta = \beta(F) = \operatorname{rank} H_*(M, F)$ for any coefficient field F. For surfaces in $\mathbb{R}^3$ we saw in Figure 7 immersions with $\mu_z(f) = \beta$, for z pointing in the vertical direction. Thus we can conclude part (i) of next theorem:

THEOREM 3.6. (i) *The total curvature of a smooth immersion $f : M^n \to \mathbb{R}^N$ is*

$$\tau \geq \gamma^{(N)} \geq \gamma^{(N+1)} = \gamma \geq \beta.$$

For surfaces, $\gamma \geq \beta$.

(ii) *The greatest lower bound of $\tau(f)$ for all immersions $f : M^n \to \mathbb{R}^N$ is*

$$\inf_f \tau = \gamma^{(N)} \geq \gamma \geq \beta.$$

For surfaces in $\mathbb{R}^N$, with $N \geq 3$, this is

$$\inf_f \tau = \beta = 4 - \chi.$$

REMARK. Sharpe [1989] proved that $\gamma^{(N)} = \gamma$ for $n > 5$.

Part (ii) of Theorem 3.6 follows from the next result:

LEMMA 3.7 [Kuiper 1959; Wilson 1965]. *Let $z \circ f$ be nondegenerate with k critical points for the smooth immersion $f : M^n \to \mathbb{R}^N$ and let $f_\varepsilon = g_\varepsilon \circ f$ where for $\varepsilon > 0$ we define g_ε by*

$$g_\varepsilon(x, z) = g(\varepsilon x, z), \quad x = (u^1, \dots, u^{N-1}) \in \mathbb{R}^{N-1}, \quad z = u^N, \quad (x, z) \in \mathbb{R}^N.$$

Then $\lim_{\varepsilon \to 0} \tau(f_\varepsilon) = k$.

Motivation: for $\varepsilon \to 0$, it seems that all curvature gets concentrated around the critical points of $z \circ f_\varepsilon$ (which correspond, one to one, to those of $z \circ f$) and they contribute k in $\tau(f_\varepsilon)$.

PROOF OF LEMMA 3.7. Any unit vector in S^{N-1} can be written

$$\cos \theta\, z^* + \sin \theta\, w^*, \quad \text{for some } w^* \in S^{N-2} \subset \mathbb{R}^{N-1} \subset \mathbb{R}^N.$$

The volume element on S^{N-1} is

$$\omega_{N-1} = \omega_{N-2} \wedge \cos^{N-1} \theta\, d\theta.$$

There exists θ_0 near $\pi/2$ such that $\mu_z(f) = k$ for all z^* with $|\theta(z^*)| > \theta_0$.

The mapping g_ε induces a map of tangent hyperplanes. A normal vector $\cos \theta\, z^* + \sin \theta\, w^*$, is replaced by a new normal vector $\cos \theta'\, z^* + \sin \theta'\, w^*$, where $\tan \theta' = \varepsilon \tan \theta$; hence

$$\frac{d\theta'}{\cos^2 \theta'} = \varepsilon \frac{d\theta}{\cos^2 \theta}$$

and

$$\cos^{N-1} \theta'\, d\theta' = \varepsilon\, \frac{\cos^{N+1} \theta'}{\cos^{N+1} \theta}\, \cos^{N-1} \theta\, d\theta.$$

For $|\theta| < \theta_0$ one finds for the volume element ω on S^{N-1}:

$$|\omega'| = \varepsilon \left(\frac{\cos \theta'}{\cos \theta} \right)^{N+1} |\omega| < \frac{\varepsilon}{(\cos \theta_0)^{N+1}} |\omega|.$$

Consider the contributions of the two parts defined by $|\theta| > \theta_0$ and $|\theta| < \theta_0$ in the integral $\tau(f)$ in (3.1). The first part is replaced by an integral of the constant k over the part $|\theta'| > \theta_0'$. As θ_0' converges to zero this converges, for $\varepsilon \to 0$, to k. The second part is replaced by an integral which is smaller than $\varepsilon / \cos^{N+1} \theta_0$ times what it was. It converges to zero for $\varepsilon \to 0$. This proves the lemma

$$\lim_{\varepsilon \to 0} \tau(f_\varepsilon) = k. \qquad \square$$

THEOREM 3.8 (MORSE INEQUALITIES). *Let $\varphi : M^n \to \mathbb{R}$ be a smooth nondegenerate function with μ_k critical points of index k and with Betti numbers $\beta_k =$ rank $H_k(M, F)$ for a field F (we mainly use $\mathbb{Z}_2 = \mathbb{Z}/2\mathbb{Z}$). Suppose the critical points have different values. Then*

$$\frac{\sum_0^n \mu_k t^k - \sum_0^n \beta_k t^k}{1 + t} = \sum_0^n K_k t^k$$

is a polynomial in t with nonnegative integer coefficients $K_k \geq 0$. In particular,

(i) $\mu_i \geq \beta_i$;
(ii) $\mu \geq \beta$ $(\mu = \sum \mu_i, \beta = \sum \beta_i)$;
(iii) $\sum(-1)^i \mu_i = \sum(-1)^i \beta_i = \chi$ $(t = -1)$;
(iv) $\sum_0^n (\mu_i - \beta_i) t^i (1 - t + t^2 - \cdots) = \sum_l K_l t^l$;
(v) $(\mu_l - \beta_l) - (\mu_{l-1} - \beta_{l-1}) + \cdots + (-1)^l (\mu_0 - \beta_0) = K_l \geq 0.$

PROOF. Let $B \supset A$ be a pair of spaces with finite-dimensional homology groups. Then we have the exact triangle

$$H_*(A) \xrightarrow{\quad i \quad} H_*(B)$$

$$\partial \qquad \qquad \alpha$$

$$H_*(B, A)$$

or, in detail,

$$H_k(A) \xrightarrow{i_k} H_k(B) \xrightarrow{\alpha_k} H_k(B, A) \xrightarrow{\partial_k} H_{k-1}(A) \xrightarrow{i_{k-1}} H_{k-1}(B)$$

where $H_*(A) = \bigoplus_k H_k(A)$, etc. Exactness means that $\text{Im } i = \text{Ker } \alpha$, etc. The commutative groups then split into direct sums

$$H_k(A) \simeq \text{Ker } i_k \oplus \text{Ker } \alpha_k,$$
$$H_k(B) \simeq \text{Ker } \alpha_k \oplus \text{Ker } \partial_k,$$
$$H_k(B, A) \simeq \text{Ker } \partial_k \oplus \text{Ker } i_{k-1}$$

Define Poincaré polynomials in t:

$$P(A) = \sum_k \dim H_k(A) t^k,$$
$$P(B) = \sum_k \dim H_k(B) t^k,$$
$$P(B, A) = \sum_k \dim H_k(B, A) t^k,$$

$$P(\operatorname{Ker} i) = \sum_k \dim(H_k(A) \cap \operatorname{Ker} i)t^k,$$

$$P(\operatorname{Ker} \alpha) = \sum_k \dim(H_k(B) \cap \operatorname{Ker} \alpha)t^k,$$

$$P(\operatorname{Ker} \partial) = \sum_k \dim(H_k(B, A) \cap \operatorname{Ker} \partial)t^k.$$

Note that ∂ decreases the dimension from k to $k - 1$. We calculate:

$$P(A) = P(\operatorname{Ker} i) + P(\operatorname{Ker} \alpha),$$
$$P(B) = P(\operatorname{Ker} \alpha) + P(\operatorname{Ker} \partial),$$
$$P(B, A) = P(\operatorname{Ker} \partial) + tP(\operatorname{Ker} i).$$

Therefore

$$P(B, A) - \big(P(B) - P(A)\big) = (1 + t)P(\operatorname{Ker} i). \tag{3.2}$$

This shows that:

LEMMA 3.9 (PRELIMINARY MORSE FORMULA). *The expression*

$$\frac{P(B, A) - \big(P(B) - P(A)\big)}{1 + t} = P(\operatorname{Ker} i)$$

is a polynomial $\sum K^k t^k$ *with integer coefficients* $K_k \geq 0$.

Next we discuss a smooth function $\varphi : M \to \mathbb{R}$ on the closed n-manifold M. Recall that the critical set is defined as $\operatorname{Cr}(\varphi) = \{p \in M : d\varphi = 0\} \subset M$. The set of critical values $\varphi(\operatorname{Cr}(\varphi))$ has measure zero on $\mathbb{R}$ by the Theorem of Sard (this is also true for a C^r-function for $r \geq n$). Assume, as in the case of Theorem 3.8, that the critical values form a finite set. This is also the case when φ is an algebraic function on an algebraic manifold.

We define the *croissant*

$$M_t = \{p \in M : \varphi(p) \leq t\}$$

and its interior

$$M_{t-} = \{p \in M : \varphi(p) < t\}.$$

Let $c_1 < c_2 < \cdots < c_L$ be the critical values of φ. Consider an arbitrary Riemannian metric on M. Then φ has as differential the one-form $d\varphi$, whose dual with respect to the metric is the gradient $\operatorname{grad} \varphi = *d\varphi$. It vanishes at p if and only if p is a critical point. Integral curves of the vector field $\operatorname{grad} \varphi$ are transversal (not tangent) to the levels of φ, except at critical points. Then we see that, *if $c_k < a < b < c_{k+1}$, the croissant M_b is diffeomorphic to M_a and*

$$H_k(M_b, M_a) = 0, \quad H_*(M_b) = H_*(M_a).$$

In other words, the diffeomorphy type of M_t can only change at critical values. There are at most $2L$ types, namely for $t < c_1$, $t = c_1$, $c_1 < t < c_2$, ..., $t = c_L$. The first type is the empty set, the last type is that of M. The open manifolds

M_{t^-} have at most $L+1$ types, namely for $t \le c_1$, $c_1 < t \le c_2, \ldots, c_{N-1} < t \le c_N$, $t \ge c_N$. They carry at most $L + 1$ homotopy types. Note that the real function $\varphi : u \to u^3$ on $M = \mathbb{R}$ has only one type for $M_{t^-} = \{p : \varphi(p) < t\}$. Let $t_0 < c_1 < t_1 < \cdots < c_L < t_L$, and consider the sequence of spaces

$$\varnothing = M_{t_0} \overset{i_1}{\subset} M_{t_1} \cdots \overset{i_L}{\subset} M_{t_L} = M.$$

By (3.2) we have

$$\frac{P(M_{t_j}, M_{t_{j-1}}) - \left(P(M_{t_j}) - P(M_{t_{j-1}})\right)}{1+t} = P(\operatorname{Ker} i_j).$$

Summation for $j = 1, \ldots, L$ yields

$$\frac{\sum_{j=1}^{L} P(M_{t_j}, M_{t_{j-1}}) - P(M)}{1+t} = \sum_{j=1}^{L} P(\operatorname{Ker} i_j). \tag{3.3}$$

We now take the assumptions of Theorem 3.8; there is exactly one nondegenerate critical point in $\{p \in M : t_{j-1} < \varphi(p) < t_j\}$. Suppose it has index k. In local coordinates near p the Morse lemma gives on an n-ball $B_\delta = \sum x_j^2 < \delta^2$ the expression

$$\varphi = c_j - \sum_1^k x_j^2 + \sum_{k+1}^n x_l^2.$$

Then

$$
\begin{aligned}
H_*(M_{t_j}, M_{t_{j-1}}) &= H_*(p \cup M_{c_j^-}, M_{c_j^-}) \\
&= H_*(p \cup \{q \in B_\delta : \varphi < c_j\}, \{q \in B_\delta : \varphi < c_j\}) \\
&= H_*\left(\left\{q \in B_\delta : \sum_{k+1}^n x_l^2 = 0\right\}, \left\{q \in \partial B_\delta : \sum_{k+1}^n x_l^2 = 0\right\}\right) \\
&= H_k\left(\left\{q \in B_\delta : \sum_{k+1}^n x_l^2 = 0\right\}, \left\{q \in \partial B_\delta : \sum_{k+1}^n x_l^2 = 0\right\}\right) \\
&= F.
\end{aligned}
$$

Thus $P(M_{t_j}, M_{t_{j-1}}) = t^k$. Counting critical points of all indices yields

$$\sum_{j=1}^{L} P(M_{t_j}, M_{t_{j-1}}) = \sum_k \mu_k t^k.$$

Since $P(M) = \sum_k \beta_k t^k$, (3.3) yields Theorem 3.8. $\square$

EDITORS' NOTE. See [Kuiper 1962; 1971] for more on Morse relations and tightness.

Minimal Total Curvature and Tightness. An embedding of topological spaces $f : A \subset B$ is called *injective in homology* over $\mathbb{Z}_2$, or H_*-*injective*, in case the induced homomorphism in $\mathbb{Z}_2$-homology

$$f_* : H_*(A) \to H_*(B)$$

is injective. It is easy to show that:

LEMMA 3.10. *If $A \subset B \subset C$, and $A \subset C$ as well as $B \subset C$ are H_*-injective, then so is $A \subset B$.*

REMARK. *Čech cohomology* theory and *Čech homology* theory have been introduced for spaces whose singular homology does not give satisfactory answers. For manifolds and CW-complexes they give the same answer. If X is a compact subset of a manifold or a CW-complex Y, then $H_*^{\text{Čech}}(X)$ is the inverse limit of $H_*^{\text{sing}}(U_i(X))$ for $U_i \supset U_{i+1} \supset \cdots \supset X$ any nested sequence of open sets of Y converging to X. For such cases this can be used as definition. In general it is known that if the compact space X is embedded in any space Y, then $H_*^{\text{Čech}}(X)$ is the inverse limit of $H_*^{\text{Čech}}(Y_i)$ for any nested sequence of subspaces $Y \supset Y_i \supset Y_{i+1} \supset \cdots \supset X$ converging to X. In our applications we will not use more than these facts from the notion of Čech homology. We will use $\mathbb{Z}_2$-Čech homology, denoted H_*. We remark also that Čech cohomology has better general properties than Čech homology and for this reason is used more.

EXERCISE. If

$$X = \{(u,v) \in \mathbb{R}^2 : u = \sin(1/v) \text{ for } v > 0 \ \text{ or } \ u = 0 \text{ for } v \leq 0 \ \text{ or } \ v = 0\}$$

then $H_0^{\text{Čech}}(X) = \mathbb{Z}_2$ but $H_0^{\text{sing}}(X) = \mathbb{Z}_2 \oplus \mathbb{Z}_2$.

In Theorem 3.6 we saw that the total curvature of a smooth immersion $f : M \to \mathbb{R}^N$ of a closed n-manifold M has total curvature $\tau(f) \geq \beta(M)$. We now study equality:

THEOREM 3.11. *The smooth immersion $f : M \to \mathbb{R}^N$ has total curvature $\tau(f) = \beta(M)$ if and only if the embeddings $f^{-1}(h) \subset M$ are homology injective for every half-space $h \subset \mathbb{R}^N$. The same statement holds if we replace half-spaces h by open half-spaces $\overset{\circ}{h}$.*

PROOF. Let $\tau(f) = \mathcal{E}_z \mu_z(f) = \beta$. Then $\mu_z(f) = \beta$ and $\mu_{kz}(f) = \beta_k$ for all indices k, of any nondegenerate height function z with $\|z\| = 1$. The croissant

$$M_t = \{p \in M : zf(p) \leq t\}$$

equals $f^{-1}(h)$ for the half-spaces $h = \{q \in \mathbb{R}^N : z(q) \leq t\}$. Assume that the initial values $c_1, \ldots, c_L$ are all different ($L = \beta(M)$) and let noncritical values t_j be chosen such that $t_0 < c_1 < t_1 < c_2 \cdots < c_L < t_L$, as in the proof of Equation (3.3). Any given noncritical value t can be assumed to be t_j for some j. As $\mu_k = \beta_k$ for all k, $K_k = 0$ for all k. So all the polynomials vanish in (3.3), $M_{t_{j-1}} \subset M_{t_j}$ is H_*-injective and so is the composition $M_t = M_{t_j} \subset M$. The space $f^{-1}(h) \subset M$ is for every half-space h the limit of a nested sequence of spaces

$$M \supset f^{-1}(h_i) \supset f^{-1}(h_{i+1}) \cdots \supset f^{-1}(h)$$

for which $f^{-1}(h_i) \subset M$ is already known to be H_*-injective. Then the same follows for $f^{-1}(h) \subset M$. In fact the limit $H_*(f^{-1}(h)) = H_*(f^{-1}(h_{i_0}))$ is already attained for some integer i_0, because $H_*(M) = \oplus_1^\beta \mathbb{Z}_2$ is finite. For any open half-space $\overset{\circ}{h}$ we find a sequence of half-spaces h_j such that the nested sequence

$$f^{-1}(h_i) \subset f^{-1}(h_{i+1}) \cdots \subset f^{-1}(\overset{\circ}{h}) \subset M$$

exhausts $f^{-1}(\overset{\circ}{h})$. The limit $H_*(f^{-1}(\overset{\circ}{h})) = H_*(f^{-1}(h_{i_0}))$ is again already attained for some integer i_0. □

DEFINITION. We call a continuous map $f : X \to \mathbb{R}^N$ of a compact connected metrizable space X *tight* or *perfect* if $f^{-1}(h) \subset X$ is H_*-injective for every half-space $h \subset \mathbb{R}^N$. We call f k-*tight* if $f^{-1}(h)$ is H_i-injective for all $i \le k$ and all h.

EXERCISE. A map $f : X \to \mathbb{R}^N$ into one point is tight.

For $k = 0$ our definition agrees with Section 1 where 0-tightness, the two-piece property, was studied. We now reformulate Theorem 3.11 with a slight extension as follows:

THEOREM 3.12. *The smooth immersion $f : M \to \mathbb{R}^N$ of a closed n-manifold has minimal total curvature equal to $\tau(f) = \beta(M)$ if and only if f is k-tight, where $n - 2 \le 2k \le n - 1$.*

PROOF. By Poincaré duality $\beta_i = \beta_{n-i}$. For any nondegenerate z one has $\mu_{n-i}(z) = \mu_i(-z)$. So if f is k-tight, then $\mu_i = \beta_i$ for $i \le k$ and for $n - i \le k$. There only remains to show for n even the equality $\mu_{k+1} = \beta_{k+1}$. This follows from $\sum_0^n (-1)^i \mu_i = \sum_0^n (-1)^i \beta_i = \chi$. □

COROLLARY. A smooth immersion $f : M^2 \to \mathbb{R}^N$ of a closed surface M^2 has minimal total curvature $\tau(f) = \beta(M^2)$ if and only if f is 0-tight.

EXAMPLE. A 0-tight smooth embedding $f : S^3 \to \mathbb{R}^4$ need not have have minimal total curvature (need not be tight). For example, consider the surface $D \subset h \subset E^3 \subset E^4$ with boundary $\partial D \subset \partial h = E^2$ shown on the right. Assume D orthogonal to ∂h along ∂D. Rotate (D, h) in E^4 around the (fixed point set) plane $\partial h = E^2$ to form the three-sphere $M \subset E^4$. See [Kuiper 1970, pp. 221–224] for more detail.

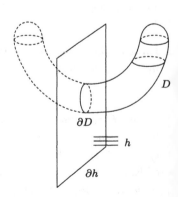

EXERCISE. Prove that the limit Swiss cheese in E^2 (page 7) is tight.

OPEN QUESTION. Is there a 0-tight smooth S^3 embedded substantially in $\mathbb{R}^5$?

For topological or PL embeddings (see also [Banchoff 1965]) the possibilities are richer:

THEOREM 3.13 [Banchoff 1971a]. *There is for every $N > n \geq 3$ a substantial 0-tight polyhedral embedding $f : S^n \to E^N$.*

4. Tight Smooth Surfaces in Codimensions 3 and Higher

This section is devoted to the proof of the following result:

THEOREM 4.1 [Kuiper 1962]. *Let $f : M \to E^N$ be a 0-tight smooth closed embedded surface substantial in E^N. Then $N \leq 5$. For $N = 5$, M is the real projective plane and f is an embedding onto a Veronese surface.*

Tightness and "high" codimension (at least 3) therefore dramatically impose uniqueness and rigidity up to projective transformation.

PROOF. We prove the theorem in a sequence of steps.

(a) $N \leq 5$.

PROOF. Let f fulfill the conditions of the theorem and let z be a height function with nondegenerate maximum at $p \in M$. Then p will be called a *nondegenerate top point*. Choose Euclidean coordinates $u_1 \ldots u_{N-2}, v_1, v_2$ vanishing at $p = 0 \in \mathbb{R}^N$, and such that $u_1, \ldots, u_{N-2}$ vanish on the tangent space $T_p(M)$, and u_1 is nondegenerate maximal at p. Then v_1 and v_2 (or more precisely $v_1 f$ and $v_2 f$) are coordinates for M near p. The height functions that vanish at p form a vector space

$$W = \{w_1 u_1 + \ldots + w_{N-2} u_{N-2} : (w_1, \ldots, w_{N-2}) \in \mathbb{R}^{N-2}\}.$$

Denote $\mathcal{J}^2 z = \mathcal{J}^2(zf)$ ($\mathcal{J}^2$ for 2-jet) the part up to degree two of the Taylor series of zf, in terms of v_1, v_2. It is a quadratic function. The induced map $\mathcal{J}^2 : W \to Z$ into the 3-space Z of quadratic functions in v_1 and v_2 is a homomorphism.

Suppose $\mathcal{J}^2 z = 0$ for some height function z with $\|z\| = 1$ and $z \neq \pm u_1$. Then $\mathcal{J}^2(u_1 + \lambda z) = \mathcal{J}^2 u_1$ is negative definite and $u_1 + \lambda z$ has a nondegenerate isolated maximum at p for all λ. If $z(p_1) \neq 0$ for some $p_1 \in M$ then the half-space

$$h = \{q \in \mathbb{R}^N : u_1(q) + \lambda z(q) \geq 0, \quad u_1(p_1) + \lambda z(p_1) = 0\}$$

meets M in an isolated point p and in another part containing p_1, contradicting tightness. So $z(p_1) = 0$ for all $p_1 \in M$, contradicting the substantiality of f. It follows that $\mathcal{J}^2$ must be injective and $\dim W \leq \dim Z = 3$, so $N \leq 5$. $\square$

REMARK. The codimension of a 0-tight smooth closed n-submanifold substantially in E^n is $N - n \leq \frac{1}{2}n(n+1)$, by the same argument [Kuiper 1970].

From now on assume $N = 5$.

(b) *Preferred affine coordinates at $p \in M$, and the topsets*

$$M_\theta = M_\theta(p) \subset E_\theta^3 \subset E_\theta^4.$$

Since $N = 5$, $\mathcal{J}^2 : W \to Z$ is bijective onto Z. Choose $u_1', u_2', u_3' \in W$ such that $\mathcal{J}^2(u_1') = -v_1^2 - v_2^2$, $\mathcal{J}^2(u_2') = -v_1^2 + v_2^2$, $\mathcal{J}^2(u_3') = 2v_1 v_2$. Denote them by u_1, u_2, u_3 and consider u_1, u_2, u_3, v_1, v_2 as an orthonormal basis of Euclidean space E. Let

$$z = w_1 u_1 + w_2 u_2 + w_3 u_3 = \sin\varphi\, u_1 + \cos\varphi \cos\theta\, u_2 + \cos\varphi \sin\theta\, u_3.$$

Then

$$\mathcal{J}^2 z = w_1(-v_1^2 - v_2^2) + w_2(-v_1^2 + v_2^2) + w_3 2 v_1 v_2,$$

with determinant

$$\begin{vmatrix} -w_1 - w_2 & w_3 \\ w_3 & -w_1 + w_2 \end{vmatrix} = w_1^2 - w_2^2 - w_3^2 = \sin^2\varphi - \cos^2\varphi = -\cos 2\varphi.$$

Therefore z is nondegenerate at p and its index is

$$\begin{array}{lll} 2 \text{ (maximum)} & \text{if and only if} & \pi/4 < \varphi \le \pi/2, \\ 1 & \text{if and only if} & -\pi/4 < \varphi < \pi/4, \\ 0 \text{ (minimum)} & \text{if and only if} & -\pi/2 < \varphi < -\pi/4. \end{array}$$

The height function z has, for $\varphi = \pi/4 - \varepsilon$ with $\varepsilon > 0$ small, a critical point of index 1 at p. Then we see that the half-space

$$h = h(\varphi, \theta) = \{q \in E^5 : z = z_{\varphi,\theta} \ge 0\}$$

contains in $h \cap M$ an essential one-cycle of M, and so does the limit for $\varepsilon = 0$, $\varphi = \pi/4$.

We now denote the half-space $h(\pi/2, \theta)$ by $h(\theta)$ and conclude that the topset $M_\theta = M \cap h(\theta)$ carries an essential 1-cycle of M. The boundary $\partial h(\theta)$ is the 4-plane with equation $u_1 + \cos\theta\, u_2 + \sin\theta\, u_3 = 0$; see Figure 17.

The half-spaces $h(\theta)$ envelop a solid cone with equations

$$u_1 \le 0, \quad u_1^2 \ge u_2^2 + u_3^2$$

with boundary the quadratic 4-dimensional cone with equation

$$u_1 \le 0, \quad u_1^2 = u_2^2 + u_3^2. \tag{4.1}$$

It contains as u_1-topset the whole tangent plane $T_p(M)$ with equation $u_1 = u_2 = u_3 = 0$. The half-space $h(\theta)$ and the 4-plane $\partial h(\theta)$ support the cone in a 3-plane E_θ^3 with equation

$$\begin{aligned} u_1 + \cos\theta\, u_2 + \sin\theta\, u_3 &= 0, \\ -\sin\theta\, u_2 + \cos\theta\, u_3 &= 0. \end{aligned} \tag{4.2}$$

The second equation is obtained from the first by differentiation with respect to θ.

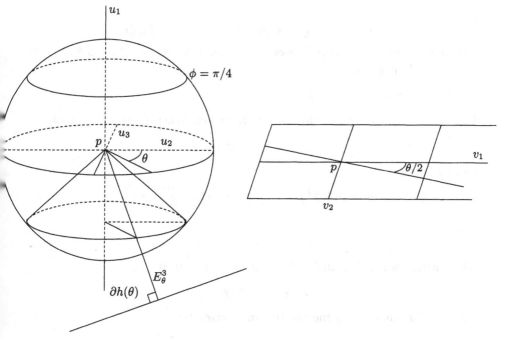

Figure 17. Left: normal space at p. The cone has equation (4.1). The 3-plane E_θ^3 has equation (4.2); dimensions v_1 and v_2 are not shown. E_θ^3 is the intersection of the cone with E_θ^4, the 4-plane generated by E_θ^3 and the u_1-axis. Moreover $E_\theta^3 = E_\theta^4 \cap \partial h(\theta)$. Right: tangent space at p.

The 4-plane with equation $-\sin\theta\, u_2 + \cos\theta\, u_3 = 0$ is denoted E_θ^4. It contains E_θ^3 as well as $E_{\theta+\pi}^3$. In summary,

$$p \in M_\theta \subset E_\theta^3 \subset E_\theta^4.$$

(c) M_θ *has a tangent line at p with equation*

$$\cos(\theta/2)\, v_1 - \sin(\theta/2)\, v_2 = 0, \qquad u_1 = u_2 = u_3 = 0. \qquad (4.3)$$

PROOF. Consider the function on M

$$u_1 + \cos\theta\, u_2 + \sin\theta\, u_3 = (-v_1^2 - v_2^2) + \cos\theta(-v_1^2 + v_2^2) + \sin\theta\, 2v_1 v_2 + O(\|v\|^3).$$

Here $O(\|v\|^3)$ means that the remainder $R(v_1, v_2)$ is such that $R(v_1, v_2)/\|v\|^3$ is bounded for small $\|v\| = \sqrt{v_1^2 + v_2^2}$. As the expression vanishes on M_θ we obtain after calculation

$$-\big(\cos(\theta/2)\, v_1 - \sin(\theta/2)\, v_2\big)^2 + O(\|v\|^3) = 0.$$

Hence

$$\cos(\theta/2)\, v_1 - \sin(\theta/2)\, v_2 = O\big(\|v\|^{\frac{3}{2}}\big)$$

and
$$\left|\cos(\theta/2)(v_1/\|v\|) - \sin(\theta/2)(v_2/\|v\|)\right| = O(\sqrt{\|v\|}).$$
The unit vector $\|v\|^{-1}(v_1, v_2)$ goes in the limit to $\pm(\sin(\theta/2), \cos(\theta/2))$, for $(v_1, v_2) \in M_\theta, \|v\| \to 0$. $\qquad\square$

(d) M_θ *is a smooth curve near* p.

PROOF. It suffices to prove this for $\theta = 0$. The equations (4.2) for $\theta = 0$ are

$$u_1 + u_2 = 0, \quad u_3 = 0,$$

and we obtain

$$v_1 = O(\|v\|^{\frac{3}{2}}), \quad \frac{v_1}{\|v\|} = O\sqrt{\|v\|}, \quad \frac{v_2}{\|v\|} \to 1,$$

and

$$2v_1 v_2 - \varphi(v_1, v_2) = 0. \tag{4.4}$$

The function φ is C^∞ with $\mathfrak{J}^2(\varphi) = 0$ at $(v_1, v_2) = (0, 0)$:

$$\varphi(v_1, v_2) = O(\|v\|^3).$$

Consider for any fixed v_2 the function in v_1 given by

$$\eta : v_1 \mapsto v_1 - (2v_2)^{-1}\varphi(v_1, v_2).$$

Since $d\eta/dv_1 \neq 0$ for small $\|v\|$, the map η is a C^∞-diffeomorphism for any small value v_2. Call its inverse η^{-1}. Then the solution of (4.4) is

$$v_1 = \eta^{-1}(0).$$

Its dependence on v_1 is expressed as

$$v_1 = \psi(v_2).$$

The inverse function theorem says that ψ is also C^∞. $\qquad\square$

We obtain for every tangent line (4.3) a smooth curve. The 4-planes E_θ^4 cut out these curves in pairs near p, M_θ and $M_{\theta+\pi}$, with tangent lines

$$\cos(\theta/2)\,v_1 - \sin(\theta/2)\,v_2 = 0, \quad \sin(\theta/2)\,v_1 + \cos(\theta/2)\,v_2 = 0.$$

(e) *Any component of* $M_\theta(p) \cap \mathcal{U}$ *is a plane convex curve.*

PROOF. If z has a nondegenerate maximum, then so has any height function z_1 near to z. The nondegenerate top points therefore form an open set $\mathcal{U}$ in M. We study the levels of the function $\theta = \arctan(u_3/u_2)$ on M in a small neighborhood $U(p) \subset U$ of p (but p is excluded as θ is not defined there). By Sard's theorem, the critical values of θ on $M \setminus E_{2,3}, (E_{2,3} : u_2 = u_3 = 0)$ have measure zero. For a regular value θ_0 the level $\{q \in M \setminus E_{2,3} : \arctan(u_3(q)/u_2(q)) = \theta_0\}$ is a smooth curve. Its intersection with $\partial h(\theta_0)$ can be completed with the point p to obtain the top set M_{θ_0}, a tight connected closed curve. By Fenchel's Theorem (Theorem 2.12) it is a convex plane curve. As θ_0 is a regular value, the function

θ is not critical on M_{θ_0}, hence on some neighborhood of M_{θ_0} in M. Then θ is also a regular value on that neighborhood, and M_θ is also a plane convex curve for $|\theta - \theta_0|$ small. So the set of θ for which the topset M_θ is plane convex is open and dense in $\mathbb{R}$ mod 2π. But inside the small neighborhood $U(p)$ of p in M the limit of a set of plane curves M_θ for $\theta \to \theta_1$ must be a plane curve as well. So every topset M_θ has a plane convex curve part in $U(p) \subset \mathcal{U}$. The same arguments apply for any $p \subset \mathcal{U}$, hence to $\mathcal{U}$. $\qquad\square$

(f) *Any component of $M_\theta(p) \cap \mathcal{U}$ is part of a conic.*

PROOF. Consider $q \in M_{\theta_1}(p)$ and $M_{\theta_2}(q) = \gamma \neq M_{\theta_1}(p)$. There is an open interval of regular values θ near to θ_1 giving rise to curves $M_\theta(p) \cap U$ that meet γ in some open interval near q. That interval lies in the intersection of the plane of γ and the quadratic cone for p. This proves statement (f) locally, hence globally. $\qquad\square$

(g) The remaining part of the proof belongs to classical projective geometry. We only indicate the main ideas. Take the above situation in some neighborhood $U(q_0)$ of

$$q_0 \in M_{\theta_1}(p), \qquad M_{\theta_2}(q_0) = \gamma \neq M_{\theta_1}(p).$$

Consider an interval of top conic sections $M_\theta(p)$, $\theta_1 - \delta < \theta < \theta_1 + \delta$, and a two-parameter family of top conic sections $M_\omega(p)$ parametrized by $q \in M_{\theta_1}(p)$, q near to q_0, and ω, $\theta_2 - \delta < \omega < \theta_2 + \delta$, such that every $M_\theta(p)$ meets every $M_\omega(q)$ in exactly one point inside $U(q_0) \subset U$. We consider $E^5 = P^5 \setminus P^4$ as the complement of a projective 4-plane P^4 in real projective space P^5. We project $U(p)$ from the center $P \in E^5$ into P^4. The point p is itself excluded from this projection, which is denoted $\hat{p}$. We see that:

(i) $\hat{p}(T_p(M) \setminus p)$ is a line $L_0 \subset P^4$.
(ii) $\hat{p}(M_\theta(p))$ is a line $L_\theta \subset P^4$, and $L_0 \cap L_\theta$ is a point.
(iii) $\hat{p}(M_\omega(q)$ is a conic $\gamma(\omega, q) \subset P^4$ in a plane $\Pi(\omega, q)$.
(iv) Every line L_θ meets every conic $\gamma(\omega, q)$ in one point in $\Pi(\omega, q)$.

These are very strong conditions on the lines L_θ and planes $\Pi(\omega, q)$.

In P^4 the family of all 2-planes that meet 4 lines in general position is a two-dimensional family of planes all meeting one more (fifth) line. However, if the four lines L_θ, for $\theta = \theta_1, \theta_2, \theta_3, \theta_4$, are in general position but for the fact that they all meet one and the same line L_0, then there is a complete one-parameter family of such lines [Kuiper 1941] and a degenerate Veronese surface $\mathcal{V}_1$, that can be described as follows.

Take three points $\theta_1, \theta_2, \theta_3$ on a projective line $L_0 \subset P^4$. Take a disjoint plane Π and in it a conic $\mathcal{E} \subset \Pi$. Let $\rho : L_0 \to \mathcal{E}$ be a birational correspondence. $\mathcal{V}_1$ is the union of all lines L_θ from $\theta \in L_0$ to $\rho(\theta) \in \mathcal{E}$.

From our description it follows that $\mathcal{V}_1$ has no projective invariants. A Veronese surface $\mathcal{V}$ can be found in E^5 which passes through p, has the same tangent plane $T_p(\mathcal{V}) = T_p(M)$, and yields the projection $\hat{p}\mathcal{V} = \mathcal{V}_1$. As other points $p' \in U(q_0)$ also give algebraic projections $\hat{p}'(U(q_0))$ in a projective surface $\mathcal{V}_1'$ we can succeed in finding a unique Veronese surface $\mathcal{V}$ which contains all of $U(p)$ and then by continuation all of U. By extension we also see that the set of nondegenerate top-points $U \subset M$ is not only open but also closed in $\mathcal{V}$. Then $f(M) = U = \mathcal{V}$, the Veronese surface. This ends the proof of Theorem 4.1. $\square$

Details of the last part can be found in [Kuiper 1980], which contains Theorem 4.3 below.

Generalization. The conclusion of Theorem 4.1 remains true if we assume $f : M \to E^5$ to be an immersion. We mention, without proof, a deep generalization of Theorem 4.1:

THEOREM 4.2 [Little and Pohl 1971]. *Let $f : M^n \to \mathbb{R}^N$ be a 0-tight smooth substantial immersion of a closed n-manifold $n \geq 2$. Then $N - n \leq \frac{1}{2}n(n + 1)$. If we assume equality, $N - n = \frac{1}{2}n(n+1)$, then M is real projective n-space and $f(M^n)$ is the standard Veronese n-manifold (up to projective transformation). Note that only 0-tightness is used.*

Hard to prove is:

THEOREM 4.3 [Kuiper 1980]. *Let $f : M^{2d} \to E^{3d+2}$ be a smooth substantial tight embedding of a manifold M like a projective plane (that is, $\beta_0 = \beta_d = \beta_{2d} = 1$, $\beta = \sum \beta_i = 3$) into E^{3d+2}, $d = 1, 2, 4$ or 8. Then $f(M^{2d})$ is an algebraic variety. For $d = 1$, we have Theorem 4.1. For $d = 2$, $f(M^4)$ is the standard model $\mathcal{V}(\mathbb{C})$ of the complex projective plane, up to real projective transformation in E^8.*

In both theorems the assumptions lead to a high degree of rigidity of the embedding.

OPEN QUESTION (perhaps not difficult). It is *conjectured* that the conclusions of Theorem 4.3 for obtaining the standard models for projective planes over $\mathbb{C}$ also hold for quaternion planes (or even the same conclusion for 3-connected 8-manifolds) and for the Cayley-plane (or even for 7-connected 16-manifolds).

EDITORS' NOTE. See also [Kuiper and Pohl 1977], in which it is shown that a TPP topological embedding of the real projective plane into E^5 is either a Veronese surface or Banchoff's piecewise linear embedding with six vertices [Banchoff 1974].

5. Tightness of Topsets

THEOREM 5.1. *If $Y \subset \mathbb{R}^N$ is compact and tight, then so is every topset X of Y, and the inclusion $X \subset Y$ is homology injective. In particular, every essential cycle in X carries an essential cycle in Y.*

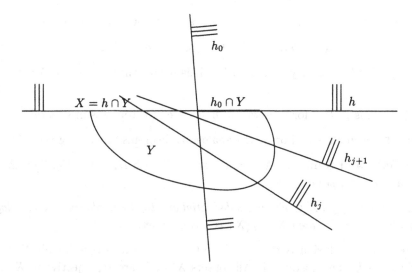

Figure 18. Proving that the topset X of a tight set Y is tight.

THEOREM 5.2. *If $f : M \subset \mathbb{R}^N$ is a smooth embedding of a closed manifold M with minimal total curvature $\tau(f) = \beta(M)$, then every topset $X \subset M$ is tight, and every essential cycle in $H_*(X)$ carries an essential cycle in $H_*(M)$.*

Note that there is no reason for X to be a manifold! This is one motivation for our generalization. Theorem 5.1 is a special case of the next result:

THEOREM 5.3. *If $f : Y \to \mathbb{R}^N$ is a tight map of a compact connected metrizable space Y and if $X \subset Y$ is any topset, then $f : X \to \mathbb{R}^N$ is tight and $X \subset Y$ is homology injective. In particular every essential cycle in X carries an essential cycle in Y.*

PROOF. (Compare the examples of tight spaces given so far). If the conclusion holds for any topset of a tight map, it holds for a topset of this topset by composition. Then, it holds for any top^{k}set for $k \geq 1$ by induction. It suffices therefore to prove the theorem for a topset $X \subset Y$, say

$$\varnothing \neq X = f^{-1}(h) = f^{-1}(\partial h) \neq Y,$$

h a supporting half-space. Let h_0 be another half-space. Say it "meets" X in $(f|_X)^{-1} = X \cap f^{-1}(h_0) = f^{-1}(h \cap h_0) \neq \varnothing$. See Figure 18. There is a sequence of half-spaces $h_1, h_2, \ldots$ in $\mathbb{R}^N$ such that

$$f^{-1}(h_1) \supset f^{-1}(h_2) \supset \cdots \supset f^{-1}(h_j) \supset \cdots \supset f^{-1}(h \cap h_0) = \bigcap_j f^{-1}(h_j) = X$$

is a nested sequence of subspaces of Y, *converging to* X.

The inclusions $X = f^{-1}(h) \subset Y$ and $f^{-1}(h_j) \subset Y$ are H_*-injective by tightness of f. The inclusion $f^{-1}(h \cap h_0) \subset Y$ is H_*-injective because $f^{-1}(h \cap h_0)$ is the inverse limit of $\{f^{-1}(h_j)\}$. Then the first inclusion in the sequence

$$(f|_X)^{-1}(h_0) = f^{-1}(h \cap h_0) \subset f^{-1}(h) = X \subset Y$$

is H_*-injective by the easy Lemma 3.10. This means that the restriction $f|_X$ is tight.   □

Our theorem is proved for topsets and holds then for top*sets in general.

EXERCISE. Formulate and prove the analogous theorems concerning k-tightness.

Tight Balls and Spheres. A compact space X with $H_*(X) = H_0(X) = \mathbb{Z}_2$ is called a *homology point space*.

THEOREM 5.4. *If $f : X \subset E^N$ is a substantial and tight embedding of a homology point space X in E^N then $X = f(X)$ is a convex set.*

PROOF. The conclusion is true for $N = 0$. Suppose it is true for all $N < k$. Let $f : X \to E^k$ be substantial. All topsets X' of X are H_*-injective in X. So they are homology points, and convex by induction. Therefore $\partial \mathcal{H} X \subset X$ is a $(k-1)$-sphere in X. It is bounding because $H_{k-1}(X) = 0$. Then, no point of $\mathcal{H} X$ can be not in X, so $X = \mathcal{H} X$. By induction the theorem follows.   □

We next prove the analogous theorem for maps.

THEOREM 5.5. *If $f : X \to E^N$ is a tight substantial map of a homology point space X, then $f(X)$ is convex.*

PROOF. Assume the theorem true for all $N < k$. It is true for $n = 0$. Let $f : X \to E^k$ be substantial tight. All topsets are homology injective in X, so they are homology point spaces, their images are convex sets by induction. Their union is contained in fX, i.e., $\partial \mathcal{H} f X \subset f X$. Denote $W = f^{-1} \partial \mathcal{H} f X \subset X$. Let $R \subset W \times S^{k-1}$ be the compact relation

$$R = \{(w, z^*) : \max_{p \in X} zf(p) = zf(w)\}.$$

Given z we see that the set of w for which $(w, z^*) \in R$ is just the topset determined by the maximal value of zf on X. It is a homology point space. Therefore the projection $p_S : W \times S^{k-1} \to S^{k-1}$ induces a projection $p_S : R \to S^{k-1}$ whose fibers are all homology-point spaces. Then by a theorem of Vietoris and Begle [Spanier 1966, p. 344], p_S induces an isomorphism in homology:

$$H_*(R) \xrightarrow{\sim} H_*(S^{k-1}).$$

Given $w \in W$, the set of $z^* \in S^{k-1}$ for which $(w, z^*) \in R$ is nonempty and convex by geometry (convexity of $\mathcal{H} f X$ containing $f(w)$). Therefore, the fibers of the projection $p_W : R \to W$ are also homology point spaces, and again the theorem of Vietoris and Begle reads that $H_*(R) \to H_*(W)$ is an isomorphism. By composition, we see that

$$H_{k-1}(W) = H_{k-1}(S^{k-1}) = \mathbb{Z}_2.$$

As $W \subset X$ and $H_*(X) = H_0(X) = \mathbb{Z}_2$, the space W is bounding in X and so is its image $\partial \mathcal{H} f X$ in $f X$. Then $f X = \mathcal{H} f X$ is convex. $\quad\square$

THEOREM 5.6 [Kuiper 1980]. *Let $f : S^n \to E^N$ be a substantial tight map. Then either $N = n + 1$ and $f(S^n)$ is a convex hypersurface or $N \leq n$ and $f(S^n)$ is a convex set.*

COROLLARY [Chern and Lashof 1957]. *Let $f : M^n \to E^N$ be a substantial smooth immersion of a closed n-manifold M^n with total curvature $\tau(f) = 2$. Then M^n is the n-sphere with standard smooth structure (not exotic), $N = n+1$, and $f(M)$ is a convex hypersurface.*

PROOF OF COROLLARY. Since $\tau(f) = \mathcal{E}_z \mu_z(f) = 2$, we have $\mu_z(f) = 2$ and M must be homeomorphic to a n-sphere. There is no immersion in R^N for $N \leq n$, so f is a smooth immersion onto a smooth convex hypersurface in $\mathbb{R}^{n+1}$. Then, it is an embedding and $f(S^n)$ has the standard smooth structure. $\quad\square$

PROOF OF THEOREM 5.6. Given f one observes that every topset $X \subset S^n$ misses at least one point of S^n and it is homology injective in S^n. Therefore $H_*(X) = H_*(\text{point})$. So by Theorem 5.5 every topset is a homology point space. As in the proof of Theorem 5.5, we get an isomorphism

$$H_*(W) \simeq H_*(\partial \mathcal{H} f S^n) \simeq H_*(S^{N-1}),$$

where $\partial \mathcal{H} f S^n \subset f S^n, W = f^{-1}(\partial \mathcal{H} f S^n)$. As $W \subset S^n$, then $N - 1 \leq n$. If $N - 1 < n$, then $W \subset S^n$ bounds in S^n, that is $H_{N-1}(W) \to H_{N-1}(S^n)$ maps into zero. Then also

$$0 \neq H_{N-1}(fW) = H_{N-1}(\partial \mathcal{H} f X) \to H_{N-1}(f X)$$

maps into zero. Then $f(X) = \mathcal{H} f(X)$ is convex.

If $N - 1 = n$ then $W \subset S^n$ induces an isomorphism $H_n(W) \to H_n(S^n)$ and so $W = S^n$ and $f(W) = \partial \mathcal{H} f S^n$, a convex hypersurface in E^{n+1}. $\quad\square$

Application: Tight Projective Planes. A closed manifold P^{2d} that contains a tame embedded d-sphere S^d so that $P^{2d} \backslash S^d$ is homeomorphic to $\mathbb{R}^{2d}$ is called a *manifold like a projective plane*. Such manifolds were studied in [Eells and Kuiper 1962]. P^{2d} is a CW-complex with three cells; $\beta = \beta_0 + \beta_d + \beta_{2d}$. Necessarily $d = 1, 2, 4, 8$. Examples are the standard projective planes over $\mathbb{R}, \mathbb{C}, \mathbb{H}$ (quaternions) and Ca (octaves or Cayley numbers), for $d = 1, 2, 4, 8$ respectively. The self intersection of the essential d-cycle in P^{2d} is one. Here we prove:

THEOREM 5.7. *Let $f : P \subset E^N$ be a tight substantial embedding of a manifold like a projective plane $P = P^{2d}$. Then $N \leq 3d + 2$.*

PROOF. Let k be the smallest number for which there exists an E^k-top*set X that contains an essential cycle in $H_*(X)$. If there is no such k, let $k = N$. Clearly, $k > 1$ because every E^0- or E^1-top*set is convex.

Every topset Y of X is homology injective in X and is therefore a homology point space. Then Y is convex. Consequently, the $(k-1)$-sphere $\partial \mathcal{H} X$, a union of such convex topsets, is in X. If $k - 1 \neq d$ or $2d$, then $\partial \mathcal{H} X$ bounds in X because by tightness if not, it also does not bound in P. Then every point inside $\partial \mathcal{H} X$ must belong to X, and $X = \mathcal{H} X$, contradicting the assumption on k.

Now assume that $k - 1 = d$. Then $\partial \mathcal{H} X$ carries the d-cycle of P^{2d}. Now we project E^N parallel to $E^k = \text{span}(X)$ into a Euclidean space E^{N-k} orthogonal to E^k. Every height function z which is constant on E^k has $\mu_z(X) = 3$. There is a minimum $\mu_{0z}(X) = 1$ for some $z \leq z(X)$, a maximum $\mu_{2dz}(X) = 1$ for some $z \geq z(X)$ and no other critical point for $z \neq z(X)$. This is so because any d-cycle in $\{p \in X : z(p) < z(X)\}$ has to meet geometrically the d-cycle in X by nonzero self intersection in homology. The space $P \setminus X$ can be exhausted by a nested sequence of half-space sections $f^{-1}(h_j) \subset f^{-1}(h_{j+1}) \ldots \subset P \setminus X$. As they do not meet X they are (open) homology point spaces all homeomorphic to $\mathbb{R}^{2d}$. The one point compactification $P/(X = \text{point})$ is then a sphere S^{2d} with a tight map into E^{n-k} :

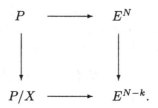

By Theorem 5.5 then $N - k = N - d - 1 \leq 2d + 1$, and $N \leq 3d + 2$.

If $k - 1 \neq d$ we are in the case $k - 1 = 2d$. Then $\partial \mathcal{H} X$ is a $2d$-sphere in $X \subset P$. This is a contradiction. $\qquad \square$

REMARKS. Equality $N = 3d + 2$ is attained for the standard smooth models of standard projective spaces. It has also been attained for polyhedral embeddings for the real projective space, for the standard complex projective space in [Kühnel and Banchoff 1983] and for some P^8 (perhaps $\mathbb{HP}(2)$) in [Brehm and Kühnel 1992]. By projection one finds tight embeddings for $N = 3d + 1$. There are no embeddings for $N \leq 3d$ by characteristic class obstructions.

By projections one finds tight maps for all $N \leq 3d + 2$.

OPEN QUESTION. Let $f : P^{2d} \to \mathbb{R}^N$ be a tight substantial *map*. Is $N \leq 3d+2$? For $N = 2d + 1$, is $f(P)$ a convex hypersurface in E^{2d+1}? For $N \leq 2d$, is $f(P)$ a convex set?

Counting Critical Points. A continuous function $\varphi : M \to \mathbb{R}$ on a compact space M is said to have a *Poincaré polynomial* in case there exists for every value t, a number $\varepsilon > 0$, such that if $t - \varepsilon < r < t \leq s < t + \varepsilon$, the following conditions are satisfied:

(a) the homomorphisms

$$H_*(M_{t-\varepsilon}) \to H_*(M_r) \quad \text{and} \quad H_*(M_s) \to H_*(M_{t+\varepsilon})$$

induced by inclusions, are bijective.

(b) the group $H_*(M_{t+\varepsilon}, M_{t-\varepsilon}) = H_*(M_t, M_{t-})$ is finitely generated.

Here $M_t = \{p \in M : \varphi(p) \le t\}$.

The value t is called *critical* in case this group is nonzero. Clearly any real algebraic function on a compact real algebraic manifold $M \subset E^N$ has a Poincaré polynomial.

Call the finitely many critical values $c_1 < c_2 \ldots < c_L$. Choose noncritical values $t_0, \ldots, t_L$ such that

$$t_0 < c_1 < t_1 < \ldots < c_L < t_L.$$

DEFINITION. The *Poincaré polynomial* of φ is

$$P(\varphi) = \sum_{j=1}^{L} P(M_{t_j}, M_{t_{j-1}}) = \sum_k \mu_k(\varphi) t^k.$$

It is independent of the choice of $t_0, \ldots, t_L$. As in equation (3.3), it obeys the Morse inequalities:

$$\frac{P(\varphi) - P(M)}{1+t} = \frac{\sum_k (\mu_k(\varphi) - \beta_k) t^k}{1+t} = \sum_k K_k t^k$$

is a polynomial with integer coefficients $K_k \ge 0$.

In particular, putting $\mu_k = \mu_k(\varphi)$, we deduce all the inequalities of Theorem 3.8 for this more general case.

The function φ is *perfect* or *tight* in case $K_k = 0$ for all k, because this holds if and only if the inclusion $M_t \subset M$ is homology-injective for all t. This is the case if and only if $\mu_k(\varphi) = \beta_k(M)$ for all k. Equivalently $M_{c-\varepsilon} \subset M_c$ is homology injective for small $\varepsilon > 0$ and every critical value $c = c_j$.

LEMMA 5.8 (THE LACUNARY PRINCIPLE). *Suppose* $\mu_{2l+1}(\varphi) - \beta_{2l+1}(M) = 0$ *for all* l. *Then* φ *is perfect. In particular* φ *is perfect in case* $\mu_{2l+1} = 0$ *for all* l.

PROOF. Write μ_i for $\mu_i(\varphi)$. By the Morse inequalities (Theorem 3.8) we have

$$(\mu_{2l} - \beta_{2l}) - (\mu_{2l-1} - \beta_{2l-1}) + \ldots + (\mu_0 - \beta_0) = K_{2l} \ge 0$$

and

$$(\mu_{2l+1} - \beta_{2l+1}) - (\mu_{2l} - \beta_{2l}) \ldots = (\mu_{2l+1} - \beta_{2l+1}) - K_{2l} = 0 - K_{2l} = K_{2l+1} \ge 0.$$

Therefore $K_{2l} = K_{2l+1} = 0$ for all l. $\qquad\square$

REMARK. The limit Swiss cheese M (page 7) is tight and its height functions are tight as well but they have no Poincaré polynomials, since $H_1(M)$ has no finite basis.

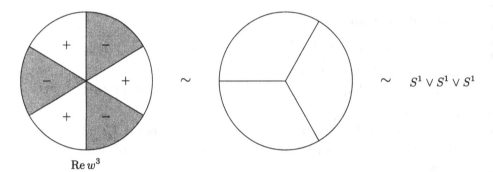

$\mathrm{Re}\,w^3$

Figure 19. There is homotopy equivalence (indicated by $\sim$) between the set of (5.1) and a bouquet of r circles.

DEFINITION. If $P(M_c, M_{c-}) = \sum_k \mu_{kc} t^k$, the critical value c counts for μ_{kc} critical points of index k, with total number

$$P(M_c, M_{c-})_{t=1} = \Sigma_k \mu_{kc} = \mu_c$$

critical points for the critical value c.

EXERCISE. Let the smooth function $\varphi : M^2 \to \mathbb{R}$ have at most one critical point p at level $\varphi^{-1}(p)$ (given by $d\varphi(p) = 0$) which in some topological chart is locally expressed by $(u, v) \to \mathrm{Re}(u + iv)^{r+1}$. Then our count is $\mu_1 = r$ for $r \geq 1$. The contribution in $P(\varphi)$ is t^r. The case $r = 1$ gives a nondegenerate critical point, $r = 2$ a monkey saddle, $r \geq 3$ an octopus saddle.

PROOF. (See Figure 19, where $r + 1 = 3$.) Look at the set of points

$$\{w = u + iv : |w| \leq 1, \ \mathrm{Re}\,w^{r+1} \leq 0 \text{ or } |w| = 1\} \tag{5.1}$$

modulo the set $\{w : |w| = 1\}$.   □

EXERCISE. Let φ be a PL-function on a PL-surface with one critical point p in the level $\varphi^{-1}(p)$. The level curve of φ near p is then in some chart either one point (minimum $\mu_0 = 1$ or maximum $\mu_2 = 1$; in both cases we have $\mu = 1$) or it is the union of $2(r + 1) \geq 4$ straight segments ending in p. In a suitable topological chart, we obtain the figure on the right. The contribution in $P(\varphi)$ at level $\varphi(p)$ is t^r [Kuiper 1971].

References

[Banchoff 1965]   T. F. Banchoff, "Tightly embedded 2-dimensional polyhedral manifolds", *Amer. J. Math.* **87** (1965), 462–472.

[Banchoff 1971a]   T. F. Banchoff, "High codimensional 0-tight maps on spheres", *Proc. Amer. Math. Soc.* **29** (1971), 133–137.

[Banchoff 1971b] T. F. Banchoff, "The two-piece property and tight n-manifolds-with-boundary in E^n", *Trans. Amer. Math. Soc.* **161** (1971), 259–267.

[Banchoff 1974] T. F. Banchoff, "Tight polyhedral Klein bottles, projective planes, and Möbius bands", *Math. Ann.* **207** (1974), 233–243.

[Borsuk 1947] K. Borsuk, "Sur la courbure totale des courbes fermées", *Ann. Polon. Math.* **20** (1947), 251–265.

[Brehm and Kühnel 1992] U. Brehm and W. Kühnel, "15-vertex triangulations of an 8-manifold", *Math. Ann.* **294**:1 (1992), 167–193.

[Casson 1986] A. J. Casson, "Three lectures on new-infinite constructions in 4-dimensional manifolds", pp. 201–244 in *À la recherche de la topologie perdue*, edited by L. Guillou and A. Marin, Progr. Math. **62**, Birkhäuser Boston, Boston, MA, 1986. With an appendix by L. Siebenmann.

[Chern and Lashof 1957] S.-s. Chern and R. K. Lashof, "On the total curvature of immersed manifolds", *Amer. J. Math.* **79** (1957), 306–318.

[Donaldson 1983] S. K. Donaldson, "An application of gauge theory to four-dimensional topology", *J. Differential Geom.* **18**:2 (1983), 279–315.

[Donaldson 1986] S. Donaldson, "Connections, cohomology, and the intersection forms of 4-manifolds", *J. Diff. Geom.* **24** (1986), 275–341.

[Eells and Kuiper 1962] J. Eells, Jr. and N. H. Kuiper, "Manifolds which are like projective planes", *Publ. Math. Inst. Hautes Études Sci.* **14** (1962), 5–46 (181–222 in year).

[Fáry 1949] I. Fáry, "Sur la courbure totale d'une courbe gauche faisant un nœud", *Bull. Soc. Math. France* **77** (1949), 128–138.

[Fenchel 1929] W. Fenchel, "Über die Krümmung und Windung geschlossener Raumkurven", *Math. Ann.* **101** (1929), 238–252.

[Freedman 1982] M. H. Freedman, "The topology of four-dimensional manifolds", *J. Diff. Geom.* **17**:3 (1982), 357–453.

[Kühnel and Banchoff 1983] W. Kühnel and T. F. Banchoff, "The 9-vertex complex projective plane", *Math. Intelligencer* **5**:3 (1983), 11–22.

[Kuiper 1941] N. H. Kuiper, "Lijnen in R_4", *Nieuw Archief voor Wiskunde* (2) **21** (1941), 124–143.

[Kuiper 1959] N. H. Kuiper, "Immersions with minimal total absolute curvature", pp. 75–88 in *Colloque Géom. Diff. Globale* (Bruxelles, 1958), Centre Belge Rech. Math., Louvain, 1959.

[Kuiper 1962] N. H. Kuiper, "On convex maps", *Nieuw Arch. Wisk.* (3) **10** (1962), 147–164.

[Kuiper 1966] N. H. Kuiper, "C^r-functions near non-degenerate critical points", mimeoraphed notes, Warwick Univ., Coventry, 1966.

[Kuiper 1970] N. H. Kuiper, "Minimal total absolute curvature for immersions", *Invent. Math.* **10** (1970), 209–238.

[Kuiper 1971] N. H. Kuiper, "Morse relations for curvature and tightness", pp. 77–89 in *Proceedings of Liverpool Singularities Symposium* (Liverpool, 1969/1970), vol. 2, edited by C. T. C. Wall, Lecture Notes in Math. **209**, Springer, Berlin, 1971.

[Kuiper 1980] N. H. Kuiper, "Tight embeddings and maps: Submanifolds of geometrical class three in E^{n}", pp. 97–145 in *The Chern Symposium* (Berkeley, 1979), edited by W.-Y. Hsiang et al., Springer, New York, 1980.

[Kuiper and Meeks 1983] N. H. Kuiper and W. H. Meeks, III, "Sur la courbure des surfaces nouées dans R^3", pp. 215–217 in *Third Schnepfenried geometry conference* (Schnepfenried, 1982), vol. 1, Astérisque **107-108**, Soc. Math. France, Paris, 1983.

[Kuiper and Meeks 1984] N. H. Kuiper and W. H. Meeks, III, "Total curvature for knotted surfaces", *Invent. Math.* **77**:1 (1984), 25–69.

[Kuiper and Meeks 1987] N. H. Kuiper and W. H. Meeks, III, "The total curvature of a knotted torus", *J. Differential Geom.* **26**:3 (1987), 371–384.

[Kuiper and Pohl 1977] N. H. Kuiper and W. F. Pohl, "Tight topological embeddings of the real projective plane in E^5", *Invent. Math.* **42** (1977), 177–199.

[Lastufka 1981] W. S. Lastufka, "Tight topological immersions of surfaces in Euclidean space", *J. Diff. Geom.* **16** (1981), 373–400.

[Little and Pohl 1971] J. A. Little and W. F. Pohl, "On tight immersions of maximal codimension", *Invent. Math.* **13** (1971), 179–204.

[Milnor 1950] J. W. Milnor, "On the total curvature of knots", *Ann. of Math.* (2) **52** (1950), 248–257.

[Milnor 1953] J. W. Milnor, "On the total curvature of closed space curves", *Math. Scand.* **1** (1953), 289–296.

[Milnor 1963] J. Milnor, *Morse theory*, Annals of Mathematical Studies **51**, Princeton U. Press, Princeton, 1963.

[Moise 1952] E. E. Moise, "Affine structures in 3-manifolds V: The triangulation theorem and Hauptvermutung", *Ann. of Math.* (2) **56** (1952), 96–114.

[Morse 1939] A. P. Morse, "The behavior of a function on its critical set", *Ann. Math.* **40** (1939), 62–70.

[Palais 1963] R. S. Palais, "Morse theory on Hilbert manifolds", *Topology* **2** (1963), 299–340.

[van Rooij 1965] A. C. M. van Rooij, "The total curvature of curves", *Duke Math. J.* **32** (1965), 313–324.

[Sharpe 1989] R. W. Sharpe, "A proof of the Chern-Lashof conjecture in dimensions greater than five", *Comment. Math. Helv.* **64**:2 (1989), 221–235.

[Spanier 1966] E. H. Spanier, *Algebraic Topology*, McGraw-Hill, New York, 1966.

[Wilson 1965] J. P. Wilson, "The total absolute curvature of an immersed manifold", *J. London Math. Soc.* **40** (1965), 362–366.

NICOLAAS H. KUIPER
⋆ JUNE 28, 1920
† DECEMBER 12, 1994

Tight and Taut Submanifolds
MSRI Publications
Volume 32, 1997

Tight Submanifolds, Smooth and Polyhedral

THOMAS F. BANCHOFF AND WOLFGANG KÜHNEL

ABSTRACT. We begin by defining and studying tightness and the two-piece property for smooth and polyhedral surfaces in three-dimensional space. These results are then generalized to surfaces with boundary and with singularities, and to surfaces in higher dimensions. Later sections deal with generalizations to the case of smooth and polyhedral submanifolds of higher dimension and codimension, in particular highly connected submanifolds. Twenty-six open questions and a number of conjectures are included.

CONTENTS

Introduction

The theory of tight submanifolds starts with attempts to generalize theorems about convex surfaces to topologically more complex surfaces such as the torus. For surfaces, it is possible to develop this generalization in terms of an elementary notion, the two-piece property, which then leads to the study of critical points of height functions and the theory of total absolute curvature. These notions can then be applied for higher-dimensional objects in higher-dimensional Euclidean spaces, producing a rich collection of examples and theorems in the global geometry of submanifolds.

An object in ordinary three-dimensional space is said to have the *two-piece property*, or TPP, if any plane cuts it into at most two pieces. Examples of surfaces with the TPP are spheres and ellipsoids and, more generally, the boundary of any bounded convex body. There are also nonconvex objects with boundaries that have the TPP: for example, a torus of revolution (Figure 1), or, more generally, a surface of revolution obtained by revolving a convex curve around an axis in the plane of the curve and not meeting the curve. If we deform a sphere into a nonconvex surface, for example a U-shaped object, or a sphere with a dent in it, the resulting surfaces (Figure 2) will not have the TPP.

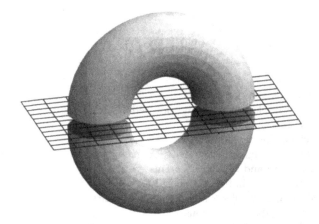

Figure 1. A torus of revolution has the two-piece property.

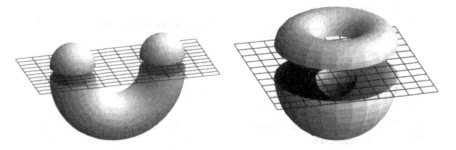

Figure 2. Smooth spheres without the two-piece property.

For closed subsets, the TPP is equivalent to the condition that the intersection of the object with every closed half-space is connected.

For compact surfaces (without boundary), the TPP is closely related to the study of *critical points of height functions*. Any plane in space can be considered a level set of a height function in a direction perpendicular to the plane. If a plane cuts a surface into more than two pieces, a height function perpendicular to this plane must have at least one maximum or minimum on each piece. It follows that if a surface does not have the TPP, there must be a height function with at least two (strict) local maxima on the surface. Conversely, if a height function has two strict local maxima on a surface, the half-space above the level set containing the lower of the two will intersect the surface in at least two pieces. It follows that a surface has the TPP if and only if no height function restricted to the surface has more than one strict local maximum. A surface with this property is called *tight*. An equivalent definition for a surface to be tight is that every local support plane be a global support plane.

The TPP is a topological condition, so it applies to any surface in space, whether it be smooth, polyhedral, or just a one-to-one continuous image of such a surface. If the surface happens to be sufficiently smooth, it is possible to characterize the tightness condition in terms of the surface's *total* or *Gaussian curvature*. Any point of *positive* curvature is a local extremum of the height function perpendicular to the tangent plane at the point. So the tightness condition implies that in any direction there is at most one point on the surface with positive curvature that is critical for that direction. For almost any direction, the maximum of the height function in that direction on the surface will occur at a point of positive curvature. It follows that, if a smooth surface is tight, the only strict local maxima of any height function must occur on the "outside", where the surface intersects its *convex hull*, the smallest convex set containing the surface.

For a smooth surface embedded in three-dimensional space, tightness can be expressed in terms of the *Gauss spherical image mapping*, which sends each point of the surface to the point of the unit sphere centered at the origin having the same outer unit normal vector. (This definition assumes that a consistent field of unit normals has been chosen over the whole surface.) For any smooth surface without boundary, almost every point of the sphere is the image of at least one point with positive curvature, so the total area of the spherical image of the part with positive curvature is at least 4π. For a tight smooth surface, almost every point of the sphere is the image of *exactly* one point of the surface with positive curvature, so the total area of the spherical image of the positive curvature part of the surface achieves the minimum value, namely 4π. Originally this property was used as the definition of tightness for smooth surfaces in ordinary space.

Although the first definition of tightness was given in terms of curvature, the critical point or TPP reformulation is much broader in scope. It applies not only to smooth surfaces in space but also to polyhedral surfaces. The critical point condition and the TPP extend naturally to surfaces embedded in higher-dimensional Euclidean spaces, and to immersions and to mappings with singularities.

It was Nicolaas Kuiper who made the first wide-ranging and systematic study of tight embeddings and immersions of surfaces, in three dimensions and higher. He produced tight embeddings of all orientable surfaces and tight immersions of all but three nonorientable surfaces in three-dimensional space. He proved that two of the remaining surfaces, the real projective plane and the Klein bottle, could not be immersed tightly in three-space even as topological surfaces, and he conjectured that the final case, a real projective plane with one handle, could not be immersed tightly into three-space. Only recently was François Haab able to prove that conjecture for smooth immersions, and even more recently Davide Cervone produced a surprising example to show that this surface can be immersed tightly as a polyhedral surface.

There are still a number of unsolved problems concerning immersions of surfaces into four-space, but thanks to the work of Kuiper the situation for five-space is better understood. First of all, Kuiper showed that any smooth immersion of a surface into Euclidean n-space for $n \geq 5$ must lie in a five-dimensional affine space; moreover, if the image does not lie in a four-dimensional subspace, the surface is the real projective plane and the immersion is affinely equivalent to the Veronese embedding, an algebraic surface. An even stronger result by Kuiper and William Pohl states that any topological tight embedding of the real projective plane into five-space whose image does not lie in a four-space must be either the smooth Veronese embedding or a simplexwise linear embedding of a triangulation with exactly six vertices.

In order to appreciate the nature of the theorems of Kuiper, it is useful to consider the TPP and tightness for closed curves in Euclidean spaces. A convex curve in the plane has the TPP, whether it is smooth or polygonal or a more general topological embedding of the circle. If an embedded curve in the plane is not convex, it does not coincide with the boundary of its convex hull, and there is a segment in the convex hull boundary containing points not in the curve. A line containing this segment bounds a half-space meeting the curve in at least two pieces, so a nonconvex plane curve does not have the TPP. Furthermore, if a curve is not planar, then there are four points on the curve, in cyclic order, not lying in a plane, and it is possible to find a plane with the first and third of these points on one side and the second and fourth on the other; this plane separates the curve into at least four pieces. Thus a TPP curve in n-space is necessarily contained in an affine two-space.

When the curve is smooth, we may recast the preceding result in terms of curvature to obtain a famous theorem in global differential geometry due to Werner Fenchel: the total curvature of a smooth closed curve in any dimension is at least 2π, and if it is exactly 2π, the curve is a convex plane curve.

For two-dimensional surfaces, the TPP restricts not only the number of maxima and minima of height functions but also the total number of critical points. This follows from the *critical point theorem* of elementary Morse theory: for almost every height function on a smooth surface, the only critical points are maxima, minima, and ordinary saddle points, and the number of maxima plus the number of minima minus the number of saddles is constant and equal to the *Euler characteristic* of the surface, also described as the number of vertices minus the number of edges plus the number of triangles in any triangulation of the surface. By "integrating" this theorem over all height functions, we obtain one of the most famous of all theorems in global differential geometry, the *Gauss–Bonnet Theorem*, relating the integral of the total curvature of a smooth surface to its Euler characteristic. We may use this fact to obtain other characterizations of tightness.

For higher-dimensional manifolds, the situation is quite different. The two-piece property no longer places such a strong restriction on the nature of the critical points of height functions. We say that an n-manifold is *tight* if the intersection of the object with any half-space is no more complicated than it has to be, that is, if the homology of the intersection is not greater than the homology of the whole object. For example, if a manifold is simply connected, we require that the intersection with every half-space also be simply connected. Thus a closed hemisphere has the TPP, but it is not tight since it is simply connected but there is a half-space that intersects it in a circle.

Morse theory gives lower bounds for the numbers of critical points of various types for almost all smooth functions defined on a higher-dimensional manifold. Tightness for higher-dimensional submanifolds of a Euclidean space requires that almost all height functions have the minimal number of critical points. Fenchel's theorem was generalized by Chern and Lashof by considering the *Lipschitz–Killing curvature* of a submanifold of a higher-dimensional Euclidean space. The total measure of the absolute value of this curvature is equal to the integral over the sphere of the number of critical points of height functions on the submanifold. Fenchel's theorem is generalized by the result that an m-dimensional sphere is immersed with minimum total absolute curvature if and only if it is a convex hypersurface in an affine subspace of dimension $m + 1$.

In this article, we will develop the theory of tight submanifolds primarily in the smooth and polyhedral situations. A related article based on the work of Nicolaas Kuiper will develop this theory for topological immersions.

1. Tight Surfaces

1.1. Definitions and notations. By a *surface*, we mean a connected two-dimensional manifold without boundary, unless stated otherwise (Section 1.5). We use the term *closed surface* to denote a compact surface without boundary.

For any closed surface M embedded in Euclidean three-space $\mathbb{E}^3$, and for any unit vector z, the height function in the direction of z achieves its absolute maximum on some subset of M. A point of M is a *local maximum* for z when a neighborhood of the point is contained in the half-space below the plane perpendicular to z through the point.

We say that the surface M is *tight* if, for almost every unit vector z, the height function in the direction of z has a unique local maximum on M. This condition is equivalent to the *Two-Piece Property*, or TPP, which states that any plane in space cuts the surface into at most two pieces. Equivalently, a surface M has the TPP if the intersection of M and any open or closed half-space is connected. Note that the property of tightness is invariant under projective transformations of three-space, since such transformations are homeomorphisms that send planes to planes.

If the closed surface satisfies some additional properties, we may find equivalent formulations of the tightness condition in terms of familiar quantities like curvature and critical points. We will be especially concerned with two important classes of surfaces: *smooth* and *polyhedral*.

For a smooth surface M in $\mathbb{E}^3$, at every point there is a well-defined tangent plane and a pair of unit normal vectors perpendicular to the tangent plane. When M is smooth and embedded, the surface is orientable and it is possible to make a choice of unit normal vector at each point over the entire surface. The *Gauss mapping* assigns each point of M to the point on the unit sphere having the same unit normal vector. For a region U, the *algebraic area* of the image of U under the Gauss mapping is given by the integral

$$\int_U K \, dA$$

of the *Gaussian curvature* function K, where the Gauss mapping preserves the orientation in a neighborhood of a point when the sign of K is positive at that point and it reverses orientation in a neighborhood of a point when the sign of K is negative at that point.

For almost all directions z on the unit sphere, only a finite number of points of M are sent to z under the Gauss mapping. At the one among those points that is highest in the direction for z, the orientation of the surface will be preserved. It follows from this observation that

$$\int_{M \cap \{K \geq 0\}} |K| \, dA \geq 4\pi.$$

Equality occurs when there is exactly one local maximum for almost every height function, so this condition gives an equivalent definition of tightness in the case of a smooth surface. This definition first appears in the work of A. D. Aleksandrov [1938].

One of the most fundamental results in elementary differential geometry is the *Gauss–Bonnet Theorem*, which states that the total algebraic area over an entire smooth closed surface M is independent of the embedding, and is given by $\int_M K\, dA = 2\pi \chi(M)$, where $\chi(M)$ denotes the *Euler characteristic* of the surface M, the number of vertices minus the number of edges plus the number of triangles in a triangulation of the surface.

It follows that

$$\int_M |K|\, dA = \int_{M \cap \{K \geq 0\}} K\, dA - \int_{M \cap \{K \leq 0\}} K\, dA$$

$$= 2 \left(\int_{M \cap \{K \geq 0\}} K\, dA \right) - 2\pi \chi(M) \geq 2\pi\big(4 - \chi(M)\big).$$

The integral $\int_M |K|$ is called the *total absolute curvature* of M, and M is *tight* when this functional achieves its minimum value.

By a fundamental result in elementary Morse theory, for almost all unit vectors z, the height function in the direction of z, when restricted to the smooth closed surface M, is *nondegenerate*, with isolated singularities that are either local maxima, local minima, or ordinary saddle points. For any such function, the number of maxima plus the number of minima minus the number of (ordinary) saddles equals the Euler characteristic $\chi(M)$.

If M is tight, almost every height function will have exactly one maximum and one minimum, so the number of saddles will be $2 - \chi(M)$ and the total number of critical points will be $4 - \chi(M)$, the minimum number of nondegenerate critical points a function can have on a closed surface. The total absolute curvature of a surface M equals the integral over the unit sphere, in the sense of Lebesgue, of the number of critical points of height functions on M corresponding to unit vectors on the sphere.

We may summarize these comments about smooth surfaces as follows:

DEFINITION 1.1.1 (SMOOTH TIGHT CLOSED SURFACES). A smooth (at least C^2) immersion of a closed surface $f : M \to \mathbb{E}^3$ is said to be *tight* if one of the following equivalent conditions is satisfied:

(i) $\frac{1}{2\pi} \int_M |K|\, dA = 4 - \chi(M)$.

(ii) Every nondegenerate height function $zf : x \mapsto \langle fx, z \rangle$ in the direction of a unit vector $z \in \mathbb{R}^3$ has exactly one local minimum and one local maximum (and, consequently, $2 - \chi(M)$ saddle points).

(iii) $\frac{1}{2\pi} \int_M |K|\, dA = \sum_i b_i(M; \mathbb{Z}_2)$. Here $b_i(M; \mathbb{Z}_2)$ denotes the i-th *Betti number* $b_i = \dim_{\mathbb{Z}_2} H_i(M; \mathbb{Z}_2)$.

(iv) f has the Two-Piece Property (TPP): For every plane $H \subset \mathbb{E}^3$, the complement $f^{-1}(M \setminus H)$ has at most two connected components.

(v) For every open half-space h bounded by a plane H, the induced morphism $H_*(f^{-1}(h)) \to H_*(M)$ is injective, where H_* denotes the singular homology with coefficients in $\mathbb{Z}_2$.

NOTE. The TPP condition is also called *0-tightness* because in this case the homomorphism $H_0(f^{-1}(h)) \to H_0(M)$ is injective for any open half-space h.

The equivalence of (i) and (ii) and (iv) has been established above. The equivalence of (ii) and (iii) follows from elementary Morse theory, and that of (iv) and (v) follows from Poincaré duality. For orientable surfaces one can replace $\mathbb{Z}_2$ by any field or by $\mathbb{Z}$ in condition (v).

In the case of a smooth immersion of a nonorientable surface M into $\mathbb{E}^3$, we will not have a field of unit normal vectors, and even in the case of an immersion of an orientable surface, it may be that the Gauss mapping is not surjective. We can take care of each of these situations by considering not just one but both unit normal vectors at each point of the surface. We define a double spherical image mapping that assigns to each point the pair of unit vectors perpendicular to the tangent plane at the point, and we get the total curvature by taking half the total integral of this double spherical image mapping.

For an immersion $f : M \to \mathbb{E}^N$ of the closed surface M for $N > 3$, the equivalences in Definition 1.1.1 still hold if we replace the word "plane" by "hyperplane" and if we no longer use the Gaussian curvature of the surface itself but rather the Lipschitz–Killing curvature of the unit normal bundle of the surface; that is, if we replace the total absolute curvature $\frac{1}{2\pi} \int_M |K|\, dA$ by

$$\mathrm{TA}(f) := \frac{1}{c_{N-1}} \int_{\perp_f} |K|\, dV,$$

where K denotes the Lipschitz–Killing curvature (the determinant of the shape operator) in a normal direction, $\perp_f$ denotes the unit normal bundle with its canonical volume element, c_{N-1} denotes the volume of the unit sphere S^{N-1}. Compare [Willmore 1971; Cecil and Ryan 1985] for more details.

For the total absolute curvature of immersions into the sphere see [Teufel 1982]. For a generalization of tightness to the case of hyperbolic space or manifolds without conjugate points other than $\mathbb{E}^N$ see [Cecil and Ryan 1979; Bolton 1982].

The Two-Piece Property is a purely topological condition, and as such it can be applied even to topological embeddings or immersions where it does not make sense to talk about a curvature measure K or the collection of nondegenerate height functions. In particular, for polyhedral surfaces embedded in $\mathbb{E}^3$ we may use the TPP as the definition of tightness. Condition (ii) still applies, since we may consider surfaces for which almost all height functions have exactly one local maximum. We can no longer assume that almost all height functions restrict to functions on the surface with only maxima, minima, and ordinary saddles, since there might be more complicated isolated critical points. Nonetheless, for each unit vector not perpendicular to any edge of a closed polyhedral surface M, it is possible to assign an index of singularity to each vertex such that maxima and minima have index 1 and such that the sum of the indices equals $\chi(M)$. The number of critical points, counted with multiplicities, once again is greater than or equal to the sum of the Betti numbers of the surface, with equality when M

has the TPP. For further discussion, see [Banchoff 1970a; Banchoff and Takens 1975].

DEFINITION 1.1.2 (POLYHEDRAL CURVATURE). In the case of a polyhedral surface, we can also find analogues for the concept of curvature. Although there is not a well-defined Gauss mapping at the vertices of a polyhedron, we may assign to each vertex v the integral of the indices of singularity at v for all height functions determined by points of the unit sphere. This average can be shown to equal the *polyhedral curvature* $K(v)$ given by

$$K(v) := 2\pi - \sum_i \alpha_i,$$

where the α_i, for $i = 1, 2, \ldots$ denote the interior angles of the faces at v.

The analogue of the Gauss–Bonnet formula for a closed polyhedral surface is then $\sum_v K(v) = 2\pi\chi(M)$.

In the case of a smooth immersion, a fundamental theorem of Gauss states that the curvature K is intrinsic, dependent only on measurements made along the surface and not depending on the way the surface sits in space. The situation is quite different for polyhedra. Although the quantity $K(v)$ depends only on intrinsic measurements, these measurements do not give as much information about the way the surface is embedded in space. For example, if two smooth surfaces are intrinsically the same and if one is convex, then the other is as well. This is not true for polyhedra: for example, a regular icosahedron is intrinsically the same as a nonconvex polyhedron where one of the vertices is pushed in.

Nevertheless, we can define a concept of the positive curvature at a vertex v of a polyhedral surface M in $\mathbb{E}^3$: Namely, K_+ is the polyhedral curvature of the local convex hull around v. If v happens to be an interior vertex of the convex hull of its neighbors, we set $K_+ := 0$. Thus K_+ represents the area of the collection of unit vectors such that the associated height function has index 1 at v. Define K_- by the condition $K = K_+ - K_-$. Then $K_* := K_+ + K_-$ is the analogue of the absolute Gaussian curvature. As in the smooth case, $\sum_v K_+(v) \geq 4\pi$, and we obtain the inequality [Brehm and Kühnel 1982]

$$\sum_v K_*(v) \geq 2\pi(4 - \chi(M)).$$

As before, the polyhedral surface M will be tight if equality is achieved.

One of the consequences of tightness for polyhedral surfaces is that a vertex v with $K(v) > 0$ (or $K_+(v) > 0$) has to lie on the boundary of its convex hull. More precisely, it has to be an *extreme vertex* of the convex hull, not contained in the interior of any segment in the convex hull.

We say that a height function in the direction of the unit vector z is *general with respect to M* if it takes distinct values at distinct vertices (so it is never constant on an edge). For an embedded polyhedral surface M, the nongeneral height functions correspond to unit vectors contained in a finite union of great circles. So almost all height functions are general for M.

We may summarize our discussion of tight polyhedral surfaces as follows:

DEFINITION 1.1.3 (POLYHEDRAL TIGHT CLOSED SURFACES). A polyhedral immersion of a closed surface $f : M \to \mathbb{E}^3$ is said to be *tight* if one of the following equivalent conditions is satisfied:

(i) $\frac{1}{2\pi} \sum_v K_*(v) = 4 - \chi(M) = \sum_{i \geq 0} b_i(M; \mathbb{Z}_2)$.

(ii) Every height function $zf : x \mapsto < fx, z >$ in the direction of a unit vector $z \in \mathbb{E}^3$ that is general for M has exactly one local minimum and one local maximum (and, consequently, $b_1 = 2 - \chi(M)$ saddle points, counted with multiplicity).

(iii) $\frac{1}{2\pi} \int_M |K| \, dA = \sum_i b_i(M; \mathbb{Z}_2)$.

(iv) f has the TPP.

(v) For every open half-space h bounded by a plane H, the induced morphism $H_*(f^{-1}(h)) \to H_*(M)$ is injective.

(The last three conditions are the same as their counterparts in Definition 1.1.1.)

In contrast to the smooth situation, where we had to modify the definitions of curvature for immersions of surfaces, or surfaces in higher codimension, these five conditions remain equivalent for any polyhedral mapping of a triangulated surface into Euclidean space of any dimension. For polyhedral surfaces in $\mathbb{E}^N$, with $N \geq 4$, condition (i) is not well defined, but the conditions (ii), (iii), (iv) still remain unchanged, and they are still equivalent.

DEFINITION 1.1.4 (TOPSETS, SUBSTANTIAL CODIMENSION). A *supporting hyperplane* for a compact subset M of $\mathbb{E}^N$ is a hyperplane that contains at least one point of M such that M is entirely contained in one of the two half-spaces determined by the hyperplane. The intersection of M with a supporting hyperplane is called an *i-topset* if it is contained in an i-dimensional linear subspace but not in any $(i - 1)$-dimensional linear subspace. A *topset* is defined as an i-topset for some i. A topset of an immersion is defined as the preimage of a supporting hyperplane. A topset is thus a subset of M where a certain height function attains its maximum (or minimum).

A closed surface M is called *substantial in* $\mathbb{E}^N$ if it is not contained in any hyperplane. An immersion f is called *substantial* if its image is substantial. If f is polyhedral and substantial in $\mathbb{E}^N$, the convex hull of $f(M)$ is an N-dimensional convex polytope, or N-polytope for short. The topsets are the preimages under $f(M)$ of the faces of this polytope. For general facts about convex polytopes see [Grünbaum 1967; Brøndsted 1983].

If f is smooth, its convex hull admits a stratification into smooth i-topsets of various dimensions. Not necessarily every i between 0 and $N - 1$ has to occur in this case: The standard sphere has only 0-topsets, while the convex hull of a torus of revolution in $\mathbb{E}^3$ and the Veronese surface in $\mathbb{E}^5$ (see Example 1.2.4 below) possess only 0-topsets and 2-topsets. A piece of cylinder capped off by two hemispheres has 0- and 1-topsets. An analytic tight closed surface can never have 1-topsets or 2-topsets.

1.2. Basic examples: smooth and polyhedral.

EXAMPLE 1.2.1 (GENUS 0). The boundary of any bounded three-dimensional convex body is a tightly embedded two-sphere. The easiest condition to verify is the TPP, which works no matter what degree of smoothness is present. It is possible to find smooth examples of any order C^k, for $k \geq 0$. The unit two-sphere is an example of a real analytic tight embedding. The boundary polyhedral surface of a Platonic solid, or of any convex three-polytope, is a polyhedral tightly embedded two-sphere.

EXAMPLE 1.2.2 (GENUS 1). A torus of revolution, obtained by revolving a circle around a disjoint axis in its plane, is a tight analytic surface in $\mathbb{E}^3$. Such a torus can be obtained by stereographic projection of the Clifford torus $S^1 \times S^1 \subset \mathbb{E}^2 \times \mathbb{E}^2 = \mathbb{E}^4$, a tight torus contained on the three-sphere S^3 of radius $\sqrt{2}$ in four-space. (Any tight surface situated on a sphere satisfies a stronger condition, called *tautness* or the *spherical TPP*; see [Cecil 1997] in this volume.) A polyhedral analogue of the Clifford torus is the Cartesian product of two square polygons in two-dimensional planes in four-space, regarded as a subcomplex of the four-cube, which it itself a Cartesian product of two square regions. The vertices of this polyhedron lie on a three-sphere, and stereographic projection from the north pole of this sphere sends the vertices of the tight "polyhedral torus of revolution" to the vertices of a Schlegel diagram of the four-cube.

An essentially different example of a tight polyhedral torus in 3-space is Császár's torus, a polyhedron without diagonals [Császár 1949; Bokowski and Eggert 1991]; see Figure 3. Any two of the seven vertices are joined by an edge. As a consequence, any simplexwise linear embedding of this particular triangulation into $\mathbb{E}^4$, $\mathbb{E}^5$ or $\mathbb{E}^6$ is tight. The abstract 7-vertex triangulation of the torus was already known to Möbius [1886], and, as the union of two disjoint triple systems on 7 points, it was mentioned by Cayley [1850].

EXAMPLE 1.2.3 (HIGHER GENUS). In order to construct smooth tight orientable surfaces of higher genus in $\mathbb{E}^3$, we start with the boundary of the convex hull of a thin torus of revolution (an ε neighborhood of a horizontal circle). This convex smooth surface contains horizontal circular discs. We may then remove smaller circular discs from these discs and attach rotationally symmetric handles given by the parts of negative curvature of other tori of revolution. This is an instance of what is known as "tight surgery" $S^0 \times B^2 \to B^1 \times S^1$. One can see that condition (i) of Definition 1.1.1 remains satisfied. Polyhedral analogues are obtained by removing pairs of small square discs from parallel faces of a cube, then connecting the square boundaries by tubes with square cross-section. We can also carry out a similar polyhedral construction to join pairs of convex polygons in disjoint planar faces of a tight polyhedral surface.

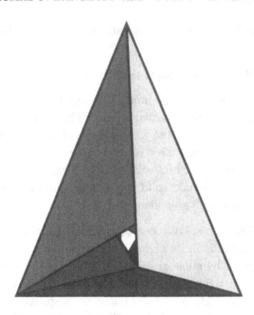

Figure 3. Császár's seven-vertex embedding of the torus.

A similar type of construction is possible in the case of a tight smooth surface in $\mathbb{E}^4$ that contains a pair of parallel planar pieces. We can obtain such a tight torus as the product of a pair of curves consisting of a pair of parallel segments capped by two semicircles. This surface contains a pair of parallel square regions lying in a three-dimensional affine subspace of $\mathbb{E}^4$, so we may add negative curvature tubes as above. Note that these examples will not be real analytic, and in fact there are no real analytic surfaces of genus greater than 1 in $\mathbb{E}^4$ not lying in any affine hyperplane [Thorbergsson 1991]. This leads to an open problem:

QUESTION 1. *For a polyhedral tight surface in $\mathbb{E}^4$, it is possible to join two polygons in different faces by a polyhedral tube, preserving tightness, even if the two faces do not lie in any three-dimensional subspace. Is an analogous construction possible for smooth surfaces, to produce examples of tight surfaces of higher genus by adding handles to nonparallel flat pieces?*

For any N it is possible to construct tight polyhedral surfaces in $\mathbb{E}^N$ not lying in any hyperplane. We can obtain such examples recursively as follows: Start with any tight polyhedral closed surface substantially embedded in $\mathbb{E}^N$, and translate a copy of it into a parallel hyperplane of $\mathbb{E}^{N+1}$. Then remove a number of open two-faces with pairwise disjoint boundaries such that every vertex of the given surface is contained in one of the boundaries. Remove the corresponding faces in the parallel copy, and join corresponding boundaries by polyhedral tubes. The resulting surface has the TPP, so this procedure gives a tight polyhedral surface substantially embedded in $\mathbb{E}^{N+1}$, of at least twice the genus of the original surface. Applying this procedure to the boundary of the three-cube gives the polyhedral torus in the four-cube constructed earlier. We may repeat the procedure to obtain a series of highly symmetric tight surfaces embedded as subcomplexes of higher-dimensional cubes. This series of examples has been developed independently by various authors [Banchoff 1965; Coxeter 1937; Kühnel and Schulz 1991; Kühnel 1995]. Surprisingly, these examples can also be realized as embedded polyhedra in $\mathbb{E}^3$ where the number of vertices can be smaller than the genus [McMullen et al. 1983].

EXAMPLE 1.2.4 (THE KLEIN BOTTLE). According to a theorem of Kuiper [1960; 1983b], it is not possible to find a tight immersion of the Klein bottle into $\mathbb{E}^3$, even as a topological surface. However, following a procedure analogous to the one used in the previous paragraph, we can construct a tight polyhedral embedding of the Klein bottle into $\mathbb{E}^4$. We start with a Möbius band in $\mathbb{E}^3$ with five vertices, all ten edges connecting pairs of vertices, and five triangular faces. Take a copy in a parallel hypersurface in $\mathbb{E}^4$ and connect corresponding vertices by segments and corresponding boundary segments by rectangles. The resulting tight polyhedral surface contains all edges of the convex hull of the ten points in $\mathbb{E}^4$. Compare [Ba7].

QUESTION 2. *Is there a tight smooth embedding or immersion of the Klein bottle substantially into $\mathbb{E}^4$?* (Haab [1994/95] has announced that the answer is negative.)

EXAMPLE 1.2.5 (THE PROJECTIVE PLANE). There is no tight immersion of the real projective plane into $\mathbb{E}^3$. This was proved by Kuiper [1960], who showed that it was not possible even for topological immersions. One of the most famous and most beautiful examples in the theory of tight surfaces is the *Veronese surface* contained substantially in a five-dimensional hyperplane in $\mathbb{E}^6$. It is defined as the immersion of the unit sphere

$$S^2 = \{(x, y, z) : x^2 + y^2 + z^2 = 1\}$$

into $\mathbb{E}^6$ by the formula

$$(x, y, z) \mapsto \left(x^2, y^2, z^2, \sqrt{2}\,xy, \sqrt{2}\,yz, \sqrt{2}\,zx\right).$$

Since antipodal points of the unit sphere are sent to the same image, and no other disjoint pairs are sent to the same point, this defines an embedding into $\mathbb{E}^6$ of the real projective plane, thought of as the unit sphere with antipodal points identified. The sum of the first three coordinates is 1, so the image is contained in a five-dimensional hyperplane, but it is not contained in any four-dimensional affine subspace. It is, however, contained in the five-sphere of radius 1, so also in the small four-sphere obtained by intersecting the unit sphere with the hyperplane. It follows that this surface is not only tight but also taut; see [Cecil 1997] in this volume.

The Veronese surface has the TPP. We can see this using projective geometry, since any hyperplane cuts the surface in a projective quadric that separates the projective plane into at most two pieces. Alternatively, one can verify condition (ii) in Definition 1.1.1 because any nondegenerate height function on the Veronese surface can be regarded as a quadratic function on S^2. Using Lagrange multipliers, we see that there are exactly six critical points on S^2, three pairs of antipodal points, giving the minimum number of critical points on the real projective plane $\mathbb{RP}^2$.

By stereographic projection, we can obtain a taut (therefore tight) embedding of the real projective plane into $\mathbb{E}^4$. The Veronese surface has a very special property called *secant-tangency*. Any secant joining two points of the Veronese surface is parallel to a line in a tangent plane to the surface, and from this it follows that any orthographic projection into four-space that leads to an immersion does in fact lead to an embedding. Such an embedding is automatically tight since almost all height functions on the projected image will still have the minimal number of critical points.

QUESTION 3. *Is it possible to attach a handle tightly to the Veronese surface projected into $\mathbb{E}^4$?*

By a result of Kuiper, it is not possible to carry out this construction smoothly in $\mathbb{E}^5$, and the construction in $\mathbb{E}^4$ would require very subtle details; compare Question 1.

A tight polyhedral embedding of the real projective plane into $\mathbb{E}^5$ is given by any simplexwise linear embedding of the unique six-vertex triangulation obtained by identifying antipodal points on a regular icosahedron (this is called the *hemi-icosahedron* in [Coxeter 1970]). As an abstract triangulation, the six-vertex $\mathbb{RP}^2$ was already known to Möbius [1886]. In this embedding the TPP follows from the fact that any two vertices are joined by an edge. As in the case of the Veronese surface, any orthogonal projection of this surface into $\mathbb{E}^4$ that results in an immersion automatically produces a tight embedding.

EXAMPLE 1.2.6. Nonorientable surfaces with even Euler characteristic $\chi \leq -2$

To construct a smooth tight immersion of a Klein bottle with one handle [Kuiper 1961], start with a tight torus smoothly embedded into $\mathbb{E}^3$ such that some line intersects the surface in four parallel flat pieces. We can then construct a tight tube as above, intersecting the original torus in a curve, thus producing an immersion of a nonorientable surface of Euler characteristic -2 (see Figure 4). Attaching additional handles produces tight immersions of all nonorientable surfaces with negative even Euler characteristic. Similar constructions can be carried out on polyhedral tori to obtain tight polyhedral immersions of such surfaces.

Figure 4. Kuiper's tight immersed Klein bottle with one handle, with a detail showing the self-intersection.

EXAMPLE 1.2.7 (NONORIENTABLE SURFACES WITH ODD EULER CHARACTER-ISTIC $\chi \leq -1$). An example with $\chi = -3$ can be obtained by attaching a certain interior part (two handles together with one cross-cap) to a convex outer part. The idea was described in [Kuiper 1961]; a more concrete description of a poly-hedral example admitting a tight smoothing was given in [Kühnel and Pinkall 1985]. Additional handles can be attached tightly.

This covers the case of odd Euler characteristic $\chi \leq -3$. The case $\chi = -1$, a projective plane with one handle, was mentioned as an open problem in [Kuiper 1961] and remained open until recently. For a long time it has been conjectured that existence or nonexistence would be the same for the smooth and the poly-hedral case. The solution was quite surprising: F. Haab [1992] proved that no smooth tight surface with $\chi = -1$ can exist, and recently D. Cervone [1994] found a polyhedral tight example (see also [Cervone 1997] in this volume).

EXAMPLE 1.2.8 (COMBINATORIAL). Assume there is given an abstract simplicial triangulation of a closed surface M with n vertices and $\binom{n}{2}$ edges (that is, any two vertices are joined by an edge). Regard the union of the triangles as a subcomplex of the $(n-1)$-dimensional simplex spanned by n vertices in $\mathbb{E}^{n-1}$. Then M is tightly and substantially embedded into $\mathbb{E}^{n-1}$ [Banchoff 1974; Kühnel 1980]. Special cases are the 7-vertex torus and the 6-vertex real projective plane.

1.3. Tight smooth surfaces. In a series of papers starting in 1958, Kuiper studied tight smooth embeddings and immersions of surfaces using a variety of approaches. He obtained existence and uniqueness results, obstructions to tight immersions, and characterizations of special examples. One of his basic observations is the following:

PROPOSITION 1.3.1 [Kuiper 1962]. *If* $f : M \to \mathbb{E}^N$ *is a smooth and substantial* 0-*tight immersion of a closed surface, then* $N \leq 5$. *In other words, a smooth closed surface immersed tightly in a Euclidean space must be contained in a three-, four-, or five-dimensional affine subspace.*

It is instructive to consider the analogue of this statement for a lower-dimensional situation, that of a smooth immersion of a closed one-dimensional manifold (a closed curve) into $\mathbb{E}^N$ possessing the TPP. We know that even for a topological immersion the TPP for a closed curve implies that the image is contained in a two-dimensional affine subspace. In the case of a smooth immersion, we observe that if the curve is not contained in the plane of the velocity and the acceleration vectors at a particular point, we can find a different plane containing the tangent line that cuts the curve away from a neighborhood of the point. This violates the TPP.

In the case of a two-dimensional surface in $\mathbb{E}^N$, at any given point all the tangent vectors to curves in the surface are contained in a two-dimensional affine subspace, and all the acceleration vectors are contained in a five-dimensional affine subspace containing that plane. If the surface is not wholly contained in that five-dimensional space, there is an affine hyperplane through the tangent plane meeting the surface away from a neighborhood of the point. This violates the TPP.

We can reformulate this argument in terms of the space of second fundamental forms as follows (compare [Cecil and Ryan 1985, p. 34]):

Let zf denote a nondegenerate height function attaining its maximum at a point $p \in M$. For any normal ξ at p, let A_ξ denote the second fundamental form in direction ξ. If the mapping $\xi \mapsto A_\xi$ is injective, the codimension $N-2$ cannot exceed the dimension of symmetric bilinear forms on the tangent plane, which for a two-dimensional surface is 3.

If this mapping were not injective, then for some ξ, the associated bilinear form would be zero. For any t, the height function in the direction of $zf + t\xi$ would have a local maximum at p; but since f is assumed to be substantial, the height function in the direction of ξ is not constant, so for some point q of M, it has a higher value at q than at p, contradicting the TPP.

PROPOSITION 1.3.2 [Chern and Lashof 1957; Kuiper 1960]. *Let* $f : S^2 \to \mathbb{E}^N$ *be a smooth and substantial tight immersion. Then* $N = 3$, *and* $f(S^2)$ *is the boundary of a convex body in* $\mathbb{E}^3$.

PROOF. Let $\mathcal{H}$ denote the convex hull of $f(S^2)$. A nonempty intersection of $\mathcal{H}$ with a hyperplane bounding a half-space containing the image of f is called a *topset*. If f is tight, every topset A of $\mathcal{H}$ is a convex set. A 0-topset is a point that must be contained in $f(S^2)$. A 1-topset is an interval that, by the TPP, is entirely contained in $f(S^2)$. A 2-topset of $\mathcal{H}$ has the property that its boundary is contained in $f(S^2)$, and the preimage of this boundary curve will separate the two-sphere into two pieces, each topologically equivalent to a disc. If the 2-topset is not contained in the image, the TPP is violated. Inductively, it follows that $f(S^2)$ would have to contain all of the topsets of $\mathcal{H}$ and finally the boundary of $\mathcal{H}$ itself. But this is impossible if $N \geq 4$. Hence $N = 3$, and $f(S^2)$ is a closed surface that contains the boundary of the convex body $\mathcal{H}$. This can only happen if $f(S^2)$ coincides with the boundary of its convex hull.   □

THEOREM 1.3.3 (EXISTENCE OF SMOOTH TIGHT SURFACES). *A tight and smooth immersion* $f : M \to \mathbb{E}^3$ *exists if*

(i) *M is orientable (and in this case, f can be chosen to be algebraic [Banchoff and Kuiper 1981]), or*

(ii) *M is nonorientable with $\chi(M) \leq -2$ (and if χ is even, f can be chosen to be algebraic [Kuiper 1983a]).*

Moreover, if M is not topologically equivalent to a two-sphere, then there exists a tight and substantial smooth immersion $f : M \to \mathbb{E}^4$ *if*

(iii) *M is orientable (but f can only be analytic in the case where M is the torus [Thorbergsson 1991]), or*

(iv) *M is nonorientable with $\chi = 1$, $\chi = -2$ or $\chi \leq -4$.*

PROOF. The existence in (i) and (ii) follows from the basic examples in Section 1.2. Case (iii) is also contained in Example 1.2.3, and the case $\chi = -1$ in (iv) is mentioned in Example 1.2.4.

For the nonorientable case $\chi = -2$ in $\mathbb{E}^4$ we start with the 4×4 torus as a subcomplex of the 4-cube $[0, 1]^4$ of Example 1.2.2, and consider the diagonally opposite squares with vertices $(0, 0, x, y)$ and $(1, 1, x, y)$. Since these squares lie in a common three-plane, it is possible to attach a polyhedral or smooth handle tightly. The outer torus can be chosen to be smooth as well; see Example 1.2.2. Additional handles can be attached tightly. This covers the case of even Euler characteristic $\chi \leq -2$. Note that these examples are embedded.

For the case of odd Euler characteristic one would like to have a starting example with $\chi = -3$ (or, even better, $\chi = -1$), which does not seem to be known. For $\chi = -5$ one can start with the torus in $\mathbb{E}^4$ as above and then attach in a certain three-plane the inner part of the surface with $\chi = -3$ in $\mathbb{E}^3$, as in Example 1.2.7. Then additional handles can be attached. This construction leads to self-intersections, which are not really necessary from the differential topological point of view; compare Question 7. □

Note that in each of cases (i)–(iv) (except possibly for $\chi = 1$) the smooth tight surface can be approximated by polyhedral tight surfaces, and it can also be obtained as the smoothing of a certain polyhedral example.

PROPOSITION 1.3.4 (TOP-CYCLES [Cecil and Ryan 1984]). *Let $f : M \to \mathbb{E}^3$ be a tight immersion of a closed surface (smooth or polyhedral), and let $\mathcal{H}$ denote the convex hull of $f(M)$. Then $\partial\mathcal{H} \setminus f(M)$ consists of a finite number of convex planar discs. Their boundaries are called top-cycles. The number $\alpha(f)$ of these top-cycles satisfies $2 \le \alpha(f) \le 2 - \chi(M)$. Moreover, if $\alpha(f) = 2 - \chi(M)$, the top-cycles are joined pairwise by cylinders of nonpositive Gaussian curvature, or by cylinders of $K_+ = 0$ in the polyhedral case. For nonorientable surfaces the sharper inequality $2 \le \alpha(f) \le 1 - \chi(M)$ holds.*

SKETCH OF PROOF. If the surface is not an immersed sphere, obviously there must exist at least one top-cycle. In a first step one has to show that the number of top-cycles is finite. Furthermore, each of the top-cycles lies in the part of the surface with vanishing Gaussian curvature. By the tightness, the part with positive Gaussian curvature is contained in the boundary of the convex hull, and the part with negative Gaussian curvature lies in the interior, connecting the various top-cycles with one another. Finally, their number can be related to the topology of the surface by a decomposition argument and the additivity of the Euler characteristic. □

In fact, any value for $\alpha(f)$ within the range of the inequalities above can be realized by a tight surface; for examples, see [Cecil and Ryan 1985, § 7.27].

EXAMPLE 1.3.5. *There exists a tight smooth immersion of the torus into $\mathbb{E}^3$ that is not an embedding (i.e., that has double points).*

The key to creating a tight smooth immersion of the torus that is not an embedding is to find a nonsingular tube with everywhere nonpositive curvature joining two convex curves in parallel planes. If $X(t)$ and $Y(t)$ are two convex closed curves both defined over the same interval $\{a \le t \le b\}$, with $X(t)$ in the horizontal plane through the origin and $Y(t)$ in the horizontal plane at height 1, then we may define a surface

$$Z(t, u) = uX(t) + (1-u)Y(t).$$

The partial derivative with respect to t will be horizontal, while the partial with respect to u is a nonzero vector from the plane at height 0 to the plane at height 1. These vectors can only be linearly dependent when the first is zero,

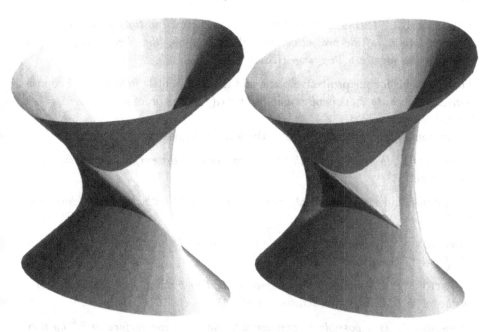

Figure 5. Construction of the inner part of a tight smooth immersion of a torus with self-intersections.

that is, when $uX'(t) + (1-u)Y'(t) = 0$ for some t and u, so the tube will be nonsingular if $X'(t)$ is never parallel to $Y'(t)$. One way to arrange this is to start with an ellipse $X(t)$ with unequal axes, then let $Y(t)$ be the parallel curve at distance r, where the distance is greater than the radius of curvature at any point of the ellipse. The tube constructed for these two curves will have four cuspidal edges meeting pairwise in a set of four swallowtail points, and there are two curves of double points. If we then rotate $Y(t)$ slightly, no $X'(t)$ will be parallel to the corresponding $Y'(t)$, so there will be no singularities. However for small enough rotations, there will still be intersection points near the original double point arcs. See Figure 5.

A similar example of a tight polyhedral immersion of a torus that is not an embedding is described in [Kuiper 1983b].

THEOREM 1.3.6 [Kuiper 1962; 1997]. *A tight and substantial smooth immersion $f : M \to \mathbb{E}^5$ exists only if M is the real projective plane, and its image is the Veronese surface (up to projective transformations of $\mathbb{E}^5$).*

This result is quite surprising, and the proof is difficult; compare the higher dimensional generalization given below as Theorem 2.4.2. It is also surprising that the tight Veronese surface $f : \mathbb{RP}^2 \to \mathbb{E}^5$ cannot be approximated by polyhedral tight surfaces [Kuiper and Pohl 1977]. Conversely, the tight polyhedral $\mathbb{RP}^2$ in 5-space cannot be approximated by smooth tight surfaces.

THEOREM 1.3.7 (NONEXISTENCE RESULTS). *There is no tight and smooth immersion into $\mathbb{E}^3$ of the projective plane* [Kuiper 1960], *the Klein bottle* [Kuiper 1960], *or the surface with $\chi = -1$* [Haab 1992].

The assertions for the projective plane and the Klein bottle follow from Proposition 1.3.4, because there is no possibility for $\alpha(f)$. The proof in the case $\chi = -1$ is much more involved.

The remaining open cases lead to the following questions.

QUESTION 4. *Is there a smooth tight immersion of the surface with $\chi = -1$ or $\chi = -3$ into $\mathbb{E}^4$?*

QUESTION 5. *Are there tight algebraic surfaces in $\mathbb{E}^3$ with odd Euler characteristic?*

QUESTION 6. *Are there tight analytic immersions of nonorientable surfaces into $\mathbb{E}^4$ if $\chi \neq 1$?* See [Haab 1994/95].

QUESTION 7. *Does there exist a smooth tight embedding into four-space of a surface with odd Euler characteristic $\chi \leq -1$?*

QUESTION 8. *Is it possible to approximate the Veronese surface in $\mathbb{E}^4$ by tight polyhedral surfaces?* A positive answer would also shed some light on Question 2.

QUESTION 9. *Is it possible to approximate Császár's seven-vertex torus in $\mathbb{E}^3$ by tight smooth surfaces?* One may start with any version of this polyhedron [Bokowski and Eggert 1991].

QUESTION 10. *Is there any difference between the case of C^2-immersions and the case of C^1-immersions, as far as existence or nonexistence of tight immersions is concerned?*

1.4. Tight polyhedral surfaces. Tight polyhedral surfaces were introduced in [Banchoff 1965]. There are a number of analogies with the smooth case, but also a number of significant differences. In particular, the structure of the topsets can be different from those in the smooth case, and the substantial codimension can be arbitrarily large (see Example 1.2.3). In [Banchoff 1974] the relationship with Heawood's map color problem was mentioned, and this was developed in a systematic way in [Kühnel 1980]. In this section we summarize the main results. For the details and proofs see [Kühnel 1995].

The 1-*skeleton* of a convex polytope, denoted by Sk_1, is defined as the set of all extreme vertices and extreme edges of the polytope. For example, the 1-skeleton of an N-dimensional simplex is the *complete graph* K_{N+1} on $N+1$ vertices.

LEMMA 1.4.1. (i) [Banchoff 1965] *Let $M \subset \mathbb{E}^d$ be a 0-tight and connected polyhedron. Then M contains the 1-skeleton of its convex hull: $\mathrm{Sk}_1(\mathcal{H}) \subset M$.*
(ii) *A polyhedral surface M with convex faces is 0-tight if and only if its 1-skeleton is 0-tight.*

PROOF. (i) Let e be an extreme edge of $\mathcal{H}$ with the extreme vertices v, w as its endpoints. By construction M contains v and w. There is a half-space h of $\mathbb{E}^d$ such that $h \cap \mathcal{H} = e$. Consequently we have $\{v, w\} \subset h \cap M \subset h \cap \mathcal{H} = e$. By the 0-tightness, $h \cap M$ must be connected. It follows that $h \cap M = e$.

(ii) If the one-skeleton is 0-tight, M is 0-tight because adding higher dimensional faces preserves the connectedness of $M \cap h$. Conversely, if $M \cap h$ is connected then $\mathrm{Sk}_1(M) \cap h$ must be connected because the faces are convex.

Note that this is not true if there are nonconvex faces. For example, if we remove two square regions from opposite faces of a cube and connect by a polyhedral tube, the resulting one-skeleton is not even connected, and if we make it connected by adding some diagonals, it is possible that the resulting one-skeleton will not be tight. □

The tightness of the one-skeleton of a polyhedron essentially means that (i) the one-skeleton of the convex hull is contained in the surface, and (ii) every vertex that is not a vertex of the convex hull lies in the relative interior of some of its neighbors. For example, a vertex might lie in the interior of a segment determined by two neighboring vertices, or in the interior of a triangle determined by three neighbors. These situations are not stable, in that tightness can be lost by small perturbations of the vertex. If, however, a vertex is in the interior of a tetrahedron spanned by four neighboring vertices, this situation is preserved under small perturbations of the vertex.

THEOREM 1.4.2 (EXISTENCE RESULTS IN SMALL CODIMENSION). *Let M be an abstract surface with Euler characteristic $\chi(M)$.*

(i) *There is a tight polyhedral embedding $M \to \mathbb{E}^3$ if M is orientable.*

(ii) *There is a tight polyhedral immersion $M \to \mathbb{E}^3$ if M is nonorientable and $\chi(M) \leq -1$ (see [Cervone 1994; 1997] for the case $\chi = -1$).*

(iii) *There are tight and substantial polyhedral embeddings $M \to \mathbb{E}^4$ and $M \to \mathbb{E}^5$ if M is not topologically equivalent to the two-sphere.*

(iv) *There is a tight and substantial polyhedral embedding $M \to \mathbb{E}^6$ if M is orientable and distinct from the two-sphere.*

In particular, any closed surface admits a tight polyhedral embedding into some Euclidean space $\mathbb{E}^N$.

PROOF. The proof consists in a series of examples, most of them already sketched in Section 1.2. The most difficult case is the nonorientable surface in $\mathbb{E}^3$ with $\chi = -1$, settled only recently by D. Cervone [1994; 1997]. The case $\chi = -3$ is also special, although it can now be obtained from the case $\chi = -1$ by attaching a handle. In [Kühnel and Pinkall 1985], a symmetric immersion of the surface with $\chi = -3$ is constructed in such a way that it can be smoothed to give a tight smooth immersion of the surface. By the nonexistence result of Haab, the Cervone immersions cannot be smoothed while maintaining tightness. □

For a tight torus in three-space, the intersection with the convex hull is a convex sphere with two open convex discs removed. In the case of a smooth torus, the closures of these two convex discs are disjoint, although in the polyhedral case they may intersect at a point or along a line segment. In any case, the complement of this outer part is an open cylinder strictly contained in the interior of the convex hull.

Although the outer part is necessarily a one-to-one image of a subset of the torus, the inner part, surprisingly, can have self-intersections. The first example of this phenomenon in the smooth case was observed by L. Rodríguez, and in the polyhedral case by Banchoff. Since this example has not been published elsewhere, we include it here.

Start with a tetrahedron inscribed in a cube, with vertices

$$v_1 = (1,1,1), \ v_2 = (1,-1,-1), \ v_3 = (-1,1,-1), \ v_4 = (-1,-1,1).$$

Shift the triangles $v_1 v_3 v_4$ and $v_2 v_3 v_4$ in the direction $(t,0,0)$ and the triangles $v_1 v_2 v_3$ and $v_1 v_2 v_4$ in the direction $(-t,0,0)$. For t between 0 and 1, the intersection of these triangles will be a skew quadrilateral with vertices $(1-t)v_1, (1-t)v_3, (1-t)v_2, (1-t)v_4$. To these four triangles, we may attach two parallelograms $v_1 + (t,0,0)$, $v_1 + (-t,0,0)$, $v_3 + (-t,0,0)$, $v_3 + (t,0,0)$ and $v_2 + (t,0,0)$, $v_2 + (-t,0,0)$, $v_4 + (-t,0,0)$, $v_4 + (t,0,0)$. This produces a self-intersecting polyhedral cylinder with boundary given by two parallelograms $v_1 + (t,0,0)$, $v_1 + (-t,0,0)$, $v_4 + (-t,0,0)$, $v_4 + (t,0,0)$ and $v_2 + (t,0,0)$, $v_2 + (-t,0,0)$, $v_3 + (-t,0,0)$, $v_3 + (t,0,0)$. We may obtain a tightly immersed polyhedral torus by attaching this cylinder to the surface of a square prism with vertices $(\pm 2, \pm 2, \pm 1)$, with the regions removed that are bounded by the parallelograms $v_1 + (t,0,0)$, $v_1 + (-t,0,0)$, $v_4 + (-t,0,0)$, $v_4 + (t,0,0)$ and $v_2 + (t,0,0)$, $v_2 + (-t,0,0)$, $v_3 + (-t,0,0)$, $v_3 + (t,0,0)$. See Figure 6.

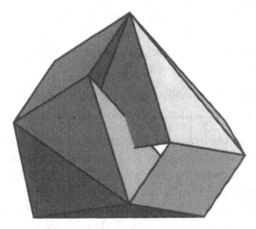

Figure 6. The inner part of a tight polyhedral torus with self-intersections.

ATTACHING LEMMA 1.4.3 [Kühnel 1980]. *Let M be a tight polyhedral surface, substantial in $\mathbb{E}^N$, for $N \geq 4$. Then there is a tight polyhedral $M \# S^1 \times S^1$ and a tight $M \# \mathbb{RP}^2$ in the same space $\mathbb{E}^N$, obtained just by local modifications of M (attaching a handle or a cross-cap).*

A handle can be attached tightly also in $\mathbb{E}^3$; see Example 1.2.3. For the construction of attaching a cross-cap one can replace the cone over a skew pentagon by a five-vertex Möbius band.

LEMMA 1.4.4 [Grünbaum 1967]. *The one-skeleton of any convex N-polytope contains the complete graph K_{N+1} as a subset (not necessarily as a subgraph).*

COROLLARY 1.4.5 [Banchoff 1971a]. *Let $f : S^2 \to \mathbb{E}^d$ be a tight and substantial polyhedral embedding or immersion. Then $d = 3$, and $f(S^2)$ is the boundary of a convex three-polytope.*

PROOF. By Lemma 1.4.1, $f(S^2)$ contains the one-skeleton of its convex hull. This in turn contains a K_{d+1} by Lemma 1.4.4. On the other hand, S^2 does not contain a K_5 because K_5 is not planar, hence it does not contain a K_{d+1} for any $d \geq 4$. Therefore $d + 1 = 4$. Assume that $f(S^2)$ is not identical with its convex hull $\mathcal{H}$. Then one of the two-dimensional faces of $\mathcal{H}$ is not contained in $f(S^2)$. On the other hand, the boundary of this two-face is certainly contained in $f(S^2)$ because this boundary contains all edges of $\mathcal{H}$. It separates S^2 into two pieces. Therefore it follows that a certain plane in $\mathbb{E}^3$, parallel to this two-face, would separate $f(S^2)$ into more than two pieces, a contradiction to 0-tightness. □

THEOREM 1.4.6 (NONEXISTENCE RESULTS IN SMALL CODIMENSION). (i) *There is no tight polyhedral immersion of the real projective plane or of the Klein bottle into $\mathbb{E}^3$ (not even topologically [Cecil and Ryan 1985; Kuiper 1983b]).*
(ii) *There is no tight and substantial polyhedral immersion of S^2 into $\mathbb{E}^N$, for $N \geq 4$.*

Assertion (i) follows from Proposition 1.3.4. For assertion (ii), see Corollary 1.4.5. For the case of higher codimension, see Theorem 1.4.8 below.

By Lemma 1.4.4, a necessary condition for tightness of a polyhedral surface into $\mathbb{E}^d$ is the embeddability of the complete graph on $d+1$ vertices in the surface. This in turn is closely related with the Heawood map color problem [Heawood 1890; Ringel 1974; White 1984]. The following result of G. Ringel and J. W. T. Youngs expresses the embeddability of the complete graph in n vertices in terms of an inequality, known as *Heawood's inequality*, between n and the genus of the surface.

THEOREM 1.4.7 [Ringel 1974]. *For every abstract surface M of genus g, apart from the Klein bottle, the following conditions are equivalent:*

(i) *There exists an embedding $K_n \to M$.*
(ii) $\chi(M) \leq n(7 - n)/6$.

(iii) $n \leq \frac{1}{2}\left(7 + \sqrt{49 - 24\chi(M)}\right)$.

(iv) $\binom{n-3}{2} \leq 3\left(2 - \chi(M)\right) = 6g$.

For the Klein bottle, condition (i) *is equivalent to* $n \leq 6$. *Moreover if equality holds in* (ii)–(iv), *the embedding of* K_n *induces an abstract triangulation of* M.

THEOREM 1.4.8 [Kühnel 1980]. *For an abstract surface* M *and a number* $n \geq 6$, *the following conditions are equivalent:*

(i) *There exists a tight and substantial polyhedral embedding* $M \to \mathbb{E}^{n-1}$.

(ii) *There exists an embedding* $K_n \to M$.

Combining this with Theorem 1.4.7, we obtain:

COROLLARY 1.4.9 [Kühnel 1978, Theorem A]. *Let* M *be an abstract surface that is not topologically equivalent to the Klein bottle. Then there exists a tight substantial polyhedral embedding* $M \to \mathbb{E}^d$ *if and only if*

$$d_0 \leq d \leq \frac{1}{2}\left(5 + \sqrt{49 - 24\chi(M)}\right),$$

where $d_0 = 3$ *if* M *is orientable and* $d_0 = 4$ *if* M *is nonorientable. The Klein bottle can be embedded tightly and substantially into* $\mathbb{E}^4$ *and* $\mathbb{E}^5$ *but not into* $\mathbb{E}^k$, $k \geq 6$ [Franklin 1934; Banchoff 1974].

SKETCH OF PROOF OF THEOREM 1.4.8.

(i) $\Rightarrow$ (ii) is just a combination of Lemmas 1.4.1 and 1.4.4:

$$K_n \subset \mathrm{Sk}_1(\mathcal{H}) \subset M \subset \mathbb{E}^{n-1}.$$

(ii) $\Rightarrow$ (i) : We start with an embedding $K_m \subset M$, where $m \leq n$ is maximal with respect to the inequality (iii) in Theorem 1.4.7. Then we extend it to a triangulation of M with those m vertices and some extra vertices. Finally the m vertices can be put into general position in $\mathbb{E}^{n-1}$, and the extra vertices have to be chosen in the relative interiors of their neighbors. This implies that the edge graph of this triangulation is 0-tight in $\mathbb{E}^{n-1}$. If the surface is embedded (i.e., without self-intersections) then it is 0-tight and tight by Lemma 1.4.1. Here it can be shown that the case of self-intersections can always be avoided by slight changes of the triangulation. $\square$

A triangulated surface (or any simplicial complex) with n vertices can always be regarded as a subcomplex of an $(n - 1)$-dimensional simplex Δ^{n-1}. We call this the *canonical embedding* of the triangulation. An abstract triangulation of a surface is called *tight* if its canonical embedding is tight.

COROLLARY 1.4.10 (TIGHT TRIANGULATIONS [Kühnel 1995]). *Let* M *be a triangulated surface of genus* g *with* n *vertices. Then the following conditions are equivalent:*

(i) *The triangulation is tight.*

(ii) *The triangulation is two-neighborly, that is, its edge graph is a complete graph K_n.*

(iii) $\binom{n-3}{2} = 3(2 - \chi(M)) = 6g.$

Conversely, given an abstract surface not topologically equivalent to the Klein bottle, and a number n satisfying (iii), *then there is a tight triangulation of M with n vertices.*

For a two-manifold M, let N_M denote the maximum dimension of Euclidean space admitting a tight and substantial polyhedral embedding of M. Let n_M denote the minimum number of vertices for any simplicial triangulation of M. Then the results of [Ringel 1955b; Jungerman and Ringel 1980; Huneke 1978] in connection with Theorem 1.4.8 can be reformulated as follows:

THEOREM 1.4.11. *For any surface M we have $N_M \leq n_M - 1 \leq N_M + 2$. Moreover, for any surface distinct from the Klein bottle, from the orientable surface of genus 2 and from the surface with $\chi = -1$, the sharper inequality $N_M \leq n_M - 1 \leq N_M + 1$ holds.*

THEOREM 1.4.12 [Banchoff 1965; Pohl 1981; Kühnel 1995]. *Let $M \subset \mathbb{E}^d$ be a tightly and substantially embedded polyhedral surface. Then:*

(i) $\binom{d-2}{2} \leq 3(2 - \chi(M)).$

(ii) *If $d \geq 4$, equality in* (i) *holds if and only if M is embedded as a subcomplex of a d-simplex Δ^d (and the induced triangulation is tight by Corollary 1.4.10).*

Equality in (i) is satisfied by the boundary of any convex three-polytope. This shows that (ii) cannot be extended to the case $d = 3$.

The case $\chi \neq 0$ was treated in [Banchoff 1974], the case $\chi = 0$ in [Pohl 1981; Kühnel 1995, § 2.17].

COROLLARY 1.4.13. *The image of a tight polyhedral real projective plane in $\mathbb{E}^5$ is affinely equivalent to the canonical embedding of the 6-vertex $\mathbb{RP}^2$, and the image of a tight polyhedral torus in $\mathbb{E}^6$ is affinely equivalent to the image of the 7-vertex torus.*

CONJECTURE 1.4.14 [Pohl 1981]. *Let $M \to \mathbb{E}^d$ be a tight and substantial topological immersion of a surface, and assume $d \geq 6$. Then:*

(i) *The convex hull of M in $\mathbb{E}^d$ is a convex polytope.*

(ii) $\binom{d-2}{2} \leq 3(2 - \chi(M)).$

(iii) *Equality in* (ii) *holds if and only if M is embedded as a subcomplex of a d-simplex Δ^d (and the induced triangulation is tight by Corollary 1.4.10).*

If (i) turns out to be true, (ii) and (iii) follow by the same arguments as in the proof of Theorem 1.4.12. A particular case of this conjecture is the uniqueness of the polyhedral model induced by the 7-vertex torus as the only tight topological torus in $\mathbb{E}^6$.

CONJECTURE 1.4.15. *Any tight torus that is centrally symmetric lies in a linear subspace of dimension at most four.* (This is true for a certain subclass of polyhedral surfaces [Kühnel 1996].)

1.5. Tight and 0-tight surfaces with boundary. For surfaces with boundary, the 0-tightness condition is much weaker than the condition of tightness. Recall that an object in $\mathbb{E}^3$ is *0-tight* (condition (iv) in Definition 1.1.1) if the intersection with every open or closed half-space is connected. For *tightness* (condition (iii) in Definition 1.1.1), we also require that every 1-chain in the intersection of an object and a half-space, that bounds in the object also bounds in the intersection of the object with the half-space. For example, a closed hemisphere satisfies the 0-tightness condition, but the plane containing the boundary circle bounds a half-space in which the circle is not a boundary of a 2-chain even though the circle bounds the hemisphere itself.

PROPOSITION 1.5.1. (i) *Assume that $f : M \to \mathbb{E}^3$ is a smooth tight immersion of a closed surface. Then for any $r \in \mathbb{N}$ there exists a smooth 0-tight immersion $\tilde{f} : M_r \to \mathbb{E}^3$ where M_r denotes the surface M with r open discs removed.*

(ii) *Assume that $f : M \to \mathbb{E}^N$ is a tight polyhedral immersion of a closed surface. Then for any $r \in \mathbb{N}$ there exists a 0-tight polyhedral immersion $\tilde{f} : M_r \to \mathbb{E}^N$.*

(iii) *The upper bound for the substantial codimension of 0-tight polyhedral surfaces M_r is the same as for the corresponding closed surfaces M.*

PROOF. To obtain (i), note that the image of a smooth tight immersion of a surface without boundary into $\mathbb{E}^3$ must have points of strictly positive curvature. By moving the tangent plane at any such point parallel to itself by an arbitrarily small amount, we can cut off a region bounded by a convex plane curve, and the resulting object will still have the TPP. This procedure can be repeated any desired finite number of times. It is not clear whether there are analogous constructions for smooth surfaces in $\mathbb{E}^4$ or higher.

In order to prove (ii), we may start with any tight polyhedral immersion of a surface M without boundary into $\mathbb{E}^N$ and then remove r disjoint convex polygonal open regions from any of the two-dimensional faces of the polyhedron to produce a TPP embedding of M_r into $\mathbb{E}^N$.

The proof of (iii) is already contained in the proof for the case of closed surfaces, Corollary 1.4.10.   □

L. Rodríguez [1976] constructed a 0-tight embedding in $\mathbb{E}^3$ of a torus with a disc removed by removing a nonconvex topological disc from the negative Gaussian curvature region of a torus of revolution. In order for the surface with boundary to remain 0-tight, it is necessary that the boundary curve consist of asymptotic curves, that is, curves where the tangent vector is directed along an asymptotic direction at the point (where an asymptotic direction is a null direction of the second fundamental form so that the principal curvature direction of the curve lies in the tangent plane of the surface at every point of the boundary curve).

PROPOSITION 1.5.2. *For any closed surface M (except $\mathbb{RP}^2$) that is known to admit a tight and substantial immersion $M \to \mathbb{E}^4$ (see Theorem 1.3.3), there is a smooth 0-tight substantial immersion $M_r \to \mathbb{E}^4$ for any $r \geq 1$. We may start with a tightly embedded surface containing a flat regions, and remove a number of disjoint open regions bounded by smooth convex curves.*

LEMMA 1.5.3. *For any immersion $f : M \to \mathbb{E}^N$ (smooth or polyhedral) of a compact surface M with boundary $\partial M \neq \varnothing$, the following two conditions are equivalent:*

(i) *f is tight.*
(ii) *f is 0-tight and $\mathcal{H}(fM) = \mathcal{H}(f(\partial M))$.*

Moreover, if f is smooth, we have $\mathrm{TA}(f|_{M \backslash \partial M}) + \frac{1}{2} \mathrm{TA}(f|_{\partial M}) \geq 2 - \chi(M)$, and condition (i) above is equivalent to either of the following conditions [Grossman 1972; Kühnel 1977; Rodríguez 1976]:

(iii) $\mathrm{TA}(f|_{M \backslash \partial M}) + \frac{1}{2} \mathrm{TA}(f|_{\partial M}) = 2 - \chi(M)$.
(iv) *For sufficiently small $\varepsilon > 0$, the boundary f_ε of the ε-tube is 0-tight.*

COROLLARY 1.5.4. *The image of a tight immersion of a two-disc into any $\mathbb{E}^N$ is a convex set contained in an affine two-dimensional subspace of $\mathbb{E}^N$.*

EXAMPLES 1.5.5. The first study of smooth tight immersions $f : M \to \mathbb{E}^2$ of surfaces with boundary into the plane was carried out by L. Rodríguez [1973]; compare [Kuiper 1997, Figures 8, 9, 10]. Except for the disc, these surfaces with boundary must have at least two boundary components: the boundary of the convex hull and one or more inner components that are "locally concave", that is, smooth curves such that, at each point, the part of the tangent line lying in a disc neighborhood of the point is contained in the image of the surface.

Smooth orientable tight immersions of surfaces with boundary $f :\to \mathbb{E}^3$ can be obtained by starting with a smooth immersion of a surface without boundary that contains a flat piece, and then removing a finite number of disjoint convex discs.

Nonorientable surfaces of this type can be obtained by cutting two convex holes into Kuiper's tight Klein bottles with a handle or into the tight surface with $\chi = -3$ described in [Kühnel and Pinkall 1985]; compare Examples 1.2.6 and 1.2.7. In any case handles can be attached and convex discs can be removed while still preserving the tightness.

CONJECTURE 1.5.6 [White 1974]. *There is no tight and substantial smooth immersion of any compact orientable surface with exactly one boundary component, except for the disc.*

The key lemma in [White 1974] states that this boundary curve would have to be planar and convex, in contradiction with Lemma 1.5.3. However, it seems that the proof of this lemma has never appeared. The conjecture may be extended to the nonorientable case as well. By a theorem of Kuiper [Kuiper 1971/72] there is no smooth tight Möbius band.

THEOREM 1.5.7 [Rodríguez 1976]. *There is no tight and substantial smooth immersion of any compact surface with nonempty boundary into* $\mathbb{E}^N$, *for* $N \geq 4$.

In contrast with Theorem 1.5.7, it is easy to construct polyhedral examples with boundary. Start with a tight polyhedral surface without boundary. Then cut out a finite number of disjoint convex holes from polyhedral faces in such a way that every vertex is contained in the boundary of one of the holes. Then the remaining surface with boundary is tight, according to Proposition 1.5.2. Another example of a polyhedral surface with the TPP is the torus with disk removed illustrated in Figure 7.

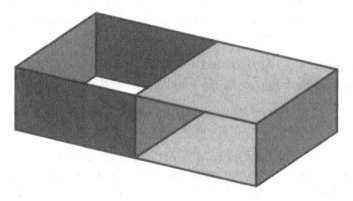

Figure 7. A tight polyhedral embedding of a torus minus a disc.

EXAMPLE 1.5.8. The 5-vertex Möbius band includes all edges joining pairs of vertices, so any simplexwise linear embedding of the 5-vertex Möbius band is tight. Similarly if we remove the open star of a vertex of the 7-vertex torus, we obtain a surface with boundary that contains every edge joining a pair of the remaining six vertices, so any simplexwise linear embedding of this surface with boundary is tight. These are examples of *tight triangulations*.

EXAMPLE 1.5.9. The product of two one-dimensional planar convex polygons γ_1, γ_2 is a tight torus in $\mathbb{E}^4$. Remove the products $\gamma_1 \times \{p\}$, where p ranges over all vertices of γ_2. This leads to a tight polyhedral torus in $\mathbb{E}^4$ with a number of holes.

EXAMPLE 1.5.10. The highest possible codimension for a tight polyhedral surface M without boundary is attained in the cases $N = \frac{1}{2}\left(5 + \sqrt{49 - 24\chi(M)}\right)$; see Corollary 1.4.10. By removing sufficiently many disjoint convex holes from one of the 2-faces, we obtain a tight surface in the same highest possible codimension. For the details see [Kühnel 1980].

EXAMPLE 1.5.11. A tight $\mathbb{RP}^2$ with three holes in $\mathbb{E}^5$ can be constructed as follows: Start with the 6-vertex triangulation, regarded as a subcomplex of the 5-simplex. Then cut three convex holes into the faces such that all 6 vertices are covered by the boundaries. Observe that two holes are not sufficient to do this: there is no tight polyhedral Möbius band with one hole, substantial in $\mathbb{E}^5$.

CONJECTURE 1.5.12 (EXTENSION OF POHL'S CONJECTURE). *If for a compact surface M with nonempty boundary $f : M \to \mathbb{E}^N$, with $N \geq 4$, is a tight topological immersion, then its convex hull $\mathcal{H}(fM)$ is a polyhedron.*

By [Kuiper 1971/72; 1980, Theorem 12], this is true for the Möbius band.

1.6. Congruence and rigidity theorems. We now turn to the question of whether or not two isometric tightly immersed surfaces are necessarily congruent.

THEOREM 1.6.1 (CONGRUENCE THEOREM FOR SMOOTH TIGHT TORI). *If two tightly immersed smooth tori in $\mathbb{E}^3$ are isometric, they are congruent if either*

(i) *they are analytic* [Aleksandrov 1938], *or*

(ii) *the immersions are at least five times differentiable,* $\operatorname{grad} K \neq 0$ *at every point for which $K = 0$, and an additional technical condition on asymptotic curves is satisfied* [Nirenberg 1963].

The situation for tight polyhedral surfaces is quite different:

EXAMPLE 1.6.2 (NONCONGRUENCE FOR POLYHEDRAL TIGHT SURFACES [Banchoff 1970b]). *There are pairs of tight polyhedral tori in $\mathbb{E}^3$ that are isometric but not congruent. By attaching handles one can obtain examples of higher genus as well.*

QUESTION 11. *Is it true that any two isometric smooth tight immersions of the torus into $\mathbb{E}^3$ are congruent?*

For smooth immersions, tightness is an intrinsic property: if two surfaces M and $\bar{M}$ are isometric and one of them is tight, the other one must be tight since the Gaussian curvature is intrinsic, so $\int |K| = \int |\bar{K}|$. Tightness of surfaces in higher codimension is definitely not an intrinsic property. Note also that for polyhedral surfaces, tightness is not an intrinsic property; in particular, a convex polyhedron can be isometric to a nonconvex polyhedron.

DEFINITION 1.6.3. A polyhedral surface in $\mathbb{E}^3$ is called *rigid* if it does not allow a globally defined continuous deformation (other than by Euclidean motions) where each edge and each face moves by a Euclidean motion (that is, moves rigidly).

A famous example of R. Connelly [1978/79] disproved the rigidity conjecture for polyhedral spheres in general. However, rigidity does hold for convex polyhedral surfaces [Connelly 1993]. In other words: *A tight polyhedral surface of genus 0 is rigid.*

For the case of higher genus this seems to be an open question:

CONJECTURE 1.6.4 [Kalai 1987]. *Any tight closed polyhedral surface in $\mathbb{E}^3$ is rigid.*

It is sufficient to consider only the case of triangulated surfaces because planar n-gons can always be subdivided into triangles. Then Conjecture 1.6.4 just says that the edge graph of any tight triangulated surface is rigid.

It is a trivial consequence of Theorem 1.4.11 that a tight surface is rigid if it is substantial in $\mathbb{E}^N$ with $N = \frac{1}{2}\left(5 + \sqrt{49 - 24\chi(M)}\right)$.

Note that a rigidity theorem is weaker than a congruence theorem: Rigidity does not a priori exclude the possibility of two noncongruent positions of the same polyhedron. It just says that there is no continuous one-parameter family of polyhedra of the same type joining them. The examples in [Banchoff 1970b] do allow such a continuous and isometric one-parameter family, but in this case the polyhedral structure is not preserved; faces are creased at continuously varying edges.

The rigidity (even infinitesimal rigidity) of smooth ovaloids in $\mathbb{E}^3$ is a classical result [Liebmann 1900]. For congruence theorems of ovaloids see [Blaschke and Leichtweiß 1973, Section 105] (smooth case) and [Pogorelov 1973] (general convex surfaces).

QUESTION 12. *Are there congruence or rigidity theorems for higher dimensional tight submanifolds, smooth or polyhedral?*

If the rank of the shape operator of a hypersurface is at least 3 everywhere, then a congruence theorem holds even locally. Besides cylinders with rank 1, there are examples of three-dimensional hypersurfaces in $\mathbb{E}^4$ with rank 2 that are isometric but not congruent [Hollard 1991]. For congruence and rigidity of convex hypersurfaces, see [Sen'kin 1972].

1.7. Isotopy, knots, and regular homotopy. Instead of studying immersions for which the minimum of the total absolute curvature is achieved, it is possible to consider lower bounds on the total absolute curvature for smooth immersions in a given isotopy class or regular homotopy class. The first examples of theorems in this area come from knot theory: By Fenchel's Theorem, a closed curve in $\mathbb{E}^N$ has total absolute curvature $\int |\kappa| \geq 2\pi$, with equality only for planar convex curves [Fenchel 1929]. However, for a knotted curve in $\mathbb{E}^3$, the total (absolute) curvature is more than twice that large: $\int |\kappa| > 4\pi$ [Fáry 1949; Fox 1950; Milnor 1950b]. This lower bound is not attained for any knot. More precisely, the infimum of $\frac{1}{2\pi} \int |\kappa|$ within a given isotopy class of embeddings is the *bridge number* of the knot, defined as the minimum number of relative maxima of any height function in this isotopy class. This infimum is attained only for the unknot (the isotopy class containing the circle). The same results hold for polygonal knots, and in fact the method of Milnor [1950b] made essential use of approximation of smooth curves by polygons in the same isotopy class.

These results have been extended to the case of knotted surfaces:

THEOREM 1.7.1 [Langevin and Rosenberg 1976; Meeks 1981; Morton 1979]. *Assume that for a smooth embedded orientable surface $M \subset \mathbb{E}^3$ the total absolute curvature satisfies*

$$\frac{1}{2\pi} \int_M |K| < 8 - \chi(M).$$

Then M is unknotted, that is, M is isotopic to the 'standard' embedding given in Example 1.2.3. The same conclusion holds for polyhedral surfaces satisfying

$$\frac{1}{2\pi} \sum_v K_*(v) < 8 - \chi(M).$$

The polyhedral case can be derived by the process of smooth approximation [Brehm and Kühnel 1982].

The bound on $\chi(M)$ cannot be improved, since there are surfaces for which the equality is achieved that are not isotopic to the standard embedding. This leads to the following notion [Kuiper and Meeks 1984]:

For a given isotopy class of embeddings, we call an embedded surface *isotopy tight* if the total absolute curvature realizes the infimum in this isotopy class. As mentioned above, a knotted curve is never isotopy tight.

THEOREM 1.7.2 [Kuiper and Meeks 1984; 1987]. (i) *A knotted torus is never isotopy tight.*

(ii) *There exist knotted surfaces of genus $g \geq 3$ that are isotopy tight and satisfy*

$$\frac{1}{2\pi} \int_M |K| = 8 - \chi(M).$$

These examples can also be made polyhedral, satisfying

$$\frac{1}{2\pi} \sum_v K_*(v) = 8 - \chi(M).$$

We can broaden the question of the existence of tight immersions of a surface by asking if there are tight mappings in a given regular homotopy class of immersions of a surface. Given two immersions f and g of a surface M, we say these immersions are *image homotopic* if there is a homeomorphism ϕ of M such that f and $g \circ \phi$ are regularly homotopic. Pinkall [Pinkall 1986b] classified all surfaces up to image homotopy, and exhibited tight immersions for a given image homotopy class in all but a finite number of cases. His examples are tight polyhedral immersions that can be smoothed preserving tightness by means of a specific algorithm. More recently, Cervone [Cervone 1996] constructed polyhedral immersions for most of the missing cases; only two cases remain unresolved. However, these models do not meet the demands of Pinkall's smoothing algorithm, and some may in fact represent tight polyhedral immersions in an image homotopy class for which there exists no tight smooth immersion. (Compare the case of the projective plane with one handle, for which there is a tight polyhedral immersion but no tight smooth immersion).

The image homotopy class of an immersion of a surface can be described in terms of the number of generators of the one-dimensional homology represented by curves with neighborhoods that are twisted cylinders. For example, if both generators of the first homology of a torus have twisted neighborhoods, the torus is not image homotopic to any embedding; such a torus is called a *twisted torus*. For the Klein bottle, one generator of the first homology has an orientable neighborhood and if this generator is represented by a twisted cylinder, then the immersion is called a *twisted Klein bottle*; such Klein bottles come in both left- and right-handed versions.

THEOREM 1.7.3 (REGULAR HOMOTOPY [Pinkall 1986b; Cervone 1996]). (i) *Apart from the previously mentioned cases of the Klein bottle and the real projective plane, there are no tight immersions of the twisted Klein bottle, the twisted torus, or the connected sum of three projective planes (all of the same handedness).*

(ii) *It is unknown whether there exist smooth or polyhedral tight immersions for the connected sum of three projective planes plus a handle, or for the twisted torus with a handle.*

(iii) *Tight polyhedral immersions exist for all the remaining image homotopy classes of surfaces* [Cervone 1996].

(iv) *There is no tight immersion of the projective plane with one handle, and no smooth examples are known for*

  (i) *the twisted Klein bottle with one, two or three handles,*

  (ii) *the connected sum of three projective planes with one, two or three handles,*

  (iii) *the Klein bottle with one twisted handle or the Klein bottle with one twisted handle and one standard handle, or*

  (iv) *the twisted torus with fewer than four handles.*

In addition to these results about mappings into $\mathbb{E}^3$, we mention another question concerning mappings into Euclidean four-space:

QUESTION 13. *Is there a tight smooth immersion of an orientable surface into $\mathbb{E}^4$ with an odd number of double points?*

The tube around such an example would be $\mathbb{Z}_2$-tight but not $\mathbb{Z}$-tight [Breuer and Kühnel 1997]. There is a polyhedral example of a tight torus in $\mathbb{E}^4$ with exactly one double point; see Example 1.8.6 below.

1.8. Tight mappings of surfaces with singularities. The title of this section has two aspects: On the one hand, it can mean that we consider ordinary nonsingular surfaces and tight mappings on them that fail to be immersions at

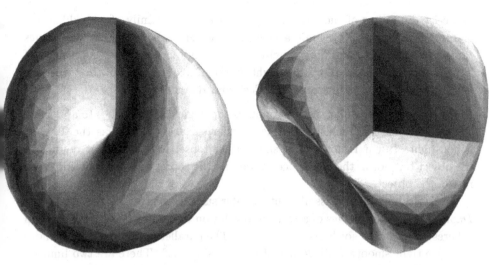

Figure 8. Tight stable mappings of the real projective plane: the cross-cap and Steiner's Roman surface.

certain points. On the other hand, it can mean that we consider surfaces with singular points, like a pinched sphere (a sphere with two points identified), and then consider tight mappings that are immersions, that is, locally one-to-one.

A smooth or polyhedral map $f : M \to \mathbb{E}^N$ from a surface is called *tight* if condition (iii) of Definition 1.1.1 is satisfied, so the preimage of any closed half-space is connected. A trivial example of a tight map on a connected topological space is the mapping that sends the entire space to a single point. More generally, the composition of a tight immersion into $\mathbb{E}^N$ with any linear mapping to a subspace of $\mathbb{E}^N$ produces a tight map.

René Thom proved that almost all differentiable mappings of nonsingular surfaces into $\mathbb{E}^3$ have at most a finite number of singularities, all equivalent to "pinch points", topologically equivalent to a cone over a figure-eight. Such mappings are called *stable*. In particular, any smooth immersion is a stable mapping.

THEOREM 1.8.1 [Kuiper 1975]. *All closed surfaces admit tight stable smooth maps into $\mathbb{E}^3$.*

In particular, Kuiper exhibited tight stable mappings into $\mathbb{E}^3$ for the real projective plane (see Figure 8) and the Klein bottle. Such mappings were analyzed further by Coghlan [1987; 1989], who also proved the following:

THEOREM 1.8.2 [Coghlan 1987; 1989]. *For any surface other than the sphere or the real projective plane, and any integer $n \geq 2$ there is a tight stable smooth map into $\mathbb{E}^3$ with exactly n top-cycles* (in contrast with Theorem 1.3.4).

EXAMPLE 1.8.3 (DUPIN CYCLIDES WITH SINGULARITIES). A torus of revolution obtained by revolving a circle around a disjoint axis in its plane is tight, and as we let the radius of the circle increase, we obtain a family of embedded tori of revolution having as a limit a singular mapping, where the circle becomes tangent to the axis and an entire circle is sent to a single point. As a limit of TPP mappings, this map is also a tight differentiable mapping of the torus into $\mathbb{E}^3$. This is called a "limit torus" in [Cecil and Ryan 1985].

The image of this singular tight mapping is topologically equivalent to a pinched sphere. The natural inclusion of this singular surface into $\mathbb{E}^3$ has the TPP so it is 0-tight. But it is not 1-tight since either topset is a circle that bounds in the object but not in the half-space containing the topset and not containing the rest of the object.

A related important class of tight singular surfaces is given by the *cyclides of Dupin*, obtained as images of a torus of revolution under inversion of $\mathbb{E}^3$ through spheres with centers not lying on the torus. The parallel surfaces of such cyclides also give tight smooth embeddings of the torus into $\mathbb{E}^3$. There are two limiting positions for such parallel surfaces, called *limit horn cyclide* and *limit spindle cyclide* [Cecil and Ryan 1985]. These are described by differentiable mappings of a torus into $\mathbb{E}^3$ that are singular along an entire circle that is mapped to a single point, and such mappings are tight. (They have the TPP, as limits of maps with the TPP).

These singular cyclides as point sets are topologically equivalent to the pinched sphere. The natural inclusions of these singular surface into $\mathbb{E}^3$ are both 0-tight but not 1-tight.

Note that conditions (iii) and (iv) in Definition 1.1.1 are not equivalent for such surfaces with singularities because this equivalence depends on the validity of Poincaré duality, which does not hold in this case. However, if we use the intersection homology [Goresky and MacPherson 1980] instead of ordinary singular homology, Poincaré duality is still valid for this type of surface with singularities, and it is possible to consider a notion of *intersection tightness*, where we consider condition (iii) of Definition 1.1.1 applied to intersection homology groups.

QUESTION 14. *Does any given closed surface with any given number of pinch points admit an embedding into some Euclidean space with intersection tightness?*

With respect to ordinary singular homology, we have the following lemma:

LEMMA 1.8.4. *Let $M \subset \mathbb{E}^N$ be a connected closed surface with a pinch point that is an isolated local maximum for a some height function. Then M is not tight.*

PROOF. For a connected closed surface with a finite number of pinch points the second homology with coefficients mod 2 is always one-dimensional. On the other hand, the maximum pinch point is a critical point of index two and multiplicity at least two, in contradiction to equality in the Morse inequalities. This argument is the same for smooth and polyhedral surfaces.   □

Note that it is easy to obtain 0-tight examples of this type, e.g., the limit horn cyclide.

COROLLARY 1.8.5. *Let $M \subset \mathbb{E}^N$ be a connected and tight closed surface with finitely many pinch points. Then each pinch point of M lies in the relative interior of the convex hull of ordinary points of M.*

The situation is altered for three-dimensional objects. There are three-manifolds with isolated singularities that have tight triangulations, and tight mappings into $\mathbb{E}^N$ for which there are height functions with an extremum at a singular point [Kühnel 1995, § 7.16].

EXAMPLE 1.8.6 [Kühnel 1995, Section 2F]. *There are projections of the 7-vertex torus into $\mathbb{E}^4$ with one double point* (by projecting along a line joining the centers of two triangles with disjoint closures). This can also be considered as a tight inclusion into $\mathbb{E}^4$ of a pinched torus, that is, a torus with two points identified.

EXAMPLE 1.8.7 [Kühnel 1992]. *There is a two-dimensional simplicial complex that is a surface with one singular cycle and a polyhedral immersion into $\mathbb{E}^4$ that is $\mathbb{Z}_3$-tight but not $\mathbb{Z}_2$-tight.*

This example is a 12-vertex triangulation containing all $\binom{12}{2}$ edges. However, as in Corollary 1.8.5, the three singularities have to lie in the interior of the 9 nonsingular vertices.

2. Tightness in Higher Dimensions

While the tightness condition and the TPP are identical for surfaces without boundary, and closely related in the case of more general two-dimensional objects like surfaces with boundary and surfaces with singularities, the situation is different when we consider higher-dimensional manifolds. No longer does control over the numbers of maxima and minima of height functions on an object enable us to conclude such strong results about the topology of the object. More subtle conditions have to be brought in, even in the case of smooth hypersurfaces. Although a number of important classes of tight examples have been discovered over the years, there are many areas where little is known about tightness in higher dimensions.

Nonetheless, in certain cases we have a good understanding of tightness, for example for manifolds topologically equivalent to spheres. The first results of this type in higher dimensions are found in [Chern and Lashof 1957; 1958].

2.1. The Chern–Lashof theorem and related results. As a first generalization of surfaces in $\mathbb{E}^3$, consider the case of a smooth hypersurface M embedded in $\mathbb{E}^N$. At each point there is an outward unit normal vector and we may determine a Gauss mapping sending each point to the point of the $(N-1)$-sphere with the same outward unit normal vector. As in the two-dimensional case, the

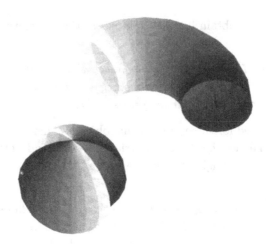

Figure 9. The spherical image of the normal tube around a space curve.

absolute measure of the image of a region U on the hypersurface, divided by the measure of the entire sphere, is called the *total absolute curvature* of the region, denoted TA(U). The total absolute curvature of the hypersurface itself can be expressed as one half the average number of critical points of height functions restricted to the hypersurface. For a strictly convex hypersurface, there will be exactly two critical points for every height function and TA(M) = 2. The theorem of Chern and Lashof shows that the condition TA(M) = 2 characterizes convex hypersurfaces even in the case where some height functions have degenerate critical points, as for example when the hypersurface contains flat pieces on which the Gauss mapping is constant.

Moreover, the Chern–Lashof results extend to immersions of surfaces of higher codimension. Let $f : M \to \mathbb{E}^N$ be an immersion of an n-dimensional submanifold M, where $n \leq N - 1$. Then, for sufficiently small r, the collection of normal vectors of length r perpendicular to the tangent space at a point forms a sphere of dimension $N - 1 - n$. If r is chosen sufficiently small, the union of all these normal spheres of radius r forms an immersed hypersurface called *the normal tube of radius r about M*. The total absolute curvature of the union of $(N-1-n)$-dimensional spheres at points of a region U is called the *total absolute curvature of U*, again denoted TA(U).

It is useful to recall the lowest dimension in which we may consider submanifolds that are not hypersurfaces, namely the case of curves in $\mathbb{E}^3$. There the normal tube of radius r about a curve M^1 is a torus of radius r, and a calculation shows that the total absolute curvature of the torus equals the integral of the absolute value of the curvature of the curve with respect to arclength (Figure 9).

Fenchel's Theorem says that, for M one-dimensional, we have TA(M) ≥ 2, with equality if and only if M is a plane convex curve. This result is generalized by the famous theorem by S.-S. Chern and R. K. Lashof, in part anticipated by Milnor [1950a]. We state it in its original form as follows:

THEOREM 2.1.1 [Chern and Lashof 1957; 1958]. *Let $f : M^n \to \mathbb{E}^N$ be a smooth immersion of a compact manifold. Then the total absolute curvature*

$$\mathrm{TA}(f) := \frac{1}{c_{N-1}} \int_{\perp_f} |K|$$

satisfies:

(i) $\mathrm{TA}(f) = (1/c_{N-1}) \int_z \sum_i \mu_i(zf)$.
(ii) $\mathrm{TA}(f) \geq \sum_i b_i(M; F)$ *for any field F.*
(iii) $\mathrm{TA}(f) < 3$ *implies that M is homeomorphic to the sphere S^n.*
(iv) $\mathrm{TA}(f) = 2$ *if and only if f is an embedding and $f(M)$ is the boundary of a convex body in an $(n+1)$-dimensional Euclidean subspace.*

If one assumes that (i) is given, then (ii) follows directly from the Morse inequalities $\mu_i(\phi) \geq b_i(M; F)$ for the number of critical points of any Morse function ϕ defined on M. In fact, almost all height functions in the sense of the Lebesgue measure satisfy this condition. Assertion (iii) follows from (i) by Reeb's theorem, which says that if a compact n-manifold that admits a Morse function with two critical points is homeomorphic to the sphere S^n.

The proof of (iv) requires more subtle geometric arguments. The main problem is how to deal with the parts of the manifold where the second fundamental form (or the Gauss mapping) degenerates. Assertion (iv) can be regarded as one of the many characterizations of convexity [Mani-Levitska 1993]. In particular, there is no immersion f of an exotic sphere with $\mathrm{TA}(f) = 2$, an observation already mentioned in [Kuiper 1959]. Generalizations for the case of less regular immersions (e.g., topological immersions) can be found in [Kuiper 1980; Lastufka 1981]. A generalization of (iii) can be obtained by using more general versions of Reeb's theorem.

Here is one of the consequences of Theorem 2.1.1:

COROLLARY 2.1.2. *If $\mathrm{TA}(f) < 4$, then M is either homeomorphic to S^n or it is a manifold with Morse number 3, a "manifold like a projective plane" in the sense of* [Eells and Kuiper 1962].

Such manifolds with Morse number 3 can occur only as the compactification of $\mathbb{E}^2$, $\mathbb{E}^4$, $\mathbb{E}^8$, or $\mathbb{E}^{16}$ by a "sphere at infinity" S^1, S^2, S^4 or S^8.

Even in the case where the total absolute curvature is not at its minimum, restrictions on the total absolute curvature can place conditions on smooth immersions of the sphere. For example, in codimension two, we have the following result of Ferus:

THEOREM 2.1.3 [Ferus 1968]. *Suppose that Σ is an n-manifold homeomorphic to the sphere S^n and admitting an immersion $f : \Sigma \to \mathbb{E}^{n+2}$ with $\mathrm{TA}(f) < 4$. Then Σ is diffeomorphic with the standard sphere.*

QUESTION 15. *Does there exist an immersion of an exotic sphere $f : \Sigma^n \to \mathbb{E}^{n+2}$ with $\mathrm{TA}(f) = 4$? Compare Example 2.7.6.*

Note that the restriction on the codimension is essential in the above theorem; if we allow higher codimension, we can always find an immersion with total absolute curvature arbitrarily close to the minimum value. Given any immersion $f : M^n \to \mathbb{E}^N$, and given any smooth function $g : M^n \to \mathbb{E}^1$, we may obtain an immersion $f = cg : M^n \to \mathbb{E}^{N+1}$, where c is a large constant. Then in $\mathbb{E}^{N+1}$, nearly all of the height functions restricted to $f = cg(M^n)$ will have the minimal number of critical points, and all the height functions with a larger number can be clustered in a region with arbitrarily small volume. In particular, if $f : S^n \to \mathbb{E}^N$ is an immersion of an exotic sphere, we can choose a function g on S^n with two critical points, and for sufficiently large c, we can make $\mathrm{TA}(f + cg) \leq 2 + \varepsilon$ for arbitrarily small positive ε.

In the case of immersions of noncompact manifolds, the Chern–Lashof inequality for the total absolute curvature is no longer valid in its original form. However, it is possible to derive an analogue by regarding the geometry and topology of the ends. We assume that $f : M \to \mathbb{E}^N$ is a proper immersion of a noncompact manifold with finitely many ends $\infty_1, \ldots \infty_k$ and finitely many limit directions in the sense of [Wintgen 1984]. A *limit direction* is a possible accumulation point of a sequence of normalized position vectors converging to an end. The number of limit directions is finite if the immersion converges to one direction at each end. In this case the Gauss–Bonnet formula remains valid [Wintgen 1984], and the following inequality holds:

THEOREM 2.1.4 [van Gemmeren 1996]. *Let $f : M \to \mathbb{R}^N$ be a proper immersion with finitely many limit directions. Then*

$$\mathrm{TA}(f) \geq \sum_i \mu_i(M) \geq \sum_i \left| b_i(M) - \tfrac{1}{2} b_i(\infty) \right|,$$

and the equality $\mathrm{TA}(f) = 1$ *for one end is possible only for convex hypersurfaces.*

The case of a cylinder shows that $\mathrm{TA}(f)=0$ is possible if there are two ends.

2.2. Tightness and k-tightness. The case of equality in Theorem 2.1.1(ii) is very special, because it forces almost all height functions zf to have the minimal number of critical points of index i, namely b_i, for all indices i. In this case every critical point is really necessary from the topological point of view, in that each critical point generates one additional homology class of the manifold. Such functions have been called "perfect functions" (of "linking type" in the sense of [Morse and Cairns 1969].) The condition of equality in Theorem 2.1.1(ii) is what we mean by *tightness* in higher dimensions.

DEFINITION 2.2.1. A smooth (at least C^2) immersion $f : M \to \mathbb{E}^N$ is said to be *tight* with respect to a field F if any of the following equivalent conditions is satisfied:

(i) $\mathrm{TA}(f) = \sum_{i \geq 0} b_i(M; F)$.

(ii) Every nondegenerate height function $zf : x \mapsto \langle fx, z \rangle$ in the direction of a unit vector $z \in \mathbb{R}^N$ has exactly b_i critical points of index i.

(iii) For every open half-space h the induced morphism $H_*(f^{-1}(h)) \to H_*(M)$ is injective, where H_* denotes the singular homology with coefficients in F.

It is also possible to obtain significant, although weaker, information about an immersion when there is a condition on the numbers of critical points of height functions of index less than or equal to some fixed number k. We have already seen that the TPP is equivalent to the condition that almost every height function restricted to the object has exactly one critical point of index 0, and this notion is known as 0-*tightness*. More generally we may define k-*tightness* for other values of k:

An immersion $f : M \to \mathbb{E}^N$ of an n-dimensional manifold is called k-*tight* with respect to F if in (ii) the number of critical points satisfies $\mu_0 = b_0$, $\mu_1 = b_1$, ..., $\mu_k = b_k$ or, equivalently, if in (iii) the induced morphism $H_i(f^{-1}(h)) \to H_i(M)$ is injective for $i = 0, 1, \ldots, k$.

In terms of critical point theory, tightness means that every critical point of a nondegenerate height function is homology-generating ("linking type"), whereas k-tightness means that this holds for critical points of index less or equal to k. By condition (iii) it follows that tightness (and k-tightness as well) is invariant under projective transformations of the ambient space. Although we have been concentrating on smooth immersions, the definition of k-tightness also makes sense for polyhedral immersions.

LEMMA 2.2.2 (DUALITY). *Let $f : M \to \mathbb{E}^N$ be an immersion (smooth or polyhedral) of a compact n-dimensional manifold satisfying Poincaré duality, so that $H_i(M; F) \cong H_{n-i}(M; F)$ for every i. Then the following conditions are equivalent:*

(i) *f is tight with respect to F.*

(ii) *f is k-tight with respect to F for one particular $k \geq \frac{1}{2}(n - 2)$.*

The proof relies on the duality for critical points $\mu_i(-zf) = \mu_{n-i}(zf)$ and the Poincaré duality $b_i = b_{n-i}$ in combination with the Euler–Poincaré equation $\sum_i (-1)^i \mu_i = \chi = \sum_i (-1)^i b_i$.

As an example, the Veronese surface is 0-tight for any field, but 1-tight and 2-tight only for fields of characteristic 2.

THEOREM 2.2.3. *The space of all smooth tight immersions of compact manifolds into Euclidean spaces is closed under the following operations:*

(i) *composition with projective transformations $P : \mathbb{E}^N \to \mathbb{E}^N$ (not sending any point of the manifold to infinity);*

(ii) *composition with linear embeddings $j : \mathbb{E}^N \to \mathbb{E}^{N+1}$;*

(iii) *cartesian products $f_1 \times f_2 : M_1 \times M_2 \to \mathbb{E}^{N_1 + N_2}$.*

Moreover, for a tight smooth immersion, the tube of sufficiently small radius r gives a tight immersion of the unit normal bundle. More precisely, if $f : M \to \mathbb{E}^N$ is tight then the ε-tube around $j \circ f : M \to \mathbb{E}^{N+1}$ is tight for any linear embedding j. If f itself is a tight embedding then the ε-tube around f itself is tight [Pinkall 1986a; Breuer and Kühnel 1997].

Note that $\mathrm{TA}(j \circ f) = \mathrm{TA}(f)$ by the choice of the normalization and that $\mathrm{TA}(f_1 \times f_2) = \mathrm{TA}(f_1)\,\mathrm{TA}(f_2)$. Formally, one follows from the other if j is regarded as the trivial embedding of the one-point space into $\mathbb{E}^1$. The proof of (iii) uses also the Künneth formula $b_k(M_1 \times M_2) = \sum_{i+j=k} b_i(M_1)b_j(M_2)$. For a more general version of (iii) see [Ozawa 1983].

2.3. Smooth examples. If we start with the tight examples in Section 1.2, Theorem 2.2.3 immediately leads to a large variety of higher-dimensional examples, just by taking products and tubes. The argument concerning tubes can also be extended to the case of submanifolds of spheres (connected with the notion of taut submanifolds).

EXAMPLE 2.3.1 (SPHERE PRODUCTS). The Cartesian product of an arbitrary number of tightly embedded spheres (smooth or polyhedral) leads to a tight embedding where the codimension equals the number of factors in the product. We can obtain examples of hypersurfaces topologically equivalent to products of spheres by iterating the tube construction. Special cases are tight n-tori $T^n \cong (S^1)^n$ recursively defined as follows: T^1 is the unit circle in the plane, and T^{n+1} is the ε-tube around $T^n \subset \mathbb{E}^{n+1} \subset \mathbb{E}^{n+2}$ with radius $\varepsilon = 2^{-n}$.

There is also a polyhedral version of the tube construction, where a circle in the normal plane is replaced by a square.

There also exists a smooth tight (and taut) immersion of a twofold quotient of the product $S^{p-1} \times S^{q-1}$ defined as the tensor product $S^{p-1} \otimes S^{q-1}$ in $\mathbb{E}^{pq}$ [Kühnel 1994b].

EXAMPLE 2.3.2 (CONNECTED SUMS OF HANDLES). As in Example 1.2.3, we can easily obtain tight hypersurfaces in $\mathbb{E}^4$ that are diffeomorphic to the standard three-manifold of Heegaard genus g, the connected sum of g copies of $S^1 \times S^2$: start with a convex hypersurface containing two flat regions in parallel hyperplanes. Then attach handles invariant under SO(3)-rotation by rotating the same curve as in Example 1.2.3 under SO(2)-rotation. Such a handle contributes $c_3 = \mathrm{Vol}(S^3)$ to the total absolute curvature. Nonorientable versions can be obtained by starting with a suitable $S^1 \times S^2$ containing two flat regions, one in the outside, the other in the inside. Then a handle joining outside and inside can be attached tightly. This extends Kuiper's construction of a tight Klein bottle with one handle (or more handles) to the case of arbitrary dimensions. Similar constructions are possible in codimension two and for connected sums of k-handles; see also Example 3.2.2.

QUESTION 16. *Is there any smooth tight immersion of a connected sum of at least two handles $S^k \times S^1$ ($k \neq 1$) in codimension at least three?*

Recall that there is a tight substantial embedding $S^1 \times S^{2k-1} \to \mathbb{E}^{4k}$, defined as the complexification of S^{2k-1} or the tensor product $S^1 \otimes S^{2k-1}$. For the case $k = 1$, compare Proposition 3.3.1.

QUESTION 17. *Is there a tight immersion of a lens space $L(p, 1)$ for $p \neq 2$? Is there any smooth tight immersion of a manifold with p-torsion for $p \neq 2$?*

There is no tight hypersurface that is a lens space, according to [Coghlan 1991].

EXAMPLE 2.3.3 (ISOPARAMETRIC SUBMANIFOLDS). Any isoparametric hypersurface in the sphere S^{N-1} is tight (and taut), regarded as a submanifold of $\mathbb{E}^N$. It is a tube around a so-called focal manifold, which is also tight (and taut) [Cecil and Ryan 1985]. Any isoparametric submanifold of arbitrary codimension is also tight (and taut) [Terng 1993]. Particular classical examples of isoparametric hypersurfaces are the tubes around the Veronese embeddings of projective planes into S^4, S^7, S^{13}, S^{25}. These are precisely the hypersurfaces of spheres with three distinct constant principal curvatures (compare also Examples 1.2.4 and 3.2.3). These Veronese embeddings are special cases of the following construction:

EXAMPLE 2.3.4 (GRASSMANNIANS). Let F denote either $\mathbb{R}$, $\mathbb{C}$, or the quaternions $\mathbb{H}$. Then the unoriented Grassmann manifold $\mathcal{G}_{p,q}(F)$ is defined as the set of all p-dimensional linear subspaces through the origin in F^{p+q}. Every such subspace A can be identified with the matrix representing the orthogonal projection from F^{p+q} onto A. This leads to the so-called *standard embedding*

$$\mathcal{G}_{p,q}(F) \to \mathbb{E}^{(p+q)^2},$$

which is substantial in a linear subspace of dimension $p + q - 1 + d\binom{p+q}{2}$, where $d = 1, 2, 4$ for $F = \mathbb{R}, \mathbb{C}, \mathbb{H}$, respectively. This standard embedding turns out to be tight with $\mathrm{TA}(f) = \binom{p+q}{p} = \sum_i b_i(\mathcal{G}_{p,q}(F); \mathbb{Z}_2)$. As an exceptional case, there is a standard embedding of the Cayley plane into $\mathbb{E}^{26}$ [Tai 1968; Kuiper 1970; Cecil and Ryan 1985, §9.4].

EXAMPLE 2.3.5 (UNITARY GROUPS AND R-SPACES). The unitary group $FU(n)$ for $F = \mathbb{R}$, $\mathbb{C}$, $\mathbb{H}$ can be regarded as a submanifold of Euclidean dn^2-space, just by regarding an element as a quadratic unitary matrix over F. This is also tight with total absolute curvature $\mathrm{TA} = 2^n = \sum_i b_i(FU(m), \mathbb{Z}_2)$. In the real case $FU(m) = O(m)$ is not connected, so one may restrict the construction to $SO(m)$.

In much more generality, the standard embedding of any symmetric R-space is tight (and taut) [Bott and Samelson 1958; Takeuchi and Kobayashi 1968]. Homogeneous Kähler manifolds can be tightly embedded [Kobayashi 1967]. For tight embeddings of other homogeneous spaces, see [Wilson 1969]. Not all homogeneous spaces can be embedded tautly [Thorbergsson 1988]. Ferus [1982] characterized the case of standard embeddings of symmetric R-spaces as the extrinsically symmetric submanifolds. For the connection between tautness and Dupin submanifolds, see [Cecil 1997] in this volume.

EXAMPLE 2.3.6 (INTERMEDIATE TIGHTNESS). Let $D^n \subset \mathbb{E}^n$ denote a tightly embedded n-ball that is rotationally symmetric with respect to the $\mathrm{SO}(n-1)$-action around a fixed axis. If we rotate it in higher-dimensional space $\mathbb{E}^{n+m}$, we obtain a tight $(n+m)$-ball D^{n+m} and a tight $(n+m-1)$-sphere as boundary. Similarly, if the D^n is only k-tight then D^{n+m} will be $(k+m)$-tight. The boundary in this case is $(k+m-1)$-tight, either by direct calculation or by applying Proposition 2.6.1 below. In particular, if $D^2 \subset \mathbb{E}^2$ is not tight (that is, not convex) and $\mathrm{SO}(1)$-symmetric (that is, congruent to its mirror image), the induced $D^3 \subset \mathbb{E}^3$ is 0-tight but not 1-tight, and its boundary is not 0-tight. The induced $D^4 \subset \mathbb{E}^4$ is 1-tight but not 2-tight, and its boundary is a 0-tight three-sphere that is not tight; compare [Kuiper 1970].

Curtin [1991] found similar examples with intermediate tautness, e.g., 0-taut three-spheres that are not taut. These examples are ellipsoids with rotational symmetry in two orthogonal planes. Inverse stereographic projection to $\mathbb{E}^5$ leads to 0-tight but not 1-tight three-spheres in codimension two, giving a positive answer to a question of Kuiper [1970].

2.4. The substantial codimension and the Little–Pohl theorem. The argument leading to the upper bound of the substantial codimension in Section 1.3 remains valid in arbitrary dimensions: The mapping $\xi \mapsto A_\xi$ is injective.

THEOREM 2.4.1 [Kuiper 1959; 1970]. *Let* $f : M^n \to \mathbb{E}^N$ *be a 0-tight smooth and substantial immersion. Then* $N - n \le \binom{n+1}{2}$.

Observe that equality is realized by the standard embeddings of $\mathcal{G}_{1,n}(\mathbb{R}) = \mathbb{RP}^n$ in Example 2.3.4.

THEOREM 2.4.2 [Little and Pohl 1971; 1985, p. 98]. *Let* $f : M^n \to \mathbb{E}^N$ *be a 0-tight smooth and substantial immersion with* $N = n + \binom{n+1}{2}$. *Then* f *is the standard embedding of* $\mathbb{RP}^n$ *(up to projective transformations of* $\mathbb{E}^N$*).*

SKETCH OF PROOF. One of the main tools is the 2-*jet* of the immersion spanning the so-called *osculating space* of second order. At an extreme point (a maximum of a nondegenerate height function) the TPP implies that the third-order derivatives stay in the same half-space as the first- and second-order derivatives. This has the consequence that $f(M)$ is contained in the osculating space at that point. Another step is to show that the dimension of the osculating space is in fact the special number $N = n + \binom{n+1}{2}$ at every extreme point. A sophisticated combination of this type of arguments with classical and very special properties of the Veronese embeddings of $\mathbb{RP}^n$ ultimately leads to the local (and global) coincidence of $f(M)$ with the Veronese embedding. $\square$

CONJECTURE 2.4.3. *The substantial codimension of any tight topological immersion of FP^n into Euclidean space is bounded by the codimension of the corresponding standard embedding in Example 2.3.3.*

This is true for $n = 1$ (tight spheres [Kuiper 1980]) and $n = 2$ (see Theorem 3.3.2). Compare [Arnoux and Marin 1991] for essentially the same bounds for the number of vertices of triangulations of FP^n.

QUESTION 18. *Are there nonsmooth tight substantial immersions of $\mathbb{RP}^3$ into $\mathbb{E}^9$?* Note that any smooth immersion is congruent to the standard embedding by Theorem 2.4.2. The "canonical" polyhedral candidate would be a 10-vertex triangulation, which, however, does not exist [Walkup 1970].

2.5. Tight polyhedra.

DEFINITION 2.5.1. For a compact polyhedron $M \subset \mathbb{E}^N$ *tightness* means that condition (iii) of Definition 2.2.1 is satisfied: For every open half-space h and for any i, the induced morphism $H_i\big(f^{-1}(h)\big) \to H_i(M)$ is injective, where H_* denotes the singular homology with coefficients in F. Naturally, k-tightness means the morphism is injective for $i = 0, 1, \ldots k$.

Equivalently, one can define tightness by the equality $\mu_i(zf) = b_i(M; F)$ for any height function in general position. In this case the number of critical points of index i is defined as

$$\mu_i(zf) := \sum_v \dim_F H_i\big(M_v, M_v \setminus \{v\}\big),$$

where the sum ranges over all vertices of the polyhedron, and where $M_v := \{p : (zf)(p) \le (zf)(v)\}$ denotes the sublevelset determined by z and v. For details of this type of critical point theory, including Morse relations and duality, see [Kuiper 1971; Banchoff 1967; Kühnel 1990; 1995]. It seems that the Chern–Lashof Theorem 2.1.1 and Corollary 2.1.2 remain valid for polyhedral immersions of manifolds. In particular, we have the following result:

PROPOSITION 2.5.2 (TIGHT SPHERES). *Any tightly embedded polyhedral sphere $\Sigma \subset \mathbb{E}^N$ of dimension n is the boundary of a convex polytope in some $(n + 1)$-space.*

PROOF. The proof of Corollary 1.4.5 can be carried over directly from dimension two to arbitrary dimension.

An alternative proof can be formulated using Corollary 3.1.3 below: Since the sphere is $(n - 1)$-connected, the tight polyhedral sphere in $\mathbb{E}^N$ must contain the n-dimensional skeleton of the convex hull $\mathcal{H}$, which is a convex N-polytope. For $N > n + 1$ this n-skeleton contains several distinct n-spheres, each as the boundary of one of the $(n + 1)$-dimensional faces. This is impossible, hence $N = n + 1$, and the image coincides with the boundary of $\mathcal{H}$.   □

For the case of even dimension $n = 2k$ one may compare this argument with the theorem of van Kampen and Flores [Grünbaum 1967, § 11.2], stating that there is no topological embedding of the underlying set of $\mathrm{Sk}_k(\triangle^{2k+2})$ into $\mathbb{E}^{2k}$ or S^{2k}.

EXAMPLE 2.5.3 (k-TIGHT SPHERES [Banchoff 1971a; Kühnel 1995]). *For any given $N \geq n + 2$ and any $k \leq (n - 3)/2$, there exists a k-tight substantial polyhedral embedding of the sphere S^n into $\mathbb{E}^N$. This inequality is the best possible, according to the polyhedral version of Lemma 2.2.2.*

Such an example can be chosen as the boundary complex of the cyclic polytope $C(N+1, n+1)$, regarded as a subcomplex of the N-dimensional simplex. Since the cyclic polytope contains all $\binom{N+1}{k+2}$ $(k + 1)$-simplices for $k + 1 \leq (n - 1)/2$ [Grünbaum 1967], it follows that the embedding into $\mathbb{E}^N$ is k-tight.

We remark also that any given compact three-manifold admits a 0-tight polyhedral embedding into $\mathbb{E}^N$ if N is large enough, just by a two-neighborly triangulation [Walkup 1970; Sarkaria 1983; Kühnel 1995].

EXAMPLE 2.5.4 (TIGHT TRIANGULATIONS). As in Example 1.2.8, one can regard every n-vertex triangulation of a manifold as a subcomplex of the $(n - 1)$-dimensional simplex in $\mathbb{E}^{n-1}$. If this embedding into $\mathbb{E}^{n-1}$ is tight, we call the triangulation a *tight triangulation*; compare Corollary 1.4.10 and [Kühnel 1995]. In this case every simplexwise linear embedding into any Euclidean space is tight. Particular cases are $(d + 1)$-dimensional 1-handlebodies with $n = 2d + 3$ vertices and their boundaries: Regard the n vertices as elements of $\mathbb{Z}_n$ and let this group act on the starting simplex $\langle 0\,1\,2\ldots d+1\rangle$. The union of this $\mathbb{Z}_n$-orbit of simplices is a tight triangulation of a one-handle, its boundary is tight as well [Kühnel 1995, Chapter 5]. The 9-vertex triangulation of the boundary of a (nonorientable) one-handle is the only triangulation of any three-manifold with 9 vertices that is not a sphere [Altshuler and Steinberg 1976].

We mention an interesting 8-vertex triangulation of a three-pseudomanifold, which is also a tight triangulation [Emch 1929; Kühnel 1995, § 7.16]. It contains $\binom{8}{2}$ edges, $\binom{8}{3}$ triangles and 28 tetrahedra, each vertex link is a 7-vertex torus. If we regard it as a subcomplex of the 7-simplex, then any slice by a hyperplane in general position is a tight polyhedral two-manifold substantial in $\mathbb{E}^6$.

EXAMPLE 2.5.5 (TIGHT SUBCOMPLEXES OF THE CUBE). We now present L. Danzer's general construction for getting tight polyhedra as subcomplexes of higher dimensional cubes [McMullen and Schulte 1989]. Let K be a simplicial complex with n vertices $1, 2, \ldots, n$. Each k-simplex of K can be identified with a subset $\triangle = \{i_0, \ldots, i_k\}$ of $\{1, \ldots, n\}$. Set

$$A_j(\triangle) := \begin{cases} [0, 1] & \text{if } j \in \{i_0, \ldots, i_k\}, \\ \{0, 1\} & \text{otherwise;} \end{cases}$$

furthermore, set $F(\triangle) := A_1(\triangle) \times \cdots \times A_n(\triangle)$ and $2^K := \bigcup_{\triangle \in K} F(\triangle)$. By definition, we may regard each $F(\triangle)$ and therefore the entire 2^K as a subcomplex of the n-dimensional cube $C^n := [0,1]^n$, as well as a subset $2^K \subset C^n \subset \mathbb{E}^n$ of the ambient Euclidean space.

A particular case is $M(n) := 2^{\{n\}}$, where $\{n\}$ denotes the boundary of an n-gon. This is a tight surface of genus $2^{d-3}(d-4)+1$ in the boundary of the n-cube [Coxeter 1937; Banchoff 1965; Ringel 1955a; Beineke and Harary 1965].

An unexpected statement is the following:

THEOREM 2.5.6 [Kühnel 1995]. *The subset 2^K is tight in $\mathbb{E}^n$ for any simplicial complex K with n vertices. This holds for any choice of a field F. If K is a triangulated sphere, 2^K is a topological manifold.*

In particular, this leads to strange examples as follows:

EXAMPLE 2.5.7 (TORSION). Let K be any simplicial sphere such that a certain subset of vertices spans a subcomplex with p-torsion in the homology. Then 2^K is a tightly embedded manifold with p-torsion in the homology.

EXAMPLE 2.5.8 (HOMOLOGY MANIFOLDS). Let K be a triangulated homology sphere that is not a sphere. Then 2^K is a tight homology manifold that is not a manifold.

EXAMPLE 2.5.9 (TOPOLOGICAL MANIFOLDS THAT ARE NOT PL). The double suspension K of a certain homology 3-sphere is a triangulated 5-sphere [Edwards 1975] and it is not PL with respect to this simplicial decomposition, since the link of some edge is the original homology sphere. Then 2^K is a tightly embedded topological 6-manifold that is not PL with respect to this induced polyhedral structure.

2.6. Manifolds with boundary and tubes around submanifolds. For manifolds with boundary M, *tightness* is again defined by condition (iii) of 2.2.1: For every open half-space h the induced morphism $H_*\big(f^{-1}(h)\big) \to H_*(M)$ is injective, where H_* denotes the singular homology with coefficients in F.

This is equivalent to the equality $\text{TA}(f) = \sum_i b_i(M; F)$. We remark that the equation $\text{TA}(M) = \text{TA}(M \setminus \partial M) + \frac{1}{2}\,\text{TA}(\partial M)$ remains valid in general.

One of the basic cases to be considered in both the smooth and the polyhedral situations is the case of n-manifolds with boundary in $\mathbb{E}^n$.

PROPOSITION 2.6.1. *Let $M \subset \mathbb{E}^n$ be a compact n-manifold with boundary ∂M. Then M is tightly embedded if and only if ∂M is tightly embedded.*

In the case $n = 2$, this proposition states that a compact tightly embedded two-manifold with boundary in $\mathbb{E}^2$ must be a closed convex set with a collection of open convex sets removed, so that the closures of these sets are disjoint from one another and from the boundary of the original convex set.

The condition that the mapping be an embedding is essential. For smooth immersions with self-intersections the proposition is not true, even for $n = 2$ (see the immersed surfaces with boundary in the plane 1.5.5, where the inner boundary curves are locally convex but not necessarily globally convex).

The proof of Proposition 2.6.1 follows from the equation $\mathrm{TA}(M) = \mathrm{TA}(M \setminus \partial M) + \frac{1}{2}\mathrm{TA}(\partial M)$ on the one hand and the equation $\sum_i b_i(\partial M) = 2\sum_i b_i(M)$ on the other hand. The latter one follows from the Alexander duality for M and its complement.

PROPOSITION 2.6.2. *For any tight immersion f of a manifold without boundary into $\mathbb{E}^N$ for $N \geq 3$, the Euclidean solid $f_{\leq \varepsilon}$ of radius ε-tube is a tight immersion. If there are no self-intersections, or if $\sum_i b_i(\partial M) = 2\sum_i b_i(M)$, the same holds also for the boundary of the tube.*

The analogous result holds for tight subcomplexes of the cube (see Section 2.5.5) if we replace the Euclidean tube by the polyhedral tube, that is, the tube with regard to the maximum norm.

THEOREM 2.6.3 [Breuer and Kühnel 1997]. *The boundary of the ε-tube around any smooth tight immersion of a compact two- or three-manifold is again tight if the codimension is at least two.*

For four-manifolds, compare Question 21 in Section 3.4 below.

EXAMPLE 2.6.4 (THE LADDER CONSTRUCTION). Let C^k denote the boundary of the $(k+1)$-dimensional unit cube in $\mathbb{E}^{k+1}$ with one vertex at the origin and with edges parallel to the coordinate axes. Take an arbitrary number of congruent copies of C^k in $\mathbb{E}^{k+1}$, attached to one another along facets in the form of a ladder. Specifically we translate each cube-boundary a certain number of units along the first coordinate axis so that the union of these cube-boundaries then consists of the boundary of a rectangular parallelepiped together with a certain number of interior copies of the k-dimensional unit cube; this is what we call a ladder. Regard $\mathbb{E}^{k+1}$ as a linear subspace of $\mathbb{E}^N$ for $N > (k+1)$. Then, for small ε, the polyhedral ε-tube around the ladder is a tightly embedded k-handlebody, and its boundary is a tightly embedded connected sum of sphere products $S^k \times S^{N-k-1}$. Similar polyhedral examples are given by 2^K for triangulated balls K.

PROPOSITION 2.6.5 [Rodríguez 1977; Kühnel 1978; Banchoff 1971b]. *For an immersion (smooth or polyhedral) $f : M^n \to \mathbb{E}^N$ with $\partial M \neq \varnothing$, the following conditions are equivalent:*

(i) *f is tight.*
(ii) *f is $(n-2)$-tight and $\mathcal{H}(fM) = \mathcal{H}(f(\partial M))$.*

COROLLARY 2.6.6. *A tight embedding (smooth or polyhedral) of a ball B^n into $\mathbb{E}^N$ is a convex embedding into some $(n+1)$-dimensional subspace.*

PROOF. Any nondegenerate height function has only one critical point on B^n, the absolute minimum. Hence it has exactly two critical points on the boundary. The tight boundary sphere is convex by Theorem 2.1.1 or Proposition 2.5.2, as the case may be. Then Theorem 2.6.3 implies the convexity of the embedding of the ball.   □

The same argument leads to this result:

COROLLARY 2.6.7. *Assume that M is a compact manifold with boundary satisfying $\sum_i b_i(\partial M) = 2\sum_i b_i(M)$. Then the tightness of a substantial smooth immersion $f : M \to \mathbb{E}^N$ implies that*

(i) *the restriction of f to ∂M is tight and substantial, and*

(ii) *the Lipschitz–Killing curvature vanishes identically in $M \setminus \partial M$.*

If the boundary has several components, the single components do not have to be substantial in $\mathbb{E}^N$.

2.7. Higher-dimensional knots. An *n-knot* is defined as a smooth or polyhedral embedding $f : S^n \to \mathbb{E}^{n+2}$ (or $f : S^n \to S^{n+2}$). Here S^n indicates the topological sphere with no particular differentiable structure, so we will also consider embeddings of spheres with exotic smooth structures.

THEOREM 2.7.1. *Let f be a smooth n-knot. If $\mathrm{TA}(f) < 4$ and n is odd, or if $\mathrm{TA}(f) < 6$ and n is even, then f is unknotted, that is, isotopic to the standard embedding.*

Note that the number of critical points of a Morse function on the sphere must be even, so if $\mathrm{TA}(f) < 4$ there must be a height function with exactly two critical points. Thus the proof in the odd-dimensional situation is analogous to the Fáry–Milnor theorem (see Section 1.7). For even n, this theorem is mentioned in [Kuiper 1984] as a consequence of hard results in topology; see [Scharlemann 1985] for the case $n = 2$.

CONJECTURE 2.7.2 [Kuiper 1984]. *The assumption in Theorem 2.7.1 can be replaced by $\mathrm{TA}(f) \le 4$ for n odd or $\mathrm{TA}(f) \le 6$ for n even.*

This is true for $n = 1$; see Section 1.7.

CONJECTURE 2.7.3. *Proposition 2.7.1 is true for polyhedral n-knots that are locally unknotted.*

EXAMPLE 2.7.4 [Wintgen 1980]. *For any $\varepsilon > 0$, there is a suspension of a polyhedral (locally unknotted) n-knot that is a polyhedral $(n+1)$-knot with $\mathrm{TA}(f) < 2 + \varepsilon$. However, this $(n+1)$-knot is locally knotted at the two additional vertices of the suspension.*

The Veronese surface $\mathbb{RP}^2 \to \mathbb{E}^4$ is unknotted in the sense that it is isotopic to the cone over an unknotted Möbius band in a 3-hyperplane. We mention the following:

THEOREM 2.7.5 [Bleiler and Scharlemann 1988]. *If $f : \mathbb{RP}^2 \to \mathbb{E}^4$ is a smooth embedding with $\mathrm{TA}(f) < 5$ then it is isotopic to the Veronese surface.*

Note that under this assumption there is a Morse height function with three critical points because the number of critical points must be odd.

Exotic spheres in codimension two can also be regarded as *n*-knots.

EXAMPLE 2.7.6 [Ferus 1968]. *For any $\varepsilon > 0$ and any $n = 4m + 1$ there is a smooth embedding of an exotic sphere (Brieskorn's sphere) $f : \Sigma^n \to \mathbb{E}^{n+2}$ with $\mathrm{TA}(f) < 4 + \varepsilon$.*

A tube around an ordinary knot in $\mathbb{E}^3$ produces a knotted torus (see Section 1.7), and there are analogous constructions for hypersurfaces arising from knots or containing knots. This tube construction leads to an unexpected phenomenon:

EXAMPLE 2.7.7 [Kuiper and Meeks 1984]. *There is a compact hypersurface embedded into four-space that satisfies $\mathrm{TA}(f) = \gamma(M) > b(M; F)$ for any F where $\gamma(M)$ denotes the Morse number, i.e., the minimum number of critical points of any Morse function defined on M. Thus it attains the minimal total absolute curvature but it is not tight with respect to any field.*

This example starts with the isotopy tight surface M of genus 3 in $\mathbb{E}^3$ given by Theorem 1.7.2(ii). Assume that M is contained in a large ball B; then M decomposes B into an interior component B_I and an exterior component B_E. Denote the closure of the exterior component by $B^\#$. The total absolute curvature of $B^\#$ is 7, and the sum of the Betti numbers is $b(B^\#) = 5$. By the knottedness, the fundamental group of $B^\#$ requires at least four generators, although the homology $H_1(M^\#)$ is only three-dimensional. Now consider the manifold $M^\#$ defined as the ε-tube around $B^\# \subset \mathbb{E}^3 \subset \mathbb{E}^4$ in four-space. This is an embedded hypersurface with $\mathrm{TA} = 14$ and $b = 10$, so it is not tight. This hypersurface is only of class C^1 along the parts arising from the boundary of $B^\#$, but it can be made smooth preserving tightness. By the same argument used for $B^\#$, we can prove that every Morse function on $M^\#$ must have at least 14 critical points.

3. Highly Connected Manifolds

An even-dimensional manifold is called *highly connected* if it has the highest degree of connectivity in the sense of homotopy theory. More precisely, a $2k$-dimensional manifold M is called highly connected if it is $(k-1)$-connected, that is, if the homotopy groups $\pi_1(M), \ldots, \pi_{k-1}(M)$ all vanish. In particular, a surface is highly connected if and only if it is simply connected, since $k = 1$ in this case. Thus any connected and highly connected surface is topologically equivalent to a two-sphere.

The theory of highly connected manifolds has a very interesting history, with some surprising developments. A classical reference is [Whitehead 1949], which proves that in dimension 4 the homotopy type is uniquely determined by a quadratic form (or the cup-product) on the two-dimensional homology (or cohomology) with integer coefficients. Another important reference is [Wall 1962] on the classification of $(k-1)$-connected $2k$-manifolds for $k > 2$. The case of simply connected four-manifolds remained quite mysterious until the spectacular results by Donaldson, Freedman and others in the 1980's [Kirby 1989].

From the geometric point of view, it seems to be natural to begin by studying the following standard examples:

(i) $S^k \times S^k$;

(ii) a nontrivial S^k-bundle over S^k [Steenrod 1944];

(iii) the projective planes FP^2 for $F = \mathbb{R}$, $\mathbb{C}$, $\mathbb{H}$, Ca (the Cayley numbers) in dimensions 2, 4, 8, 16;

(iv) "manifolds like projective planes" in dimension 8 and 16 [Eells and Kuiper 1962] (by definition, they are manifolds admitting a Morse function with 3 critical points);

(v) exotic spheres (compare Theorem 2.1.3 and Example 2.7.6).

If $k = 1$, the first example is the torus, the second is the Klein bottle, and the third is the real projective plane. Taking the connected sum of a surface with one of these examples amounts to adding a handle, adding a twisted handle, or adding a cross-cap.

The unimodular quadratic forms corresponding to the manifolds in the list above are respectively $(+1)$, (-1) (only possible for $k = 1, 2, 4, 8$) and the matrix

$$\begin{pmatrix} 0 & 1 \\ (-1)^k & 0 \end{pmatrix}.$$

This last unimodular quadratic form represents the only indecomposable case, for k odd. The connected sum of α copies of the first, β of the second and γ of the third will then have quadratic form

$$\alpha(+1) \oplus \beta(-1) \oplus \gamma \begin{pmatrix} 0 & 1 \\ (-1)^k & 0 \end{pmatrix}.$$

The Euler characteristic of such a connected sum is $\chi(M) = 2 + (-1)^k b_k(M) = 2 + (-1)^k(\alpha + \beta + 2\gamma)$. For even k, the *signature* is defined to be $\sigma(M) = \alpha - \beta$. Note, however, that other definite quadratic forms may correspond to manifolds. For example, the form E_8 occurs as part of the quadratic form of a K3 surface.

Unless stated otherwise, in this section a *manifold* will always be compact, connected, without boundary, and of dimension $2k \geq 4$.

3.1. Tightness and the highly connected two-piece property.

LEMMA 3.1.1. *Let M be a highly connected manifold of dimension $2k$ and let $f : M \to \mathbb{E}^N$ be a smooth or polyhedral immersion. Then the following conditions are equivalent:*

(i) *f is tight.*

(ii) *f is $(k-1)$-tight.*

(iii) *Every nondegenerate height function (or height function in general position) has exactly one minimum and one maximum and $(-1)^k(\chi(M) - 2)$ critical points of index k (counted with multiplicity in the polyhedral case), and no critical points of any other index.*

(iv) *For any hyperplane H of $\mathbb{E}^N$, the preimage $f^{-1}(\mathbb{E}^N \setminus H)$ has at most two connected components, each of them being $(k-1)$-connected.*

We call (iv) the highly connected two-piece property (HTTP).

PROOF. (i) $\Leftrightarrow$ (ii) holds for manifolds in general by Lemma 2.2.2.

(i) $\Leftrightarrow$ (iii) follows directly from the Morse relations.

In order to see (ii) $\Leftrightarrow$ (iv), consider an open half-space $h \subset \mathbb{E}^N$ and the inclusion $j : f^{-1}(h) \to M$. The $(k-1)$-tightness means that $\tilde{H}_i(j) : \tilde{H}_i(f^{-1}(h)) \to \tilde{H}_i(M) = 0$ is injective for $i = 1, \ldots, k-1$, where $\tilde{H}$ denotes the i-th reduced homology. By the Hurewicz isomorphism theorem [Dold 1972], this is equivalent to the injectivity of $\pi_i(j) : \pi_i(f^{-1}(h)) \to \pi_i(M) = 0$. This in turn just says that $f^{-1}(h)$ is $(k-1)$-connected.   □

COROLLARY 3.1.2. *Let A be a k-topset of the convex hull of a tightly embedded highly connected $2k$-manifold $M \subset \mathbb{E}^N$. Then A is convex and contained in M. The same holds for i-topsets for $0 \le i \le k$.*

PROOF. Induction on i.   □

COROLLARY 3.1.3. *For a polyhedral highly connected $2k$-manifold M the tightness condition implies that M contains the k-dimensional skeleton of its convex hull. Conversely, for a subcomplex of a convex polytope, this necessary condition for tightness is also sufficient.*

For the sufficiency we observe that on the one hand every $(k-1)$-cycle in the manifold can be deformed into the k-skeleton, and that on the other hand the k-skeleton of a simplex and its intersections with arbitrary half spaces is $(k-1)$-connected. This implies the HTPP.

COROLLARY 3.1.4. *Let B be a $(k+1)$-topset of a tightly embedded highly connected $2k$-manifold M. Then either B is convex or B is a convex set minus a number of convex open sets in its interior, in any case the boundary ∂B^* of the induced topset B^* of the convex hull is contained in M.*

If B itself is not convex then a generator of $H_k(B)$ is called a *top-cycle* or a *convex cycle* [Thorbergsson 1983]. By the tightness it certainly represents a nonvanishing element of $H_k(M)$; compare Proposition 1.3.4.

3.2. Examples, smooth and polyhedral.

EXAMPLE 3.2.1 (THE SPHERE S^n). The boundary of any convex body in $(n+1)$-space is a tightly embedded sphere S^n. Such convex hypersurfaces can have any degree of differentiability, including C^∞ and C^ω. Examples of tight polyhedral spheres are given by boundaries of convex polytopes. With respect to the topsets, the main difference is that a polyhedral example necessarily has i-topsets for any $i = 0, 1, \ldots, n$, while in the differentiable case there may be gaps. For example, a strictly convex body of class C^2 or any convex body of class C^ω has only 0-topsets.

EXAMPLE 3.2.2 (CONNECTED SUM OF "HANDLES" $S^k \times S^k$). Examples of tight embeddings of $S^k \times S^k$ into $\mathbb{E}^{2k+1}$ and $\mathbb{E}^{2k}$ can be obtained by constructions analogous to the ones in Example 1.2.2: The Cartesian product of two standard spheres $S^k(1) \times S^k(1) \subset S^{2k+1}(\sqrt{2}) \subset \mathbb{E}^{2k+2}$ is tight. Stereographic projection to $\mathbb{E}^{2k+1}$ gives examples similar to the torus of revolution in $\mathbb{E}^3$, and its conformal images (generalized Dupin cyclides). Polyhedral examples are given by the product of two $(k+1)$-cubes as a subcomplex of the $(2k+2)$-cube. A Schlegel diagram of a hypercube in $(2k+1)$-space includes cubical versions of the Dupin cyclides.

In order to attach handles tightly to these examples, we observe first that we have to replace an $S^{k-1} \times B^{k+1}$ by $B^k \times S^k$, an instance of ordinary surgery. In the case of $k = 1$ (Example 1.2.3) we replace an $S^0 \times B^2$ (where B^2 lies in a flat region) by a rotationally symmetric cylinder $B^1 \times S^1$ and smooth this out along the boundary $S^0 \times S^1$. This can be done by a suitable choice of a concave radius function r of the (warped product) cylinder $B^1 \times_r S^1$ depending on the radius ρ in $B^1 = [-1, +1]$. We use the same function $r(\rho)$ in general where ρ is the polar radius of B^k (in ordinary polar coordinates) and r is the scaling factor of the fibre S^k in the warped product $B^k \times_r S^k$. To get started we have to find a product $S^{k-1} \times B^{k+1}$ where S^{k-1} is a standard round sphere and B^{k+1} lies in a flat region. There are certainly convex hypersurfaces in $\mathbb{E}^{2k+1}$ containing such regions $S^{k-1} \times B^{k+1}$ (rotationally symmetric ovaloids containing flat regions).

To obtain examples in $\mathbb{E}^{2k+2}$ we may start with the Cartesian product of an S^{k-1}-rotationally symmetric ovaloid A^k containing a 1-flat with another ovaloid B^k containing a k-flat. We can then attach a handle $S^k \times S^k$ by surgery. Without loss of generality, we can assume that the original example contains arbitrarily many regions of this type, so we can attach arbitrarily many handles tightly. The tightness can be seen from the HTPP because the handles are rotationally symmetric warped products with a concave radius function and because the fibre S^k is $(k-1)$-connected. These smooth examples are essentially due to J. Hebda [1984].

Polyhedral examples $(S^k \times S^k) \# \ldots \# (S^k \times S^k)$ in $\mathbb{E}^{2k+1}$ can be obtained by the polyhedral ε-tube around the ladder construction in Section 2.6, just by setting $N = 2k + 1$.

In high codimension we can apply the 2^K-construction from Section 2.5.5: Let K denote the boundary complex of a cyclic polytope $C(N+1, 2k)$. This is a simplicial sphere containing all $\binom{N+1}{k}$ simplices of dimension $k - 1$. Therefore $2^{\partial C(N+1, 2k)}$ is a $2k$-manifold containing every k-dimensional face of the N-dimensional cube. It is tight and substantial in $\mathbb{E}^N$ by 2.5.5, and it is homeomorphic to a connected sum of copies of $S^k \times S^k$ [Kühnel and Schulz 1991].

EXAMPLE 3.2.3 (PROJECTIVE PLANES). For $F = \mathbb{C}$ or $F = \mathbb{H}$, the projective plane FP^2 has a standard embedding, sending F^3 to $R^3 \oplus F^3$ in such a way that

(x, y, z) and (ux, uy, uz) have the same image for any nonzero element u of F:

$$(x, y, z) \mapsto (x\bar{x}, y\bar{y}, z\bar{z}, \sqrt{2}x\bar{y}, \sqrt{2}y\bar{z}, \sqrt{2}z\bar{x}.)$$

If we restrict to the unit sphere, where $x\bar{x} + y\bar{y} + z\bar{z} = 1$, this defines a mapping $S^5 \to S^7 \subset \mathbb{E}^8$ or $S^{11} \to S^{13} \subset \mathbb{E}^{14}$ so that (x, y, z) and (ux, uy, uz) have the same image for any unit element u. It follows that this mapping gives an embedding $\mathbb{CP}^2 \to S^7 \subset \mathbb{E}^8$ or $\mathbb{HP}^2 \to S^{13} \subset \mathbb{E}^{14}$. These embeddings are tight because any nondegenerate height function has exactly three critical points (it is sufficient to show here that the index of each critical point is even). For the Cayley plane $\mathbb{CaP}^2$ there is a similar tight embedding into $S^{25} \subset \mathbb{E}^{26}$ but the formulas are different [Tai 1968; Kuiper 1970; Kuiper 1980].

Under this embedding, the images of each of these projective planes lies on a sphere of the appropriate dimension. Stereographic projection of these three embeddings from the north pole on the sphere leads to tight embeddings of these projective planes into $\mathbb{E}^7$, $\mathbb{E}^{13}$, and $\mathbb{E}^{25}$ respectively. By an argument involving characteristic classes, there are no topological embeddings of these manifolds into lower dimensional Euclidean spaces [Cecil and Ryan 1985].

As in the case of the real Veronese surface, we may ask:

QUESTION 19. *Is it possible to attach a handle $S^2 \times S^2$ tightly to the projected Veronese type embedding $\mathbb{CP}^2 \to \mathbb{E}^7$?*

A remarkable polyhedral analogue of the standard smooth tight embedding of $\mathbb{CP}^2$ is given by the unique 9-vertex triangulation $\mathbb{CP}^2_9$ [Kühnel and Banchoff 1983] regarded as a subcomplex of the 8-dimensional simplex $\triangle^8$:

$$\mathrm{Sk}_2(\triangle^8) \subset \mathbb{CP}^2_9 \subset \mathrm{Sk}_4(\triangle^8) \subset \mathbb{E}^8.$$

This triangulation contains every edge determined by a pair of vertices, and moreover it contains every two-simplex determined by any triple of vertices (so the triangulation is *three-neighborly* [Kühnel and Lassmann 1983]). From this the HTPP follows directly. Condition (iv) of Lemma 3.1.1 is satisfied because any 1-cycle can be deformed homotopically into the 2-skeleton, and because $\mathrm{Sk}_2(\triangle^8) \cap h$ is simply connected for any half-space $h \subset \mathbb{E}^8$.

This situation is analogous to the tight polyhedral embedding of the real projective plane of Example 1.2.4:

$$\mathrm{Sk}_1(\triangle^5) \subset \mathbb{RP}^2_6 \subset \mathrm{Sk}_2(\triangle^5) \subset \mathbb{E}^5,$$

where the TPP is satisfied because $\mathrm{Sk}_1(\triangle^5) \cap h$ is connected for any half-space h.

EXAMPLE 3.2.4 (POLYHEDRAL MANIFOLDS WITH ODD INTERSECTION FORM). Examples 3.2.2 and 3.2.3 leave open the question whether we can combine "handles" of type $S^2 \times S^2$ with the complex projective plane. From the point of view of intersection forms, a connected sum of handles represents the case of an even intersection form and signature $\sigma = 0$, whereas the complex projective plane represents an odd intersection form.

A tight polyhedral embedding $\mathbb{CP}^2 \# (-\mathbb{CP}^2) \to \mathbb{E}^8$ can be constructed from the 9-vertex triangulation of $\mathbb{CP}^2$, just as a tight Klein bottle was constructed from the 6-vertex triangulation of $\mathbb{RP}^2$ in Example 1.2.5: Cut out the open star of one vertex, take two parallel copies of the remaining part in a 7-simplex (or 4-simplex, respectively), and then join the two boundaries by a straight cylinder. The tightness is easily verified since the TPP and HTPP are satisfied. By this cylinder construction, the two copies of $\mathbb{CP}^2$ have opposite orientations. The case of the same orientation seems to be open:

QUESTION 20. *Does there exist a tight polyhedral (or topological) embedding of* $\mathbb{CP}^2 \# \mathbb{CP}^2$ *into any* $\mathbb{E}^N$?

Tight polyhedral connected sums $\mathbb{CP}^2 \# (-\mathbb{CP}^2) \# \cdots \# (-\mathbb{CP}^2)$ can be constructed by truncation of the tight $\mathbb{CP}^2 \# (-\mathbb{CP}^2)$; see Theorem 3.5.7.

EXAMPLE 3.2.5 (COMBINATORIAL). As a generalization of Example 1.2.8 and $\mathbb{CP}^2_q$, we observe the following: when an n-vertex triangulation of a $2k$-manifold M that contains all $\binom{n}{k+1}$ k-dimensional simplices, M is highly connected and the natural inclusion of M as a subcomplex of the $(n-1)$-dimensional simple is a tight embedding into $\mathbb{E}^{n-1}$:

$$\mathrm{Sk}_k(\Delta^{n-1}) \subset M \subset \mathrm{Sk}_{2k}(\Delta^{n-1}) \subset \mathbb{E}^{n-1}.$$

Again, the tightness follows from the HTPP, condition (iv) in Lemma 3.1.1.

3.3. The substantial codimension of a tight immersion. For any tight smooth immersion $f : M \to \mathbb{E}^N$ of a $2k$-dimensional manifold M, the substantial codimension $N - 2k$ is less than or equal to $\binom{2k+1}{2}$ (see 2.4.1). If M is highly connected, this upper bound can be improved by using the fact that only quadratic forms of indices $0, k$ and $2k$ can occur in the image space of the mapping

$$\xi \mapsto A_\xi.$$

We have the following generalization of 1.3.1:

PROPOSITION 3.3.1 [Kuiper 1970]. *The substantial codimension of a smooth tight immersion of a highly connected $2k$-manifold in $\mathbb{E}^N$ satisfies*

$$N - 2k - 2 \le \begin{cases} k & \text{if } k \in \{1, 2, 4, 8\}, \\ 0 & \text{otherwise.} \end{cases}$$

Note that equality $N = 3k + 2$ is attained for the Veronese embeddings of the projective planes FP^2; see Example 3.2.3. In these exceptional cases the same upper bound is valid even without the assumption of smoothness:

THEOREM 3.3.2 [Kuiper 1980]. *If $f : M \to \mathbb{E}^N$ is a tight substantial continuous embedding of a $2k$-dimensional "manifold like a projective plane", then $N \le 3k + 2$.*

SKETCH OF PROOF. We first must show that there is a k-dimensional top-cycle given by the boundary of a $(k + 1)$-topset; see Corollary 3.1.4. This is a convex hypersurface in some $\mathbb{E}^{k+1}$. The orthogonal projection onto the subspace perpendicular to this space leads to a mapping $f^* : M \to \mathbb{E}^{N-k-1}$. Since now the k-th homology is killed, the image is a homotopy sphere of dimension $2k$, therefore the boundary of a convex set in $(2k+1)$-space. This implies $N - k - 1 \leq 2k + 1$.   □

A sharp upper bound for the substantial codimension of a tight $S^k \times S^k$ does not seem to be known in general. In the smooth case, the upper bound is 2, but for polyhedra it may be considerably larger, as indicated by the case $k = 1$, where the tight 7-vertex embedding of the torus in $\mathbb{E}^6$ has substantial codimension 4.

CONJECTURE 3.3.3. *Any continuous and substantial tight embedding $S^k \times S^k \to \mathbb{E}^N$ that is centrally-symmetric satisfies $N \leq 2k + 2$.*

This extends Conjecture 1.4.15.

CONJECTURE 3.3.4 [Kühnel 1995]. *Let $M \subset \mathbb{E}^N$ be a tight and substantial polyhedral embedding of a $(k - 1)$-connected $2k$-manifold. Then*

$$\binom{N - k - 1}{k + 1} \leq (-1)^k \binom{2k + 1}{k + 1}(\chi(M) - 2) = \binom{2k + 1}{k + 1}b_k(M),$$

with equality for $N \geq 2k + 2$ only for subcomplexes of the N-simplex containing the k-skeleton of the N-simplex.

This would generalize Theorem 1.4.11. The inequality can be regarded as a generalization of the classical Heawood inequality of Theorem 1.4.7 in the case $k = 1$. In the case of $M = S^k \times S^k$, Conjecture 3.3.4 states that $N - k - 1 \leq 2k + 2$.

THEOREM 3.3.5 [Kühnel 1994a]. *Conjecture 3.3.4 is true under the additional assumption that M is a subcomplex of the boundary complex of a simplicial convex n-polytope P that contains all vertices of P.*

For centrally symmetric versions, see [Sparla 1997a]. In this case the upper bound for N in terms of $\chi(M)$ can be improved; compare Conjectures 3.3.3 and 1.4.15. There is an example of a 12-vertex triangulation of $S^2 \times S^2$ as a tightly embedded subcomplex of the 6-dimensional cross-polytope. See [Sparla 1997b].

3.4. The smooth case. According to Proposition 3.3.1 tight smooth and substantial immersions of simply connected 4-manifolds can exist only in $\mathbb{E}^5$, $\mathbb{E}^6$, $\mathbb{E}^7$, and $\mathbb{E}^8$. Unfortunately, in codimension greater than two, no construction principle seems to be known for smooth tight immersions. Therefore this Section contains more negative results on restrictions and obstructions than positive results and examples.

THEOREM 3.4.1 [Kuiper 1980]. *Let $f : \mathbb{CP}^2 \to \mathbb{E}^8$ be a smooth tight and substantial immersion. Then, up to projective transformations of $\mathbb{E}^8$, the image is* *congruent to the image of the standard (Veronese type) embedding in Example 3.2.2.*

For tight embeddings of the complex projective plane into $\mathbb{E}^7$ no geometric uniqueness result can be expected, just as there is no uniqueness in the case of tight embeddings of the real projective plane into $\mathbb{E}^4$.

THEOREM 3.4.2 [Thorbergsson 1983]. *Let f be a substantial tight and smooth immersion of a simply connected four-manifold M into $\mathbb{E}^N$, for $N = 6$ or $N = 7$. Then, for a suitable choice of an orientation:*

(i) *If $N = 6$ then M splits diffeomorphically as a connected sum $(S^2 \times S^2) \# M^*$ and the middle Betti number $b_2(M)$ is even. Moreover, if the intersection form is odd, then $b_2(M) \geq 4$.*

(ii) *If $N = 7$ then M splits diffeomorphically as a connected sum $\mathbb{CP}^2 \# M^*$. In particular, the intersection form is odd.*

The proof is quite involved and relies on a careful study of the intersections of various top-cycles (see the end of Section 3.1). This can be considered as an obstruction to the existence of tight immersions:

COROLLARY 3.4.3 [Thorbergsson 1983]. *Infinitely many distinct simply connected differentiable four-manifolds do not admit a tight immersion into any $\mathbb{E}^N$.*

Particular examples are the K3-surfaces and algebraic surfaces in $\mathbb{CP}^3$ of even degree $d \geq 4$.

QUESTION 21 [Thorbergsson 1983]. *Is there a smooth tight immersion $f : M \to \mathbb{E}^6$ of a simply connected four-manifold with odd intersection form?*

According to Theorem 3.4.2, candidates would be $\mathbb{CP}^2 \# \mathbb{CP}^2 \# (-\mathbb{CP}^2) \# (-\mathbb{CP}^2)$ or connected sums with more copies of $\pm \mathbb{CP}^2$. Such a tight example would have quite unexpected behavior with respect to tubes: The ε-tube around it (regarded as an immersion of an S^1-bundle over the manifold) would not be tight [Breuer and Kühnel 1997]. So far there does not seem to be any example of a tight smooth immersion in codimension at least two for which ε-tube is not tight. (Compare Proposition 2.6.2 and Question 13 at the end of Section 1.7.)

THEOREM 3.4.4 [Thorbergsson 1983]. *Let $f : M \to \mathbb{E}^N$ be a substantial tight and smooth immersion of a highly connected 2k-manifold with $k = 4$ or $k = 8$. Then:*

(i) *If the intersection form of M is even then $N \leq 2k + 2$.*

(ii) *If the intersection form of M is odd then $N = 3k + 1$ or $N = 3k + 2$.*

For the case $k \notin \{1, 2, 4, 8\}$ see Proposition 3.3.1.

   Any such substantial immersion with $N = 3k + 2$ is projectively equivalent to the Veronese-type embedding of the projective plane over the complex, quaternion, or Cayley numbers [Niebergall and Thorbergsson 1996].

THEOREM 3.4.5. *For any given natural number $m \geq 1$ there exists a tight smooth embedding of a connected sum of m copies of $S^k \times S^k$ into $\mathbb{E}^{2k+1}$ and $\mathbb{E}^{2k+2}$ [Hebda 1984], but there is no tight analytic embedding into $\mathbb{E}^{2k+2}$ for $m \geq 2$ [Niebergall 1994].*

See Example 3.2.2.

QUESTION 22. *Is there a smooth tight immersion of any simply connected four-manifold with even intersection form that is not diffeomorphic to a connected sum of copies of $S^2 \times S^2$?*

3.5. The polyhedral case. In the polyhedral case we have the opposite situation to the smooth case in Section 3.4. There are many examples and various construction principles but only a few restrictions. Recall first the construction principle mentioned in Corollary 3.1.3: If a $2k$-dimensional M is a subcomplex of the boundary complex of a convex polytope, tightness is satisfied if M contains the whole k-skeleton of this polytope.

THEOREM 3.5.1 [Kühnel 1995]. *For arbitrary given numbers k, N satisfying $N \geq 2k + 1$ there is a tight and substantial polyhedral embedding of a $(k-1)$-connected $2k$-manifold into $\mathbb{E}^N$. Particular examples are PL homeomorphic to a connected sum of copies of $S^k \times S^k$.*

PROOF. Let K be a k-neighborly triangulation of S^{2k-1} with N vertices, and define M to be $2^K \subset C^N \subset \mathbb{E}^N$, as in Example 3.2.2. Then M is $(k-1)$-connected and tight. In the particular case of the cyclic polytope $K = \partial C(d, 2k)$, the manifold 2^K is PL homeomorphic to a connected sum of $(-1)^k \frac{1}{2}(\chi(d, k) - 2)$ copies of $S^k \times S^k$ where, by definition,

$$\chi(d, k) = 2\chi\big(\operatorname{Sk}_k(C^{d-k-1})\big).$$

The number $\chi(d, k)$ is the Euler characteristic of any k-Hamiltonian submanifold of C^d, that is, a submanifold containing k-dimensional faces of C^d [Kühnel and Schulz 1991]. For any such K the intersection form of 2^K on $H_k(2^K)$ is a sum of copies of

$$\begin{pmatrix} 0 & 1 \\ (-1)^k & 0 \end{pmatrix}. \qquad \square$$

It does not seem to be known whether there are distinct topological types of such examples in the skeleton of C^N for $k \geq 2$.

THEOREM 3.5.2 [Kühnel and Banchoff 1983; Morin and Yoshida 1991]. *There exists a unique tight 9-vertex triangulation of the complex projective plane $\mathbb{CP}^2$. The canonical embedding of this complex into the 8-simplex determines a tight polyhedral embedding $\mathbb{CP}^2 \to \mathbb{E}^8$ and following this embedding by projection into almost any 7-dimensional linear subspace gives a tight embedding $\mathbb{CP}^2 \to \mathbb{E}^7$.*

PROOF. Since all vertices, edges, and two-dimensional faces of Δ^8 must be contained in any such tight embedding of a simply connected four-manifold, we may determine the number f_i of i-dimensional simplices as $f_0 = 9$, $f_1 = \binom{9}{2} = 36$, $f_2 = \binom{9}{3} = 84$. From the Dehn–Sommerville relations, it follows that $f_3 = 90$ and $f_4 = 36$. To construct the 9-vertex triangulation, called $\mathbb{CP}^2_9$, we denote the nine vertices by $1, 2, 3, \ldots, 9$ and we take the union of the two orbits of the four-dimensional simplices $\langle 12456 \rangle$ and $\langle 12459 \rangle$ under the action of a group H_{54} on $\{1, 2, \ldots, 9\}$ generated by

$$\alpha = (147)(258)(369), \quad \beta = (123)(465), \quad \gamma = (12)(45)(78).$$

The generator γ corresponds to the action of complex conjugation; in fact its fixed point set is combinatorially isomorphic to an $\mathbb{RP}^2_6$. The triangulation is unique (up to relabelling of the vertices) [Kühnel and Lassmann 1983; Arnoux and Marin 1991; Bagchi and Datta 1994]. The link of each vertex is combinatorially isomorphic to the so-called *Brückner–Grünbaum sphere* $\mathfrak{M}$ [Grünbaum and Sreedharan 1967], a triangulation of the three-sphere with unusual properties. $\square$

QUESTION 23 [Kuiper 1980]. *Are there tight and substantial topological embeddings* $\mathbb{CP}^2 \to \mathbb{E}^8$ *other than the standard algebraic embedding and the canonical polyhedral embedding of* $\mathbb{CP}^2_9$ *(up to projective transformations)?*

By Theorem 3.3.5 there is no such example in the boundary complex of a simplicial polytope. Conjecture 3.3.4 together with Theorem 3.5.2 would imply the uniqueness in the polyhedral case.

We remark that the combinatorial formula for the first Pontrjagin number of a four-manifold has been explicitly evaluated for $\mathbb{CP}^2_9$ by L. Milin [1994]. The flattenings of the (nonpolytopal) Brückner–Grünbaum sphere play a particular role in Milin's work.

EXAMPLE 3.5.3. *For any* m, $1 \le m \le 256$, *there is a tight polyhedral embedding into* $\mathbb{E}^8$ *of a simply connected four-manifold with* $\mathrm{rank}(H_2) = 62 + m$ *whose intersection form on* H_2 *is odd.*

To construct this example, we take $2^{\mathfrak{M}}$, where $\mathfrak{M}$ is the Brückner–Grünbaum sphere mentioned above. The link of each vertex is combinatorially equivalent to $\mathfrak{M}$. Therefore we can truncate at each vertex by a hyperplane section and attach in this hyperplane a copy of $\mathbb{CP}^2_9$ minus an open vertex star. The tightness follows from Corollary 3.1.3. The intersection form of this manifold is the one of $2^{\mathfrak{M}}$ plus m direct summands (± 1).

PROPOSITION 3.5.4 [Banchoff and Kühnel 1992]. *There is a tight polyhedral embedding* $\mathbb{CP}^2 \to \mathbb{E}^7$ *that is essentially different from a linear projection of the one in Theorem 3.5.2. It is a simplexwise linear embedding of a 10-vertex triangulation, denoted by* $\mathbb{CP}^2_{10}$.

This triangulation is based on the decomposition of the complex projective plane into three 4-balls as "zones of influence" of three points $X = [1, 0, 0], Y = [0, 1, 0]$ and $Z = [0, 0, 1]$, given in homogeneous coordinates. The *equilibrium torus* is the set of points $[z_0, z_1, z_2]$ with the same absolute value of each coordinate z_i. We take the 7-vertex triangulation of this equilibrium torus and then introduce X, Y, Z as extra vertices. Each of the three 4-balls is triangulated as a cone over the boundary complex of the cyclic polytope $C(7,4)$, which occurs in three different, and combinatorially equivalent, versions. For the tight embedding the 7 vertices of the torus are chosen in general position in an $\mathbb{E}^6$, then X and Y as vertices of a double cone over the seven ones, and finally Z at the centre.

THEOREM 3.5.5 [Casella and Kühnel 1996]. *There is a tight 16-vertex triangulation of a K3 surface, leading to a tight polyhedral embedding into $\mathbb{E}^{15}$.*

The construction is algebraic and combinatorial. The triangulation has 16 vertices and 288 four-simplices. One can regard the vertices as the elements of a field with 16 elements. Then the triangulation is invariant under the group of all invertible affine transformations $x \mapsto ax + b$ of this field.

THEOREM 3.5.6 [Brehm and Kühnel 1992]. *There are at least three combinatorially distinct tight 15-vertex triangulations of an 8-manifold "like the quaternionic projective plane". These triangulations induce tight polyhedral embeddings into $\mathbb{E}^{14}$ and $\mathbb{E}^{13}$.*

The construction of such a triangulated 8-manifold M_{15}^8, which is quite complicated, generalizes the construction of $\mathbb{CP}_9^2$ above, according to Example 3.2.5:

$$\mathrm{Sk}_4(\Delta^{14}) \subset M_{15}^8 \subset \mathrm{Sk}_8(\Delta^{14}) \subset \mathbb{E}^{14}.$$

By a straightforward computation the numbers f_i of i-dimensional simplices are $f_0 = n = 15$, $f_1 = \binom{15}{2} = 105$, $f_2 = \binom{15}{3} = 455$, $f_3 = \binom{15}{4} = 1365$, $f_4 = \binom{15}{5} = 3003$, $f_5 = 4515$, $f_6 = 4230$, $f_7 = 2205$, $f_8 = 490$. The actual example M_{15}^8 is presumably a triangulated quaternionic projective plane; for some evidence of this conjecture see [Brehm and Kühnel 1992].

QUESTION 24. *Is there a tight polyhedral embedding of a 16-dimensional manifold "like the Cayley plane" into $\mathbb{E}^{26}$, possibly as a tight 27-vertex triangulation?*

A tight 27-vertex triangulation would have exactly 100386 16-dimensional simplices and would contain all $\binom{27}{9}$ 8-dimensional subsimplices:

$$\mathrm{Sk}_8(\Delta^{26}) \subset M_{27}^{16} \subset \mathrm{Sk}_{16}(\Delta^{26}) \subset \mathbb{E}^{26}.$$

THEOREM 3.5.7 [Kühnel 1995]. *Let M be a tight triangulation of a $(k-1)$-connected $2k$-manifold with n vertices. Then for an arbitrary integer $m \geq 0$ there is a tight and substantial polyhedral embedding $M \# m(-M) \to \mathbb{E}^{n-1}$.*

The proof uses the construction by iterated truncation. Start with the n-vertex triangulation, regarded as a subcomplex of the $(n-1)$-dimensional simplex. Then truncate the simplex at a certain vertex and glue in a small copy of the same triangulation minus an open vertex star. Then repeat this procedure, either at (old) vertices of the $(n-1)$-simplex or at (new) vertices of the truncated simplex.

COROLLARY 3.5.8. *For an arbitrary integer $m \geq 0$ there is a tight polyhedral embedding $\mathbb{CP}^2 \# m(-\mathbb{CP}^2) \to \mathbb{E}^8$, and a tight polyhedral embedding $M^8 \# m(-M^8) \to \mathbb{E}^{14}$, where M^8 is the 8-manifold "like the quaternionic projective plane" from Theorem 3.5.6.*

We mention the following sharper form of Conjecture 3.3.4: For a highly connected $2k$-manifold M let N_M denote the maximum dimension of a Euclidean space admitting a tight and substantial polyhedral embedding of M. Let n_M denote the minimum number of vertices for any simplicial triangulation of M. The approximate size of N_M and n_M should satisfy the relations $N_M \approx n_M - 1$ and

$$\binom{N_M - k - 1}{k+1} \approx (-1)^k \binom{2k+1}{k+1}(\chi(M) - 2) = \binom{2k+1}{k+1} b_k(M);$$

see Conjecture 3.3.4.

QUESTION 25. *Is there a universal constant C such that $N_M \leq n_M - 1 \leq N_M + C$ for any highly connected manifold M that is a connected sum of the standard examples 1, 2, and 3 of page 99?*

This is true for $k = 1$ with $C = 2$ ($C = 1$ with only a few exceptions); see Lemma 1.4.11. For higher dimensions it is a very general form of a Heawood problem, compare the generalized Heawood inequalities in [Kühnel 1994b; Kühnel 1995].

PROPOSITION 3.5.9. *If a highly connected manifold M admits a tight polyhedral embedding into $\mathbb{E}^N$, there is an embedding of the k-dimensional skeleton of the N-simplex into M.*

This follows from Corollary 3.1.3 and a lemma of Grünbaum [Grünbaum 1967, §11.1] saying that the k-skeleton of any convex N-polytope contains the k-skeleton of the N-simplex as a subset. Compare Example 3.2.5 and Conjecture 3.3.4.

The question remains whether the converse of Proposition 3.5.9 is true. More precisely: If M admits an embedding of the k-skeleton of the N-simplex $\triangle^N$ (N sufficiently large, tame embedding in the topological sense), does there exist a tight polyhedral embedding into $\mathbb{E}^N$?

QUESTION 26. *Given a $(k-1)$-connected $2k$-manifold M and a number $N \geq 4k$, are the following conditions equivalent?*

(i) *There exists a tight and substantial polyhedral embedding $M \to \mathbb{E}^N$.*
(ii) *There exists a (topologically tame) embedding $\mathrm{Sk}_k(\triangle^N) \to M$.*

This is true for $k = 1$ by Theorem 1.4.8. The implication (i) $\Rightarrow$ (ii) holds in general by Proposition 3.5.9. One strategy for a proof of the converse could be the construction of a suitable triangulation from the embedding $\mathrm{Sk}_k(\triangle^N) \to M$ as a kind of starting data; compare the construction in Theorem 1.4.8.

References

[Aleksandrov 1938] A. D. Aleksandrov, "On a class of closed surfaces", *Mat. Sbornik* **4** (1938), 69–77. In Russian.

[Altshuler and Steinberg 1976] A. Altshuler and L. Steinberg, "An enumeration of combinatorial 3-manifolds with nine vertices", *Discrete Math.* **16**:2 (1976), 91–108.

[Arnoux and Marin 1991] P. Arnoux and A. Marin, "The Kühnel triangulation of the complex projective plane from the view point of complex crystallography II", *Mem. Fac. Sci. Kyushu Univ. Ser. A* **45**:2 (1991), 167–244.

[Bagchi and Datta 1994] B. Bagchi and B. Datta, "On Kühnel's 9-vertex complex projective plane", *Geom. Dedicata* **50**:1 (1994), 1–13.

[Banchoff 1965] T. F. Banchoff, "Tightly embedded 2-dimensional polyhedral manifolds", *Amer. J. Math.* **87** (1965), 462–472.

[Banchoff 1967] T. F. Banchoff, "Critical points and curvature for embedded polyhedra", *J. Diff. Geom.* **1** (1967), 245–256.

[Banchoff 1970a] T. F. Banchoff, "Critical points and curvature for embedded polyhedral surfaces", *Amer. Math. Monthly* **77** (1970), 475–485.

[Banchoff 1970b] T. F. Banchoff, "Non-rigidity theorems for tight polyhedra", *Arch. Math. (Basel)* **21** (1970), 416–423.

[Banchoff 1971a] T. F. Banchoff, "High codimensional 0-tight maps on spheres", *Proc. Amer. Math. Soc.* **29** (1971), 133–137.

[Banchoff 1971b] T. F. Banchoff, "The two-piece property and tight n-manifolds-with-boundary in E^n", *Trans. Amer. Math. Soc.* **161** (1971), 259–267.

[Banchoff 1974] T. F. Banchoff, "Tight polyhedral Klein bottles, projective planes, and Möbius bands", *Math. Ann.* **207** (1974), 233–243.

[Banchoff and Kühnel 1992] T. F. Banchoff and W. Kühnel, "Equilibrium triangulations of the complex projective plane", *Geom. Dedicata* **44**:3 (1992), 313–333.

[Banchoff and Kuiper 1981] T. F. Banchoff and N. H. Kuiper, "Geometrical class and degree for surfaces in three-space", *J. Diff. Geom.* **16** (1981), 559–576.

[Banchoff and Takens 1975] T. F. Banchoff and F. Takens, "Height functions on surfaces with three critical points", *Illinois J. Math.* **19** (1975), 325–335.

[Beineke and Harary 1965] L. W. Beineke and F. Harary, "The genus of the n-cube", *Canad. J. Math.* **17** (1965), 494–496.

[Blaschke and Leichtweiß 1973] W. Blaschke and K. Leichtweiß, *Elementare Differentialgeometrie*, 5th ed., Die Grundlehren der mathematischen Wissenschaften **1**, Springer, Berlin, 1973.

[Bleiler and Scharlemann 1988] S. Bleiler and M. Scharlemann, "A projective plane in $\mathbb{R}^4$ with three critical points is standard. Strongly invertible knots have property P", *Topology* **27**:4 (1988), 519–540.

[Bokowski and Eggert 1991] J. Bokowski and A. Eggert, "All realizations of the Möbius torus with seven vertices", *Structural Topology* **17** (1991), 59–78.

[Bolton 1982] J. Bolton, "Tight immersions into manifolds without conjugate points", *Quart. J. Math. Oxford (2)* **33** (1982), 159–167.

[Bott and Samelson 1958] R. Bott and H. Samelson, "Applications of the theory of Morse to symmetric spaces", *Amer. J. Math.* **80** (1958), 964–1029. Corrections in **83** (1961), pp. 207–208.

[Brehm and Kühnel 1982] U. Brehm and W. Kühnel, "Smooth approximation of polyhedral surfaces regarding curvatures", *Geom. Dedicata* **12** (1982), 61–85.

[Brehm and Kühnel 1992] U. Brehm and W. Kühnel, "15-vertex triangulations of an 8-manifold", *Math. Ann.* **294**:1 (1992), 167–193.

[Breuer and Kühnel 1997] P. Breuer and W. Kühnel, "The tightness of tubes", 1997. To appear in *Forum Math.*

[Brøndsted 1983] A. Brøndsted, *An introduction to convex polytopes*, Graduate Texts in Math. **90**, Springer, New York, 1983.

[Casella and Kühnel 1996] M. Casella and W. Kühnel, "A triangulated K3 surface with the minimum number of vertices", preprint, 1996.

[Cayley 1850] A. Cayley, "On the triadic arrangement of seven and fifteen things", *Phil. Mag.* **37** (1850), 50–53. Reprinted as pp. 481–484 in *Collected mathematical papers*, vol. 1, Cambridge U. Press, 1896.

[Cecil 1997] T. E. Cecil, "Taut and Dupin submanifolds", pp. 135–180 in *Tight and Taut Submanifolds*, edited by T. E. Cecil and S.-s. Chern, Cambridge U. Press, 1997.

[Cecil and Ryan 1979] T. E. Cecil and P. J. Ryan, "Tight and taut immersions into hyperbolic space", *J. London Math. Soc.* (2) **19** (1979), 561–572.

[Cecil and Ryan 1984] T. E. Cecil and P. J. Ryan, "On the number of top-cycles of a tight surface in 3-space", *J. London Math. Soc.* (2) **30**:2 (1984), 335–341.

[Cecil and Ryan 1985] T. E. Cecil and P. J. Ryan, *Tight and taut immersions of manifolds*, Research Notes in Mathematics **107**, Pitman, Boston, 1985.

[Cervone 1994] D. P. Cervone, "A tight polyhedral immersion of the real projective plane with one handle", 1994. Available at http://www.geom.umn.edu/locate/rp2-handle.

[Cervone 1996] D. P. Cervone, "Tight immersions of simplicial surfaces in three space", *Topology* **35**:4 (1996), 863–873.

[Cervone 1997] D. P. Cervone, "On tight immersions of the real projective plane with one handle", pp. 119–133 in *Tight and Taut Submanifolds*, edited by T. E. Cecil and S.-s. Chern, Cambridge U. Press, 1997.

[Chern and Lashof 1957] S.-s. Chern and R. K. Lashof, "On the total curvature of immersed manifolds", *Amer. J. Math.* **79** (1957), 306–318.

[Chern and Lashof 1958] S.-s. Chern and R. K. Lashof, "On the total curvature of immersed manifolds II", *Michigan Math. J.* **5** (1958), 5–12.

[Coghlan 1987] L. Coghlan, "Tight stable surfaces I", *Proc. Roy. Soc. Edinburgh* Sect. A **107**:3-4 (1987), 213–232.

[Coghlan 1989] L. Coghlan, "Tight stable surfaces. II", *Proc. Roy. Soc. Edinburgh* Sect. A **111** (1989), 213–229.

[Coghlan 1991] L. Coghlan, "Some remarks on tight hypersurfaces", *Proc. Roy. Soc. Edinburgh* Sect. A **119**:3-4 (1991), 279–285.

[Connelly 1978/79] R. Connelly, "A flexible sphere", *Math. Intelligencer* **1**:3 (1978/79), 130–131.

[Connelly 1993] R. Connelly, "Rigidity", pp. 223–271 in *Handbook of convex geometry*, Vol. A, edited by P. M. Gruber and J. M. Wills, North-Holland, Amsterdam, 1993.

[Coxeter 1937] H. S. M. Coxeter, "Regular skew polyhedra in three and four dimensions, and their topological analogues", *Proc. London Math. Soc.* (2) **43** (1937), 33–62. Reprinted as pp. 76–105 in *Twelve Geometric Essays*, Southern Illinois U. Press, Carbondale, IL and Feffer & Simons, London and Amsterdam, 1968.

[Coxeter 1970] H. S. M. Coxeter, *Twisted honeycombs*, CBMS Reg. Conf. Series in Math. 4, Amer. Math. Soc., Providence, 1970.

[Császár 1949] Á. Császár, "A polyhedron without diagonals", *Acta Univ. Szeged. Sect. Sci. Math.* **13** (1949), 140–142.

[Curtin 1991] E. Curtin, "Manifolds with intermediate tautness", *Geom. Dedicata* **38**:3 (1991), 245–255.

[Dold 1972] A. Dold, *Lectures on algebraic topology*, Grundlehren der mathematischen Wissenschaften **200**, Springer, Berlin, 1972.

[Edwards 1975] R. D. Edwards, "The double suspension of a certain homology 3-sphere is S^5", *Notices Amer. Math. Soc.* **22** (1975), A–334.

[Eells and Kuiper 1962] J. Eells, Jr. and N. H. Kuiper, "Manifolds which are like projective planes", *Publ. Math. Inst. Hautes Études Sci.* **14** (1962), 5–46 (181–222 in year).

[Emch 1929] A. Emch, "Triple and multiple systems, their geometric configurations and groups", *Trans. Amer. Math. Soc.* **31** (1929), 25–42.

[Fáry 1949] I. Fáry, "Sur la courbure totale d'une courbe gauche faisant un nœud", *Bull. Soc. Math. France* **77** (1949), 128–138.

[Fenchel 1929] W. Fenchel, "Über die Krümmung und Windung geschlossener Raumkurven", *Math. Ann.* **101** (1929), 238–252.

[Ferus 1968] D. Ferus, *Totale Absolutkrümmung in Differentialgeometrie und -topologie*, Lecture Notes in Math. **66**, Springer, Berlin, 1968.

[Ferus 1982] D. Ferus, "The tightness of extrinsic symmetric submanifolds", *Math. Z.* **181** (1982), 563–565.

[Fox 1950] R. H. Fox, "On the total curvature of some tame knots", *Ann. of Math.* (2) **52** (1950), 258–260.

[Franklin 1934] P. Franklin, "A six colour problem", *J. of Mathematics and Physics* **13** (1934), 363–369.

[van Gemmeren 1996] M. van Gemmeren, "Total absolute curvature and tightness of noncompact manifolds", *Trans. Amer. Math. Soc.* **348**:6 (1996), 2413–2426.

[Goresky and MacPherson 1980] M. Goresky and R. MacPherson, "Intersection homology theory", *Topology* **19** (1980), 135–165.

[Grossman 1972] N. Grossman, "Relative Chern-Lashof theorems", *J. Differential Geometry* **7** (1972), 607–614.

[Grünbaum 1967] B. Grünbaum, *Convex Polytopes*, Interscience, New York, 1967.

[Grünbaum and Sreedharan 1967] B. Grünbaum and V. P. Sreedharan, "An enumeration of simplicial 4-polytopes with 8 vertices", *J. Combinatorial Theory* **2** (1967), 437–465.

[Haab 1992] F. Haab, "Immersions tendues de surfaces dans E^{3}", *Comment. Math. Helv.* **67**:2 (1992), 182–202.

[Haab 1994/95] F. Haab, "Surfaces tendues dans E^{4}", preprint, Depto. de Matemática, Universidade Federal de Minas Gerais, Belo Horizonte, MG, Brazil, 1994/95.

[Heawood 1890] P. J. Heawood, "Map colour theorem", *Quart. J. Math.* **24** (1890), 332–338.

[Hebda 1984] J. J. Hebda, "Some new tight embeddings which cannot be made taut", *Geom. Dedicata* **17**:1 (1984), 49–60.

[Hollard 1991] A. Hollard, "Plongements isométriques locaux de variétés proprement riemanniennes de dimension 3 dans un espace plat de dimension 4. Hypersurfaces rigides", pp. viii, 79–97 in *Séminaire Gaston Darboux de Géometrie et Topologie Différentielle* (Montpellier, 1989/1990), Univ. Montpellier II, Montpellier, 1991.

[Huneke 1978] J. P. Huneke, "A minimum-vertex triangulation", *J. Combin. Theory Ser. B* **24**:3 (1978), 258–266.

[Jungerman and Ringel 1980] M. Jungerman and G. Ringel, "Minimal triangulations on orientable surfaces", *Acta Math.* **145** (1980), 121–154.

[Kalai 1987] G. Kalai, "Rigidity and the lower bound theorem I", *Invent. Math.* **88**:1 (1987), 125–151.

[Kirby 1989] R. C. Kirby, *The topology of 4-manifolds*, Lecture Notes in Mathematics **1374**, Springer-Verlag, Berlin, 1989.

[Kobayashi 1967] S. Kobayashi, "Imbeddings of homogeneous spaces with minimum total curvature", *Tôhoku Math. J.* (2) **19** (1967), 63–70.

[Kühnel 1977] W. Kühnel, "Total curvature of manifolds with boundary in E^{n}", *J. London Math. Soc.* (2) **15**:1 (1977), 173–182.

[Kühnel 1978] W. Kühnel, "$(n-2)$-tightness and curvature of submanifolds with boundary", *Intern. J. Math. Math. Sci.* **1** (1978), 421–431.

[Kühnel 1980] W. Kühnel, "Tight and 0-tight polyhedral embeddings of surfaces", *Invent. Math.* **58** (1980), 161–177.

[Kühnel 1990] W. Kühnel, "Triangulations of manifolds with few vertices", pp. 59–114 in *Advances in differential geometry and topology*, edited by F. Tricerri, World Sci., Teaneck, NJ, 1990.

[Kühnel 1992] W. Kühnel, "Tightness, torsion, and tubes", *Ann. Global Anal. Geom.* **10**:3 (1992), 227–236.

[Kühnel 1994a] W. Kühnel, "Manifolds in the skeletons of convex polytopes, tightness, and generalized Heawood inequalities", pp. 241–247 in *Polytopes: abstract, convex and computational* (Scarborough, ON, 1993), edited by T. Bisztriczky et al., NATO Adv. Sci. Inst. Ser. C Math. Phys. Sci. **440**, Kluwer, Dordrecht, 1994.

[Kühnel 1994b] W. Kühnel, "Tensor products of spheres", pp. 106–109 in *Geometry and topology of submanifolds, VI* (Leuven/Brussels, 1993), edited by F. Dillen et al., World Scientific, River Edge, NJ, 1994.

[Kühnel 1995]   W. Kühnel, *Tight polyhedral submanifolds and tight triangulations*, Lecture Notes in Math. **1612**, Springer, Berlin, 1995.

[Kühnel 1996]   W. Kühnel, "Centrally-symmetric tight surfaces and graph embeddings", *Beiträge zur Algebra und Geometrie* **37** (1996), 347–354.

[Kühnel and Banchoff 1983]   W. Kühnel and T. F. Banchoff, "The 9-vertex complex projective plane", *Math. Intelligencer* **5**:3 (1983), 11–22.

[Kühnel and Lassmann 1983]   W. Kühnel and G. Lassmann, "The unique 3-neighborly 4-manifold with few vertices", *J. Combin. Theory Ser. A* **35**:2 (1983), 173–184.

[Kühnel and Pinkall 1985]   W. Kühnel and U. Pinkall, "Tight smoothing of some polyhedral surfaces", pp. 227–239 in *Global differential geometry and global analysis* (Berlin, 1984), edited by D. Ferus et al., Lecture Notes in Math. **1156**, Springer, Berlin, 1985.

[Kühnel and Schulz 1991]   W. Kühnel and C. Schulz, "Submanifolds of the cube", pp. 423–432 in *Applied geometry and discrete mathematics*, edited by P. Gritzmann and B. Sturmfels, DIMACS Ser. Discrete Math. Theoret. Comput. Sci. **4**, Amer. Math. Soc., Providence, RI, 1991.

[Kuiper 1959]   N. H. Kuiper, "Immersions with minimal total absolute curvature", pp. 75–88 in *Colloque Géom. Diff. Globale* (Bruxelles, 1958), Centre Belge Rech. Math., Louvain, 1959.

[Kuiper 1960]   N. H. Kuiper, "On surfaces in euclidean three-space", *Bull. Soc. Math. Belg.* **12** (1960), 5–22.

[Kuiper 1961]   N. H. Kuiper, "Convex immersions of closed surfaces in E^3. Nonorientable closed surfaces in E^3 with minimal total absolute Gauss-curvature", *Comment. Math. Helv.* **35** (1961), 85–92.

[Kuiper 1962]   N. H. Kuiper, "On convex maps", *Nieuw Arch. Wisk.* (3) **10** (1962), 147–164.

[Kuiper 1970]   N. H. Kuiper, "Minimal total absolute curvature for immersions", *Invent. Math.* **10** (1970), 209–238.

[Kuiper 1971]   N. H. Kuiper, "Morse relations for curvature and tightness", pp. 77–89 in *Proceedings of Liverpool Singularities Symposium* (Liverpool, 1969/1970), vol. 2, edited by C. T. C. Wall, Lecture Notes in Math. **209**, Springer, Berlin, 1971.

[Kuiper 1971/72]   N. H. Kuiper, "Tight topological embeddings of the Moebius band", *J. Differential Geometry* **6** (1971/72), 271–283.

[Kuiper 1975]   N. H. Kuiper, "Stable surfaces in euclidean three space", *Math. Scand.* **36** (1975), 83–96.

[Kuiper 1980]   N. H. Kuiper, "Tight embeddings and maps: Submanifolds of geometrical class three in E^{nn}", pp. 97–145 in *The Chern Symposium* (Berkeley, 1979), edited by W.-Y. Hsiang et al., Springer, New York, 1980.

[Kuiper 1983a]   N. H. Kuiper, "Polynomial equations for tight surfaces", *Geom. Dedicata* **15**:2 (1983), 107–113.

[Kuiper 1983b]   N. H. Kuiper, "There is no tight continuous immersion of the Klein bottle into $\mathbb{R}^3$", preprint, Inst. Hautes Études Sci., 1983.

[Kuiper 1984] N. H. Kuiper, "Geometry in total absolute curvature theory", pp. 377–392 in *Perspectives in mathematics* (Oberwolfach, 1984), edited by W. Jäger et al., Birkhäuser, Basel, 1984.

[Kuiper 1997] N. H. Kuiper, "Geometry in curvature theory", pp. 1–50 in *Tight and Taut Submanifolds*, edited by T. E. Cecil and S.-s. Chern, Cambridge U. Press, 1997.

[Kuiper and Meeks 1984] N. H. Kuiper and W. H. Meeks, III, "Total curvature for knotted surfaces", *Invent. Math.* **77**:1 (1984), 25–69.

[Kuiper and Meeks 1987] N. H. Kuiper and W. H. Meeks, III, "The total curvature of a knotted torus", *J. Differential Geom.* **26**:3 (1987), 371–384.

[Kuiper and Pohl 1977] N. H. Kuiper and W. F. Pohl, "Tight topological embeddings of the real projective plane in E^5", *Invent. Math.* **42** (1977), 177–199.

[Langevin and Rosenberg 1976] R. Langevin and H. Rosenberg, "On curvature integrals and knots", *Topology* **15**:4 (1976), 405–416.

[Lastufka 1981] W. S. Lastufka, "Tight topological immersions of surfaces in Euclidean space", *J. Diff. Geom.* **16** (1981), 373–400.

[Liebmann 1900] H. Liebmann, "Über die Verbiegung der geschlossenen Flächen positiver Krümmung", *Math. Ann.* **53** (1900), 81–112.

[Little and Pohl 1971] J. A. Little and W. F. Pohl, "On tight immersions of maximal codimension", *Invent. Math.* **13** (1971), 179–204.

[Mani-Levitska 1993] P. Mani-Levitska, "Characterizations of convex sets", pp. 19–41 in *Handbook of convex geometry*, Vol. A, edited by P. M. Gruber and J. M. Wills, North-Holland, Amsterdam, 1993.

[McMullen and Schulte 1989] P. McMullen and E. Schulte, "Regular polytopes from twisted Coxeter groups", *Math. Z.* **201**:2 (1989), 209–226.

[McMullen et al. 1983] P. McMullen, C. Schulz, and J. M. Wills, "Polyhedral 2-manifolds in $\mathbb{E}^3$ with unusually large genus", *Israel J. Math.* **46** (1983), 127–144.

[Meeks 1981] W. H. Meeks, III, "The topological uniqueness of minimal surfaces in three-dimensional Euclidean space", *Topology* **20** (1981), 389–410.

[Milin 1994] L. Milin, "A combinatorial computation of the first Pontryagin class of the complex projective plane", *Geom. Dedicata* **49**:3 (1994), 253–291.

[Milnor 1950a] J. W. Milnor, *On the relationship between the Betti numbers of a hypersurface and an integral of its Gaussian curvature*, Junior thesis, Princeton University, 1950. Reprinted as pp. 15–26 in *Collected papers*, Vol. I, Publish or Perish, Houston, TX, 1994.

[Milnor 1950b] J. W. Milnor, "On the total curvature of knots", *Ann. of Math.* (2) **52** (1950), 248–257.

[Möbius 1886] A. F. Möbius, "Zur Theorie der Polyëder und der Elementarverwandtschaft", pp. 519–559 in *Gesammelte Werke*, vol. 2, edited by F. Klein, Hirzel, Leipzig, 1886.

[Morin and Yoshida 1991] B. Morin and M. Yoshida, "The Kühnel triangulation of the complex projective plane from the view point of complex crystallography I", *Mem. Fac. Sci. Kyushu Univ. Ser. A* **45**:1 (1991), 55–142.

[Morse and Cairns 1969] M. Morse and S. S. Cairns, *Critical point theory in global analysis and differential topology: An introduction*, Pure and Applied Mathematics **33**, Academic Press, New York, 1969.

[Morton 1979] H. R. Morton, "A criterion for an embedded surface in $\mathbb{R}^n$ to be unknotted", pp. 93–98 in *Topology of low-dimensional manifolds* (Chelwood Gate, Sussex, 1977), edited by R. Fenn, Lecture Notes in Math. **722**, Springer, Berlin, 1979.

[Niebergall 1994] R. Niebergall, "Tight analytic immersions of highly connected manifolds", *Proc. Amer. Math. Soc.* **120**:3 (1994), 907–916.

[Niebergall and Thorbergsson 1996] R. Niebergall and G. Thorbergsson, "Tight immersions and local differential geometry", preprint, University of Northern British Columbia and Universität Köln, 1996.

[Nirenberg 1963] L. Nirenberg, "Rigidity of a class of closed surfaces", pp. 177–193 in *Nonlinear Problems* (Madison, WI, 1962), edited by R. E. Langer, U. of Wisconsin Press, Madison, 1963.

[Ozawa 1983] T. Ozawa, "Products of tight continuous functions", *Geom. Dedicata* **14**:3 (1983), 209–213.

[Pinkall 1986a] U. Pinkall, "Curvature properties of taut submanifolds", *Geom. Dedicata* **20**:1 (1986), 79–83.

[Pinkall 1986b] U. Pinkall, "Tight surfaces and regular homotopy", *Topology* **25**:4 (1986), 475–481.

[Pogorelov 1973] A. V. Pogorelov, *Extrinsic geometry of convex surfaces*, Translations of Math. Monographs **35**, Amer. Math. Soc., Providence, 1973.

[Pohl 1981] W. F. Pohl, "Tight topological immersions of surfaces in higher dimensions", 1981. Manuscript.

[Ringel 1955a] G. Ringel, "Über drei Probleme am n-dimensionalen Würfel und Würfelgitter", *Abh. Math. Sem. Univ. Hamburg* **20** (1955), 10–19.

[Ringel 1955b] G. Ringel, "Wie man die geschlossenen nichtorientierbaren Flächen in möglichst wenig Dreiecke zerlegen kann", *Math. Ann.* **130** (1955), 317–326.

[Ringel 1974] G. Ringel, *Map color theorem*, Die Grundlehren der mathematischen Wissenschaften **209**, Springer-Verlag, New York, 1974.

[Rodríguez 1973] L. L. Rodríguez, *The two-piece property and relative tightness for surfaces with boundary*, Ph.D. thesis, Brown University, 1973.

[Rodríguez 1976] L. L. Rodríguez, "The two-piece-property and convexity for surfaces with boundary", *J. Differential Geometry* **11**:2 (1976), 235–250.

[Rodríguez 1977] L. L. Rodríguez, "Convexity and tightness of manifolds with boundary", pp. 510–541 in *Geometry and topology* (Rio de Janeiro, 1976), edited by J. Palis and M. do Carmo, Lecture Notes in Math. **597**, Springer, Berlin, 1977.

[Sarkaria 1983] K. S. Sarkaria, "On neighbourly triangulations", *Trans. Amer. Math. Soc.* **227** (1983), 213–239.

[Scharlemann 1985] M. Scharlemann, "Smooth spheres in $\mathbb{R}^4$ with four critical points are standard", *Invent. Math.* **79**:1 (1985), 125–141.

[Sen'kin 1972] E. P. Sen'kin, "Rigidity of convex hypersurfaces", *Ukrain. Geometr. Sb.* **12** (1972), 131–152, 170. Supplement in **17** (1975), 132–134.

[Sparla 1997a] E. Sparla, "A new lower bound theorem for combinatorial $2k$-manifolds", 1997. To appear in *Graphs and Comb.*

[Sparla 1997b] E. Sparla, "An upper and a lower bound theorem for combinatorial 4-manifolds", 1997. To appear in *Disc. Comput. Geom.*

[Steenrod 1944] N. E. Steenrod, "The classification of sphere bundles", *Ann. of Math.* (2) **45** (1944), 294–311.

[Tai 1968] S.-S. Tai, "Minimum imbeddings of compact symmetric spaces of rank one", *J. Diff. Geom.* **2** (1968), 55–66.

[Takeuchi and Kobayashi 1968] M. Takeuchi and S. Kobayashi, "Minimal imbeddings of R-spaces", *J. Differential Geometry* **2** (1968), 203–215.

[Terng 1993] C.-L. Terng, "Recent progress in submanifold geometry", pp. 439–484 in *Differential geometry: partial differential equations on manifolds* (Los Angeles, 1990), edited by R. Greene and S. T. Yau, Proc. Sympos. Pure Math. **54** (part 1), Amer. Math. Soc., Providence, RI, 1993.

[Teufel 1982] E. Teufel, "Differential topology and the computation of total absolute curvature", *Math. Ann.* **258** (1982), 471–480.

[Thorbergsson 1983] G. Thorbergsson, "Dupin hypersurfaces", *Bull. London Math. Soc.* (2) **15**:5 (1983), 493–498.

[Thorbergsson 1988] G. Thorbergsson, "Homogeneous spaces without taut embeddings", *Duke Math. J.* **57**:1 (1988), 347–355.

[Thorbergsson 1991] G. Thorbergsson, "Tight analytic surfaces", *Topology* **30**:3 (1991), 423–428.

[Walkup 1970] D. W. Walkup, "The lower bound conjecture for 3- and 4-manifolds", *Acta Math.* **125** (1970), 75–107.

[Wall 1962] C. T. C. Wall, "Classification of $(n-1)$-connected $2n$-manifolds", *Ann. of Math. (2)* **75** (1962), 163–189.

[White 1974] J. H. White, "Minimal total absolute curvature for orientable surfaces with boundary", *Bull. Amer. Math. Soc.* **80** (1974), 361–362.

[White 1984] A. T. White, *Graphs, groups and surfaces*, 2nd ed., North-Holland Math. Studies **8**, North-Holland Publishing Co., Amsterdam, 1984.

[Whitehead 1949] J. H. C. Whitehead, "On simply connected, 4-dimensional polyhedra", *Comment. Math. Helv.* **22** (1949), 48–92.

[Willmore 1971] T. J. Willmore, "Tight immersions and total absolute curvature", *Bull. London Math. Soc.* (2) **3** (1971), 129–151.

[Wilson 1969] J. P. Wilson, "Some minimal imbeddings of homogenous spaces", *J. London Math. Soc.* (2) **1** (1969), 335–340.

[Wintgen 1980] P. Wintgen, "Über die totale Absolutkrümmung verknoteter Sphären", *Beiträge zur Algebra und Geometrie* **9** (1980), 131–147.

[Wintgen 1984] P. Wintgen, "On total absolute curvature of nonclosed submanifolds", *Ann. Global Anal. Geom.* **2**:1 (1984), 55–87.

THOMAS F. BANCHOFF
MATHEMATICS DEPARTMENT
BROWN UNIVERSITY
PROVIDENCE, RI 02912
USA

WOLFGANG KÜHNEL
MATHEMATISCHES INSTITUT B
UNIVERSITÄT STUTTGART
70550 STUTTGART
GERMANY

Tight and Taut Submanifolds
MSRI Publications
Volume **32**, 1997

Tightness for Smooth and Polyhedral Immersions of the Projective Plane with One Handle

DAVIDE P. CERVONE

ABSTRACT. The recent discovery that there is a tight polyhedral immersion of the projective plane with one handle, while there is no smooth tight immersion of the same surface, provides a rare example in low dimensions of a significant difference between smooth and polyhedral surfaces. In this paper the author shows that the obstruction to smoothing the polyhedral model is not local in nature, and describes some of the ways in which the proof of the nonexistence of the smooth tight surface does not carry over to the polyhedral case.

1. Introduction

A longstanding open problem in the study of tight surfaces centered around a question posed by Nicolaas Kuiper [1961] asking whether the surface with Euler characteristic -1 (a real projective plane with one handle) could be tightly immersed in three-space. Kuiper had established that all other surfaces admitted tight immersions in space except for the Klein bottle and the real projective plane, which do not. More than thirty years passed before François Haab [1992] proved that, for smooth surfaces, no such immersion exists. In light of this result and the failure of the attempts to find a polyhedral counterexample, it seemed only a matter of time before a corresponding proof would be found for the polyhedral case as well. Surprisingly, a polyhedral tight immersion of this surface *does* exist, as shown recently in [Cervone 1994]. Although the smooth and polyhedral theories differ substantially for surfaces in high-dimensional spaces, they correspond quite closely in low dimensions; the case of the real projective plane with one handle is important in that it represents one of only a handful of low-dimensional examples where the theories differ in a significant way (see Section 3).

In this article, we compare the smooth and polyhedral behaviors of the projective plane with one handle, and try to illuminate some of the reasons why they differ. The subject is approached from two different directions: first, we analyze the polyhedral example in detail, especially the potential smoothability of specific configurations within it, and find that the obstruction to smoothing this model is not local in nature. Second, we outline the basic components of Haab's proof, and discuss why this proof does not carry over directly to the polyhedral case.

2. The Polyhedral Tight Immersion

We begin by presenting a tight polyhedral immersion of the real projective plane with one handle [Cervone 1994]. The model has 13 vertices, and their mapping into space is given in Figure 1 along with the 28 triangular faces, and a view of the surface from above. The self-intersection can be seen where faces meet without a heavy black line.

Faces:

$$
\begin{array}{lllll}
a\,b\,k & b\,g\,k & b\,g\,j & b\,i\,j & g\,f\,j \\
a\,d\,l & a\,k\,l & c\,d\,l & f\,l\,m & c\,l\,m \\
h\,i\,j & e\,h\,i & b\,c\,i & c\,i\,m & f\,i\,m \\
e\,f\,i & h\,k\,l & h\,j\,l & f\,j\,l & g\,h\,k \\
a\,b\,e & b\,e\,f & b\,c\,f & c\,f\,g & \\
c\,g\,h & c\,d\,h & a\,d\,h & a\,e\,h &
\end{array}
$$

Vertices:

$a = (-2, 0, 0)$

$b = (0, 0, 0)$

$c = (1, 0, 0)$

$d = (0, 1, 0)$

$e = (-2, -1, 2)$

$f = (1, -1, 2)$

$g = (1, 1, 2)$

$h = (0, 3, 2)$

$i = (-3/8, 0, 1/2)$

$j = (1/2, 1/4, 1)$

$k = (-1/4, 7/12, 7/6)$

$l = (0, 3/4, 7/6)$

$m = (1/4, 0, 1/2)$

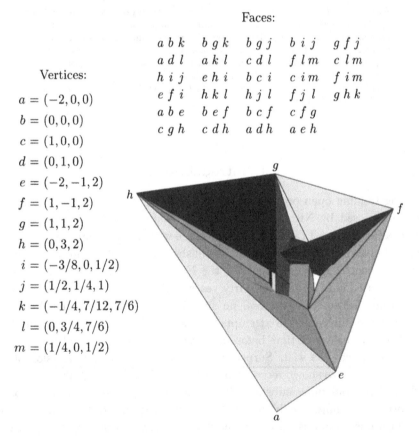

Figure 1. The tight polyhedral projective plane with one handle, including its vertex mapping and face list. The point of view is approximately $(-30, -10, 60)$. The x-axis and y-axis are parallel to the edges ef and fg. See http://www.geom.umn.edu/locate/rp2-handle for interactive three-dimensional viewing.

The surface is composed of a central core (an immersed real projective plane with two disks removed) surrounded by a cylinder formed by removing two disks from the convex envelope (the surface of the convex hull) of the central core. The core is based on the projective plane described by Kuiper [1961], who gave level sets for an immersed smooth surface with one maximum, one minimum, and one saddle point. The level sets of the polyhedron given here correspond closely to those described by Kuiper. In the complete surface, the maxima and minima are removed and replaced by a tube connecting the top to the bottom (the outer cylinder). The two curves where the core joins the tube are called *top cycles* (for a complete definition and more information about top cycles, see [Banchoff and Kühnel 1997] in this volume).

Any immersion of a surface of odd Euler characteristic must have a triple point [Banchoff 1974]; this model has exactly one, and it is visible at the center of the figure. Since the central core is essentially a projective plane with one triple point, we would expect the double curve to form three loops meeting at the triple point: six doubly covered lines emanate from the triple point, and since in an immersed surface the double locus forms closed curves, these six lines must be joined pairwise by the double curve, thus forming three loops (see Figure 2). This is indeed the case in our polyhedral model.

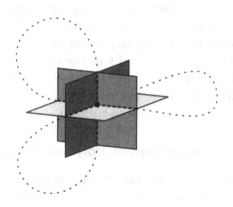

Figure 2. A triple point is formed by the intersection of three sheets, where six arcs of double points (dashed lines) meet. In an immersed projective plane with only one triple point, these arcs are joined pairwise by the double curve (dotted lines).

To verify that this is in fact a projective plane with one handle, we compute its Euler characteristic: the object has 13 vertices, 42 edges, and 28 faces, so its Euler characteristic is $V - E + F = -1$ as expected. A polyhedral surface is immersed if, and only if, the star of each of its vertices is embedded. An explicit check of the vertices of this model reveals that each star is embedded (the self-intersection curve does not pass through any vertex), so it is indeed an immersion.

There are several equivalent definitions of tightness (see [Banchoff and Kühnel 1997] in this volume). A geometric definition that applies to both smooth and polyhedral surfaces is that an immersion $f : M \to \mathbb{R}^3$ is *tight* if it has the *two-piece property*, namely that the preimage of any half-space is connected in M, or in other words any plane cuts the surface into at most two parts. For polyhedral surfaces, this provides a simple characterization of tight immersions [Banchoff and Kühnel 1997, Lemma 1.4.1]:

LEMMA 2.1. *A polyhedral immersion is tight if, and only if,*

(i) *the one-skeleton of the convex hull is contained in the surface;*
(ii) *every vertex that is not a vertex of the convex hull lies in the relative interior of some subset of its adjacent vertices; and*
(iii) *the self-intersection set does not include any vertex of the convex hull of the surface.*

The final condition is required for immersions since such a vertex could be cut off, thereby dividing the surface into at least three pieces (two copies of the vertex, plus the rest of the surface).

We can use this lemma to check that the polyhedron given above is, in fact, tight: first, note that the convex envelope has seven vertices: a, c, d, e, f, g, and h (although b is on the convex envelope, it is not a vertex of it) and that the last eight faces listed in Figure 1 contain all the edges of the convex envelope. Of the remaining six vertices, two lie on the straight line segment between two neighbors (b lies on the segment ac, and k on the segment ag), two lie within a triangle formed by three neighbors (i lies within triangle beh, and j within bfg), and two lie inside tetrahedra formed by four neighbors (l lies within $acfh$, and m within $cfil$). Thus the surface is tight, as claimed.

3. Tight Smoothings of Polyhedral Surfaces

In the previous section we presented a tight polyhedral model of the real projective plane with one handle, and Haab [1992] provided a proof that no smooth immersion of this surface exists. An important question to ask is: Why can't this polyhedral model be smoothed tightly? Given a polyhedral surface, a *tight smoothing* is a tight smooth surface of the same topological type lying within an ε-neighborhood of the polyhedron, for some small ε. Frequently, there exists such a surface for arbitrary ε and these form a continuous deformation from the polyhedral to the smooth surface that is tight at every step. Such a deformation need not always exist, however. For example, in 5-space, there are essentially only two substantially embedded tight immersions of the projective plane: the Veronese surface (a smooth embedding), and the canonical embedding of the six-vertex polyhedral real projective plane (see [Banchoff and Kühnel 1997, Theorem 1.3.6 and Corollary 1.4.12] in this volume). In one sense, the Veronese surface is the obvious tight smoothing of the polyhedral projective plane, but there is no continuous deformation by tight surfaces from one to the other, since no intermediate tight surfaces exist. Although no examples like this are known in three-space, we will not require a deformation by tight surfaces for our purposes, but only a smooth tight surface sufficiently near the polyhedral one.

One feature that might interfere with smoothability would be the presence of exotic saddle points. These are saddles where there is no direction for which the orthogonal projection of a neighborhood of the saddle is one-to-one, as in Figure 3. Such a saddle is not possible in a smooth surface, since the implicit function theorem guarantees that such a direction always exists. Exotic saddles play a crucial role in one of the few other low-dimensional examples of a difference between the smooth and polyhedral theories, namely the existence of an embedded polyhedral torus with a height function having exactly three critical points, but no such smooth embedding [Banchoff and Takens 1975]. Taking our cue from this, we look for exotic saddles in the polyhedral surface given in Section 2. Checking each of its vertices, however, we find that it contains no exotic saddles, since every vertex has a direction where the orthogonal projection of its star in that direction is a one-to-one mapping.

Figure 3. A polyhedral monkey saddle of the exotic type. There is no direction for which the orthogonal projection is a one-to-one mapping.

A second feature that might interfere with smoothability would be vertices that are not "generic" enough, i.e., ones that can not be perturbed without destroying tightness. For example, vertex k lies on the boundary of the convex hull of its neighbors (it is on the segment ag), and if it were raised in the z direction (for example) it no longer would be in the relative interior of any subset of its neighbors, and the model would fail to be tight (see lemma 2.1). Vertex b has a similar problem, and vertices i and j, which lie on faces of the convex hulls of their neighbors, have directions in which they can not be perturbed without losing tightness. The basic observation is that if a vertex lies on the convex envelope of its neighbors, then it can not be perturbed in some directions, nor can some of its neighbors, without losing tightness. On the other hand, if a vertex lies in the strict interior of the convex hull of its neighbors, then it can be moved slightly in all directions, and so can its neighbors, without destroying tightness at that vertex. Ideally, then, we would like the model to be in *general position* (i.e., no three vertices lie on a line, and no four vertices lie in a plane).

It is possible to move vertices i, j, and k slightly so that they *do* lie in the interior of the convex hull of their neighbors while maintaining the tightness of the model (move i in the positive x direction, j in the negative x direction, and k in the negative z direction slightly). Since b lies on the convex envelope of the surface, it must be treated with a bit more care. The solution in this case is to move b in the negative y direction; this makes it a vertex of the convex hull of the surface, and so condition *ii* of lemma 2.1 no longer applies. Once this is done, the vertices that form the convex hull of the surface also can be perturbed, provided they stay within the planes of the top cycles to which they belong. Thus all the vertices of the model can be "jiggled" slightly without losing tightness. This is not quite general position, however, since the top cycles must remain in a plane; but these planes can be jiggled slightly as well, provided that all the vertices in the plane are moved together. We will call this *nearly general position.*

One may ask whether the model can be put into truly general position, and indeed it can. The crucial condition is that, to be in general position, the top cycles must be triangular since each top cycle must lie in a plane. The top cycles in this model are formed by quadrilaterals; however, the polyhedron can be modified so that its top cycles *are* triangular. One way to do this is to remove the convex envelope (leaving only the core) and then place large triangles in the planes of the top cycles so that the top cycles lie inside them. Triangulate the annular regions between each triangle and the top cycle contained within it, then move each triangle slightly in the direction perpendicular to its plane but away from the central core (the annular region will become a funnel-shaped "flange"). Finally, add a new cylinder connecting the two large triangles, to replace the convex envelope that was removed at the outset. Provided the new triangles are large enough and the distance they were moved is small enough, the resulting surface will be a tight immersion of the real projective plane with one handle, and its top cycles will be triangular. The vertices that previously formed the top cycles will now be inside the convex hull of the surface, and can be moved slightly to put the entire surface into general position.

Once in general position, every vertex can be perturbed slightly without damaging tightness, including those that form the top cycles. Note that this is in sharp contrast to the smooth situation, where the top cycles are less stable: small changes to a top cycle can easily destroy tightness.

We have seen that the model presented in Section 2 is not initially in general or even nearly general position. The reason for this is twofold: first, it shows that such unusual configurations are possible and not just contrived, and second, it makes checking tightness simpler, since it is easier to check that a vertex lies on a line segment or in a plane than it is too see that it lies inside a tetrahedron. It is important to note, however, that the presence of such configurations is *not* an obstacle to smoothing the polyhedral surface, since the surface can be put into nearly general position with only minor adjustments, or general position with more extensive changes, without losing tightness.

For additional features that might interfere with smoothability, we turn to an algorithm developed in [Kühnel and Pinkall 1985] that will smooth a tight polyhedral immersion to produce a tight smooth immersion of the same surface that agrees with the polyhedron everywhere but in an ε-neighborhood of the edges. The algorithm does not apply to all polyhedra, however, but only to ones with certain properties, and the conditions that the polyhedral surface must meet are rather strict. Given a vertex of a polyhedral immersion, consider a small sphere centered at the vertex; its intersection with the surface forms a spherical polygon, and a neighborhood of the vertex is the cone over such a polygon. If this cone is convex, then the vertex is called *convex*. A vertex is a *standard saddle* if (i) it is 4-valent; (ii) all the angles of the faces at the vertex are strictly less than π; and (iii) there are no local support planes through the vertex.

The conditions of the smoothing algorithm require that the nonconvex vertices of the surface be either 3-valent or standard saddles. In particular, no vertex interior to the convex hull can be more than 4-valent, and all the vertices that lie on the convex envelope either must be convex or as simple as possible, i.e., only 3-valent. On the other hand, the algorithm also allows nontriangular faces, and indeed, nonconvex and even non-simply-connected faces, so these restrictions aren't so severe as they at first appear.

The polyhedral model presented in the previous section does not satisfy these conditions, nor can it be modified to satisfy them, otherwise its smoothing would represent a counterexample to Haab's theorem. This leads us to ask: Which vertices cause trouble, and what is the obstruction to smoothing them? All the vertices except m fail the valence conditions, and vertices b, k, i, and j have facets with angles of π or greater (note: coplanar faces that share an edge form a single facet, so abk and kbg form one facet $abgk$ that has an angle of π at k since k lies on the segment between a and g). The latter problem can be resolved by putting the vertices into nearly general position as discussed above, since then all the interior facets are triangles, and so have angles strictly less than π. The remaining difficulty is the large number of edges at each vertex.

One approach to the valence problem is to try to split up a vertex of high valence into several vertices of lesser valence. The trick is to do it while still maintaining tightness. Since the model is in nearly general position, if the new vertices all are in a small neighborhood of the original vertex, this will not disrupt the tightness at its neighbors, so it is only necessary to check that the new vertices satisfy condition *ii* of lemma 2.1.

As an example, consider vertex a, a vertex of the convex hull of the surface that is nonconvex and 7-valent. It can be broken into three vertices, as shown in Figure 4. One of these, a_1, is convex, and the other two, a_2 and a_3, are 3-valent. Note that the introduction of nontriangular faces is allowed, and indeed

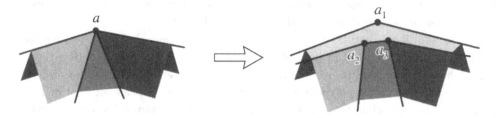

Figure 4. The nonconvex vertex a can be broken into three vertices, one convex, the other two 3-valent, all lying in the plane of the top cycle containing a.

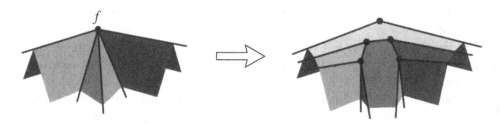

Figure 5. The nonconvex vertex f can be broken into five vertices, one convex, two 3-valent, and the other two standard saddles.

is crucial to obtaining the correct valences at a_2 and a_3. This modification maintains tightness, since a_1 is now a vertex of the convex hull, while a_2 and a_3 are in the relative interiors of their neighbors (a_2 is interior to the triangle formed by a_1, a_3 and the vertex to the left of the original vertex a, while a_3 is interior to a_1, a_2 and the vertex to the right of a).

We will call a vertex *locally smoothable* if it can be replaced by a collection of new vertices such that the resulting polyhedral model is still tight, and all the new vertices satisfy the smoothing criterion of Kühnel and Pinkall. Thus vertex a is locally smoothable. Vertices c, d and e likewise are locally smoothable using a similar decomposition. A local smoothing configuration for vertex f is given in Figure 5. Again, nontriangular faces are crucial to the construction, and care must be taken at the 4-valent vertices to assure that the angles are less than π. Vertex g can be handled similarly.

Vertex b presents a more difficult challenge. It can be modified as shown in Figure 6. Here, rather than pushing b outward so that it becomes a vertex of the convex envelope, we pull it into the interior and subdivide it. It is important that b_1, b_2, b_3 and g lie in a plane, with the angle at b_3 less than π. Vertex b_1 should be placed on the back side of the plane containing a, c and g, while b_2 and b_3 should be placed in the interior of the tetrahedron $b_1 g i j$. A similar, though slightly more complex, construction is possible at vertex h.

Vertices i and k are interior to the convex hull and are 5-valent; they can be treated in much the same way that b was above. Vertex m already satisfies the smoothing conditions. Vertex j can be split into two vertices, one 5-valent, the

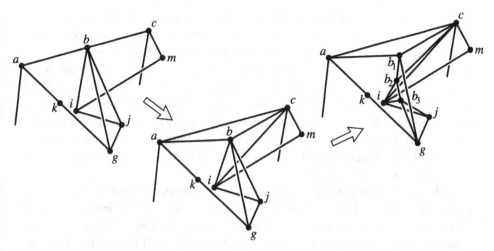

Figure 6. Vertex b can be pulled into the interior of the convex hull (middle) and then split into three vertices (right), all of which are standard saddles.

other 4-valent (Figure 7, middle) provided j_1, j_2, b and g are coplanar; j_1 can then be split in a way similar to how b was modified above (Figure 7, right). An extra vertex j_5 is added between j_2 and h to divert from j_2 an edge generated while subdividing j_1. This requires j_1, j_2, j_3 and j_5 to be coplanar, as well as j_1, j_3, j_4, and f. One can arrange that these planarity conditions are satisfied while still placing each new vertex in the interior of the tetrahedron formed by its four neighbors. Thus j is locally smoothable. Finally, l can be split into two 5-valent vertices that can be handled in a similar fashion.

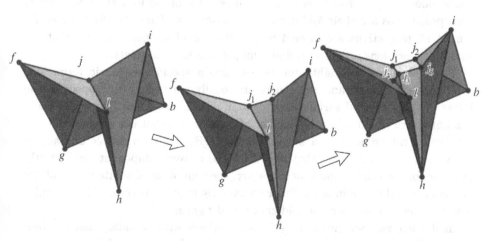

Figure 7. The 6-valent vertex j can be broken into two vertices (middle), one a standard saddle, the other 5-valent; the latter can in turn be broken into three standard saddles (right). An extra vertex is added near j_2 in order to divert an edge that would have increased the valence of j_2.

We see, then, that *every* vertex of the polyhedral model is locally smooth-able; but this does *not* mean that the model itself is smoothable. The reason is that we can not carry out all the modifications described above simultane-ously without them interfering with each other. For example, the modification to b adds edges to a and c, which will disrupt the carefully planned valence configurations at those vertices. Some coordination among these modifications may be possible, however. For example, two new edges generated at neighboring vertices may form triangles that can be combined into a single planar quadrilat-eral, thus maintaining the proper valence at each vertex (the author currently is investigating such possibilities).

Although we can not carry out all the modifications concurrently, the fact that each vertex is locally smoothable has an important implication, namely that the obstruction to smoothing this polyhedral surface is not a local one, isolated at some unusual vertex (since neighborhoods of all the vertices look like smoothable patches that could become part of a smooth tight surface). Rather the obstruction is a global one, intrinsic to the surface itself. What this global property is remains an open question at this point.

4. Haab's Proof for Smooth Surfaces

Haab's proof [1992] of the fact that there is no smooth immersion of the real projective plane with one handle requires considerable machinery, of which we outline some of the highlights.

His basic idea is to consider mappings of surfaces into the plane, and to determine when these can be "factored" into an immersion of the surface in space followed by a projection into the plane. One of the important features of such projections are their fold curves. For smooth surfaces, the fold curves for almost all projections are formed by a collection of images of circles that are smooth immersions except at a finite number of cuspidal points.

The fold curves for tight immersions have a specific geometric form: one component is convex, and the others are locally concave (with respect to the image of the projected surface) and contain all the cusps (see Figure 8). Haab computes strict bounds on the number of components that can exist in the fold set for a tight immersion of a given surface. He defines a degree on each of the fold curves, and shows that it is 2 for the convex component, and strictly negative for the others, and that moreover, the sum of the absolute values of the degrees is equal to 4 minus the Euler characteristic of the surface. This provides a key connection between the fold curves and tightness.

Haab then considers height functions on surfaces with boundary and classifies their saddle points according to whether passing the saddle point changes the number of components in the level set of the height function. He uses a gener-alization of the Morse inequalities to show that the number of saddles that do not change the number of components is equivalent mod 2 to the genus of the

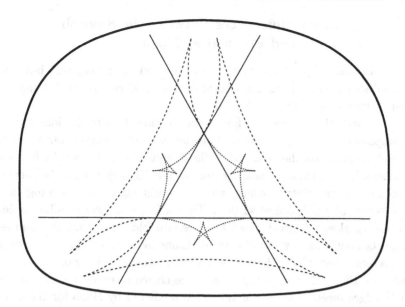

Figure 8. The fold curves for a tight immersion of the surface with Euler characteristic −3, after [Haab 1992]. The saddles that change the number of components in the level set are shown by dashed lines, and those that don't are shown
by dotted lines. Note that the type of saddle changes only at points where there
is a doubly tangent line to the fold set.

surface. This allows him to prove an actual bound (in terms of the genus of the
surface) on the number of such saddle points for any height function. Haab then
considers the fold sets of projections of tight surfaces, and determines that the
type of saddle that occurs (for height functions in directions perpendicular to
the direction of projection) on the fold set for a given projection can change only
at points where the tangent line to the fold curve is tangent to the fold set at
more than one point (see Figure 8).

He applies this information to the case of the projective plane with one handle,
and concludes that, for a tight immersion, the type of saddle is constant on each
component of the fold set, and that the fold set has exactly three components.
He uses the fold curves and the top cycles to decompose the surface into disjoint
regions, and shows that one of these contains a cycle that separates it, but
on which the Gauss map is constant. He considers a height function in the
direction of the Gauss map, and, after showing that the curve bounds a region
on which the Gauss map is *not* constant, deduces that the height function has
a local extremum inside that region. The initial decomposition of the surface
into regions ensures that this point is not a global extremum for the height
function, which contradicts the fact that the immersion is tight, since a plane
perpendicular to the direction of the height function and just below the lower
extremum will cut the surface into three disjoint pieces.

5. Some Differences Between the Smooth
and Polyhedral Theories

Haab relies heavily on the smoothness of the surface in his proof, but not all of his results carry over to the polyhedral situation. There are several important differences that arise in the polyhedral case.

First, for smooth surfaces, the fold curves of almost all projections form disjoint components, so it is possible to count the number of components accurately. In the polyhedral case, the analogue of the fold curves are formed by fold edges (ones where the two triangles sharing that edge both project onto the same side of the edge), but an arbitrarily large number of fold edges may come together at a single vertex of a polyhedral surface. Thus it is not always possible to determine a canonical way to divide the fold edges into fold curves, and the number of components may change with different divisions into curves. In the polyhedral model presented here the projection onto the xy-plane has either one or two components (see Figure 9) depending on how the choice is made at vertex h where four fold edges meet. This is not the three predicted by Haab for the smooth case, so already a difference has emerged between the two types of surfaces. It is interesting to note that the conditions of Kühnel and Pinkall's smoothing algorithm [1985] guarantee that there are at most two fold edges at each vertex, so for these polyhedra, the fold curves can be separated into components without ambiguity.

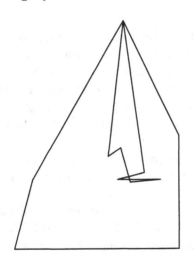 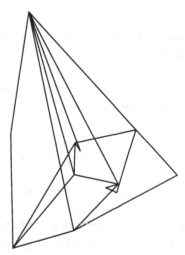

Figure 9. The fold curves for two projections of the tight polyhedral projective plane with a handle; each can be broken into fold curves in more than one way. The projection onto the xy plane (left) can be broken into one or two components depending on the decision made at the topmost vertex, h. The other projection (right) can be broken into as many as six components; it has three interior vertices where more than two fold edges meet. There are projections for which the fold curves are even more complicated.

Second, the idea of a cusp and of locally convex curves is harder to formulate in the polyhedral case. One might begin by identifying analogous polyhedral structures, such as those shown in Figure 10, but this becomes more complicated when more than two fold edges meet at one vertex.

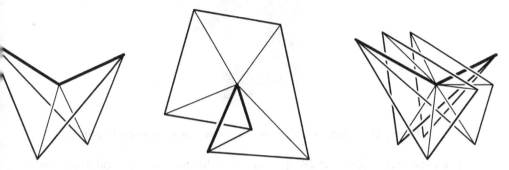

Figure 10. Polyhedral analogues of the fold and cusp (left). The neighborhood of a vertex can wrap around the vertex an arbitrary number of times (right), though only a single revolution is shown here. The fold edges themselves do not contain enough information to distinguish between the two folds shown.

The problem is compounded by the fact that a star can wind around a vertex an arbitrary number of times with no visible effect on the angle between the fold edges. This makes computing the degree of a fold curve more complicated than in the smooth case (where small loops would be present). In the polyhedral case, the fold curve itself does not contain enough information to distinguish between the two situations shown in Figure 10, so any analog of Haab's fold-curve degree would have to be more complex, and would probably involve looking at the strip neighborhoods of the fold curves. (Again, note that the conditions of Kühnel and Pinkall's smoothing algorithm do not allow for such aberrant behavior.)

For smooth tight surfaces, almost every point of the (nonconvex) fold curves is a saddle for some direction, but this is no longer the case for polyhedral surfaces, where only vertices can be saddles. Moreover, these saddles can easily be of higher degree (i.e., monkey saddles or worse), and can be the exotic type that have no smooth counterpart (Figure 3). Such saddles are stable under small changes of direction, which is not the case for smooth surfaces. This complicates the issue of determining the "type" of each saddle, and thus the type of each fold component, which is crucial to Haab's argument. Haab's results concerning when the type of a fold curve can change would require additional work in the polyhedral case, since the idea of bitangent lines to the fold curves does not have a direct analog.

Finally, Haab uses the fact that the top cycles in a smooth tight immersion lie in distinct planes to break the central projective plane into regions, one of which is an annulus. In the polyhedral case, the top cycles can share vertices or even edges, making such a decomposition harder to formulate. For example,

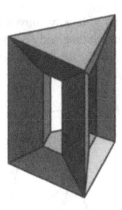

Figure 11. A surface with three top cycles that share edges in pairs.

the Császár torus [Császár 1949], which is tight, has two triangular top cycles that share a vertex. To generate an example where the top cycles share edges, begin with a hexagonal prism and remove every other rectangular face. Place the prism within a large triangular prism whose faces are parallel to the ones removed from the hexagonal prism. Finally, replace each rectangular face of the triangular prism with four trapezoidal faces that connect its boundary to the boundary of the corresponding removed face of the hexagonal prism. The resulting embedded surface is tight (see Figure 11); it has three rectangular top cycles (the boundaries of the rectangular faces of the outer prism) that share edges in pairs: each edge joining the top and bottom triangles of the prism belongs to two distinct top cycles.

6. Conclusion

The existence of a tight polyhedral immersion of the real projective plane with one handle, but no smooth one, provides an opportunity to study some of the details of how these two types of surfaces differ even in low dimensions. One question that remains is how unique is this situation? Are there other tight polyhedral immersions of surfaces that can not be tightly smoothed? The paper [Cervone 1996] presents several tight polyhedral immersions to which Kühnel and Pinkall's smoothing algorithm does not apply, and for which no corresponding smooth tight immersion is known; are these models also examples of this same phenomena? Or is there a more general smoothing algorithm that will handle these cases? The example presented here shows that no smoothing algorithm will work for *every* tight polyhedral surface, and an understanding of just why that is so should provide insight into both the smooth and polyhedral worlds.

References

[Banchoff 1974]   T. F. Banchoff, "Triple points and surgery of immersed surfaces", *Proc. Amer. Math. Soc.* **46** (1974), 407–413.

[Banchoff and Kühnel 1997]   T. F. Banchoff and W. Kühnel, "Tight submanifolds, smooth and polyhedral", pp. 51–118 in *Tight and Taut Submanifolds*, edited by T. E. Cecil and S.-s. Chern, Cambridge U. Press, 1997.

[Banchoff and Takens 1975]   T. F. Banchoff and F. Takens, "Height functions on surfaces with three critical points", *Illinois J. Math.* **19** (1975), 325–335.

[Cervone 1994]   D. P. Cervone, "A tight polyhedral immersion of the real projective plane with one handle", 1994. Available at http://www.geom.umn.edu/locate/rp2-handle .

[Cervone 1996] D. P. Cervone, "Tight immersions of simplicial surfaces in three space", *Topology* **35**:4 (1996), 863–873.

[Császár 1949] Á. Császár, "A polyhedron without diagonals", *Acta Univ. Szeged. Sect. Sci. Math.* **13** (1949), 140–142.

[Haab 1992]   F. Haab, "Immersions tendues de surfaces dans E^3", *Comment. Math. Helv.* **67**:2 (1992), 182–202.

[Kühnel and Pinkall 1985]   W. Kühnel and U. Pinkall, "Tight smoothing of some polyhedral surfaces", pp. 227–239 in *Global differential geometry and global analysis* (Berlin, 1984), edited by D. Ferus et al., Lecture Notes in Math. **1156**, Springer, Berlin, 1985.

[Kuiper 1961]   N. H. Kuiper, "Convex immersions of closed surfaces in E^3. Nonorientable closed surfaces in E^3 with minimal total absolute Gauss-curvature", *Comment. Math. Helv.* **35** (1961), 85–92.

DAVIDE P. CERVONE
MATHEMATICS DEPARTMENT
UNION COLLEGE
SCHENECTADY, NY 12308
dpvc@union.edu

Tight and Taut Submanifolds
MSRI Publications
Volume **32**, 1997

Taut and Dupin Submanifolds

THOMAS E. CECIL

ABSTRACT. This is a survey of the closely related fields of taut submanifolds and Dupin submanifolds of Euclidean space. The emphasis is on stating results in their proper context and noting areas for future research; relatively few proofs are given. The important class of isoparametric submanifolds is surveyed in detail, as is the relationship between the two concepts of taut and Dupin. Also included is a brief introduction to submanifold theory in Lie sphere geometry, which is needed to state many known results on Dupin submanifolds accurately. The paper concludes with detailed descriptions of the main known classification results for both Dupin and taut submanifolds.

Dupin [1822] determined which surfaces M embedded in Euclidean three-space $\mathbb{R}^3$ can be obtained as the envelope of the family of spheres tangent to three fixed spheres. These surfaces, known as the *cyclides of Dupin*, can all be constructed by inverting a torus of revolution, a circular cylinder or a circular cone in a sphere. The cyclides of Dupin were studied extensively in the nineteenth century (see, for example, [Cayley 1873; Liouville 1847; Maxwell 1867]). They have several other important characterizations. They are the only surfaces M in $\mathbb{R}^3$ whose focal set consists of two curves, which must, in fact, be a pair of focal conics. This is equivalent to requiring that M have two distinct principal curvatures at every point, each of which is constant along each of its corresponding lines of curvature. It is also equivalent to the condition that all lines of curvature in both families are circles or straight lines.

The cyclides reappeared in modern differential geometry in a paper by Banchoff [1970]. He considered compact surfaces M embedded in $\mathbb{R}^3$ with the property that every metric sphere in $\mathbb{R}^3$ cuts M into at most two pieces; this is called the *spherical two-piece property*, or STPP. For surfaces, the STPP is equivalent to requiring that M be *taut*, i.e., that every nondegenerate Euclidean distance function $L_p(x) = |p - x|^2$, where $p \in \mathbb{R}^3$, has the minimum number of critical points allowed by the Morse inequalities. Banchoff showed that tautness implies

Research supported by NSF Grant DMS-9504535.

that M must be a metric sphere or a cyclide of Dupin, and the close link between these notions was established.

A hypersurface M in $\mathbb{R}^n$ is said to be *Dupin* if, along each curvature surface, the corresponding principal curvature is constant. A Dupin hypersurface M is called *proper Dupin* if the number of distinct principal curvatures is constant on M. These concepts can both be generalized in a natural way to submanifolds of codimension greater than one in $\mathbb{R}^n$. A fundamental result in the theory due to Pinkall [1986] is that a taut submanifold must be Dupin. Conversely, the work of Thorbergsson [1983a] and Pinkall [1986] shows that a compact proper Dupin submanifold embedded in $\mathbb{R}^n$ must be taut. A major open question is whether the condition that the number of distinct principal curvatures is constant can be dropped; in other words, does Dupin imply taut? These results are discussed in Section 4.

This paper is a survey of the major results on taut and Dupin submanifolds. We concentrate on stating the results in their proper context and noting areas for future research and give very few proofs. In particular, we do not repeat proofs of many fundamental results in the field that can be found in [Cecil and Ryan 1985], and we will concentrate on work done since that reference appeared. We will not attempt to cover the related field of tight immersions, since that is done in the paper by Banchoff and Kühnel [1997] in this volume.

Important examples of taut submanifolds are the isoparametric submanifolds. These will be reviewed in Section 2, but the reader is referred to the excellent article [Terng 1993] for an in-depth survey of that field. There is also extensive research on real hypersurfaces with constant principal curvatures in complex space forms, which is covered by the paper by Niebergall and Ryan [1997] in this volume, and which will not be discussed here. We now give a brief overview of the contents of this article.

In Section 1 we review the critical point theory and submanifold theory needed to formulate the definition of a taut submanifold, and we list some basic results and methods for constructing taut embeddings. In Section 2 we list the primary known results for isoparametric hypersurfaces in spheres, which play an important role in the theory of Dupin hypersurfaces.

In Section 3 we give the definition of a Dupin submanifold and review Pinkall's standard local constructions of proper Dupin hypersurfaces with an arbitrary number of distinct principal curvatures and respective multiplicities. In Section 4 we discuss the relationship between the taut and Dupin conditions in detail.

Many of the main classifications of proper Dupin submanifolds are done in the context of Lie sphere geometry. In Section 5 we give a brief introduction to this theory in order to be able to explain these classifications accurately.

Section 6 is a survey of the known results on compact proper Dupin hypersurfaces. Thorbergsson [1983a] applied the work of Münzner [1980; 1981] to show that the number g of distinct principal curvatures of such a hypersurface must be 1, 2, 3, 4, or 6, the same as for an isoparametric hypersurface in a sphere. For some time, it was conjectured that every compact proper Dupin hypersurface is equivalent by a Lie sphere transformation to an isoparametric hypersurface. However, this is not the case, as examples constructed by Pinkall and Thorbergsson [1989a] and by Miyaoka and Ozawa [1989] demonstrate. We describe these examples in detail.

In Section 7 we study the local classifications of proper Dupin hypersurfaces that have been obtained using Lie sphere geometry. We describe the known results and mention several areas for further research.

Finally in Section 8 we survey the known classifications of taut embeddings. To some extent, this section can be read independent of the rest of the paper, although some references to the previous sections are necessary.

1. Taut Submanifolds

We begin with a brief review of the critical point theory and submanifold theory needed to formulate the definition of tautness. In this paper, all manifolds are assumed to be connected unless explicitly stated otherwise. Let M be a smooth, connected n-dimensional manifold, and let ϕ be a smooth real-valued function defined on M. A point $x \in M$ is a *critical point* of ϕ if the differential ϕ_* is zero at x. The critical point x is *nondegenerate* if the Hessian H of ϕ is a nondegenerate bilinear form at x, and otherwise it is said to be *degenerate*. The *index* of a nondegenerate critical point x is equal to the index of H as a bilinear form, that is, the dimension of a maximal subspace on which H is negative-definite. The function ϕ is called a *Morse function* or *nondegenerate function* if it has only nondegenerate critical points on M.

Let ϕ be a Morse function on M such that the set

$$M_r(\phi) = \{x \in M : \phi(x) \leq r\}$$

is compact for all $r \in \mathbb{R}$. Of course, this is true for any Morse function on a compact manifold M. Let $\mu_k(\phi, r)$ be the number of critical points of ϕ of index k on $M_r(\phi)$. If M is compact, let $\mu_k(\phi)$ denote the number of critical points of index k on M. For a field F, let

$$\beta_k(\phi, r, F) = \dim_F H_k(M_r(\phi); F)$$

be the k-th F-Betti number of $M_r(\phi)$, and let $\beta_k(M; F)$ be the k-th F-Betti number of a compact M. Then the *Morse inequalities* (see [Morse and Cairns 1969, p. 270], for example) state that

$$\mu_k(\phi, r) \geq \beta_k(\phi, r, F)$$

for all F, k, r, and for a compact M,

$$\mu_k(\phi) \geq \beta_k(M; F)$$

for all F, k. A Morse function ϕ on M is said to be *perfect* if ϕ has the minimum number of critical points possible by the Morse inequalities, that is, if there exists a field F such that

$$\mu_k(\phi, r) = \beta_k(\phi, r, F)$$

for all k, r. For a compact manifold M, this is equivalent to the condition $\mu_k(\phi) = \beta_k(M; F)$ for all k. Equivalently, it can be shown (see [Morse and Cairns 1969, p. 260], for example) that a Morse function ϕ on a compact manifold M is perfect if there exists a field F such that, for all k, r, the map on homology

$$H_k(M_r(\phi); F) \to H_k(M; F) \tag{1.1}$$

induced by the inclusion of $M_r(\phi)$ in M is injective. This formulation has proved to be quite useful in the theory of tight and taut immersions.

Let $f : M \to \mathbb{R}^n$ be a smooth immersion of a manifold M into n-dimensional Euclidean space. Since f is an immersion, it is an embedding on a suitably small neighborhood of any point $x \in M$. Thus, for local calculations, we often identify the tangent space $T_x M$ with its image $f_*(T_x M)$ under the differential f_* of f. Suppose that $X \in T_x M$ and that ξ is a field of unit normal vectors to $f(M)$ defined on a neighborhood of x. Then we have the fundamental equation

$$D_X \xi = -A_\xi X + \nabla_X^\perp \xi,$$

where $-A_\xi X$ is the component of $D_X \xi$ tangent to M, and $\nabla_X^\perp \xi$ is the component normal to M. Here A_ξ is a symmetric tensor of type $(1,1)$ on M called the *shape operator* determined by ξ, and $\nabla^\perp$ is a covariant derivative operator in the normal bundle of M called the *normal connection*. The eigenvalues of A_ξ are called the *principal curvatures* of A_ξ. When $f(M)$ is a hypersurface, a local field of unit normal vectors ξ is determined up to a sign. In this case, the shape operator A_ξ is often denoted simply by A, and the eigenvalues of A_ξ are determined up to a sign, depending on the choice of ξ. In that case, these eigenvalues are called the principal curvatures of M or of f.

The *normal exponential map* F from the normal bundle $N(M)$ to $\mathbb{R}^n$ is defined by

$$F(x, \eta) = f(x) + \eta,$$

where η is a normal vector to $f(M)$ at $f(x)$. A point $p \in \mathbb{R}^n$ is called a *focal point of multiplicity m of (M, x)* if $p = F(x, \xi)$ and the differential F_* has nullity m at (x, ξ). A point $p \in \mathbb{R}^n$ is called a focal point of M if p is a focal point of (M, x) for some $x \in M$. The set of all focal points of M is called the *focal set* of M. Since $N(M)$ and $\mathbb{R}^n$ have the same dimension, Sard's Theorem implies that the focal set of M has measure zero in $\mathbb{R}^n$. A direct computation (see [Milnor 1963, p. 34], for example) shows that if $p = F(x, t\xi)$, where $|\xi| = 1$, then p is a focal point of (M, x) of multiplicity m if and only if $1/t$ is a principal curvature of A_ξ of multiplicity m. In this paper, we often consider an immersion $f : M \to S^n$ into the unit sphere S^n in $\mathbb{R}^{n+1}$. In that case, one can also define the notions of normal exponential map and focal point in a manner analogous to the definitions given here for submanifolds of Euclidean space.

A *Euclidean distance function* is a function $L_p : \mathbb{R}^n \to \mathbb{R}$ given by the formula $L_p(q) = |p - q|^2$, where $p \in \mathbb{R}^n$. The level sets of L_p are spheres centered at the point p. Let f be an immersion of a smooth manifold M into $\mathbb{R}^n$. We consider the restriction of L_p to M defined by $L_p(x) = |p - f(x)|^2$. It is well-known (see [Milnor 1963, pp. 33–38], for example) that L_p has a critical point at $x \in M$ if and only if p lies along the normal line to $f(M)$ at $f(x)$. The critical point x is degenerate precisely when p is a focal point of (M, x). If L_p has a nondegenerate critical point at x, its index is the number of focal points of (M, x) on the line segment from p to $f(x)$, taking into account multiplicities. Since the set of focal points of f has measure zero in $\mathbb{R}^n$, L_p is a Morse function for almost all $p \in \mathbb{R}^n$. The immersion f is said to be *taut* if every Morse function of the form L_p is perfect, that is, there exists a field F such that

$$\mu_k(L_p, r) = \beta_k(M_r(L_p); F)$$

for every Morse function of the form L_p and for every k, r. This definition makes sense for noncompact manifolds, and there do exist taut immersions of noncompact manifolds—for example, a circular cylinder in $\mathbb{R}^3$—but most results deal with compact manifolds.

As in the theory of tight immersions, there is a formulation of tautness due to Kuiper in terms of Čech homology that has proved to be very useful in establishing certain fundamental results (see [Cecil and Ryan 1985, Section 2.1] for more detail). So far the field $F = \mathbb{Z}_2$ has been sufficient for almost all considerations, so we will use it exclusively here. Recall that a map $f : M \to \mathbb{R}^n$ is *proper* if $f^{-1}K$ is compact for every compact subset K of $\mathbb{R}^n$. Using 1.1 and the Čech theory, one can show that a proper immersion f of a manifold M into $\mathbb{R}^n$ is taut if and only if, for every closed ball B in $\mathbb{R}^n$, the induced homomorphism

$$H_i(f^{-1}B) \to H_i(M) \tag{1.2}$$

in Čech homology with $\mathbb{Z}_2$-coefficients is injective for every i. The use of Čech homology allows one to use all closed balls in $\mathbb{R}^n$ rather than only those determined by level sets of nondegenerate distance functions. Note that this formulation of tautness makes sense even if f is only assumed to be a proper continuous map and M a topological space. In that case, f is called a *taut map*.

A few key facts follow quickly from the definition. First, a taut immersion must be an embedding. In fact, this is true even if f is only assumed to be *0-taut*, i.e., the induced homomorphism 1.2 is injective for $i = 0$. For a compact manifold M, 0-tautness is equivalent to the *spherical two-piece property* (STPP) of Banchoff [1970], which requires that $f^{-1}\Omega$ be connected whenever Ω is a closed ball, the complement of an open ball, or a closed half-space. This is also equivalent to the condition that every Morse function of the form L_p have exactly one local maximum and one local minimum, which is equivalent to tautness for compact manifolds of dimension two by the Morse inequalities. The STPP is quite strong, and for a long time, every known STPP embedding was actually taut, but Curtin [1991] showed that there exist STPP embeddings that are not taut. Specifically, an embedding $f : M \to \mathbb{R}^n$ is said to be *k-taut* if the induced homomorphism 1.2 is injective for all $i \leq k$. Curtin found substantial embeddings of S^n into S^{n+d}, for $d \geq 1$, that are k-taut but not $(k+1)$-taut for every $n \geq 3$ and every $k \geq 0$ provided that $(d+1)(k+2) \leq n+1$. He also produced k-taut embeddings of manifolds other than spheres. Later [Curtin 1994] he also introduced a notion of tautness for manifolds with boundary.

As noted above, tautness can be studied for maps defined on spaces that are not manifolds. In fact, the first paper on tautness [Banchoff 1970] determined all STPP subsets of the plane. Later, Kuiper [1984] determined all taut subsets of $\mathbb{R}^2$ and all compact taut ANR (absolute neighborhood retract) subsets of $\mathbb{R}^3$.

Next, a taut embedding f of a compact manifold M into $\mathbb{R}^n$ must be *tight*, that is, every nondegenerate linear height function $l_p(x) = \langle p, f(x) \rangle$, for p a unit vector in $\mathbb{R}^n$, must be perfect. This is easily shown using Čech homology, since a closed half-space can be obtained as the limit of closed balls. A map f of a compact topological space into $\mathbb{R}^n$ is said to have the *two-piece property*, or TPP, if $f^{-1}h$ is connected for every closed half-space h. Of course, the STPP implies the TPP. As it turns out, tautness is a much stronger condition than tightness.

It is sometimes said that taut is equivalent to the combination of tight and spherical. This is true in the following sense. First, suppose that f is an embedding of a compact manifold M into $\mathbb{R}^{n+1}$ that lies in the unit sphere S^n in $\mathbb{R}^{n+1}$, in which case we say that f is *spherical*. Then, if f is a tight immersion into $\mathbb{R}^{n+1}$, it must also be taut, because the intersection of any closed ball B with S^n can be realized as the intersection of a closed half-space with S^n. Note also that the distance in S^n from p to $f(x)$ is given by the *spherical distance function* $d_p(x) = \arccos l_p(x)$, which has the same critical points as l_p. Thus, for simplicity, we usually use linear height functions rather than spherical distance functions in treating taut submanifolds of S^n. Next, if $P_q : S^n - \{q\} \to \mathbb{R}^n$ is stereographic projection with pole q not in $f(M)$, then $P_q \circ f$ is a taut embedding of M into $\mathbb{R}^n$, since P_q maps a metric ball in S^n to either a closed ball, the complement of an open ball, or a closed half-space in $\mathbb{R}^n$. Thus, f is tight and spherical if and only if $P_q \circ f$ is taut. This was first observed in [Banchoff 1970].

Hence, the theory of taut embeddings of compact manifolds is essentially the same whether one maps into $\mathbb{R}^n$ or S^n, and we work with whichever ambient space is more convenient for the problem at hand. By the same type of reasoning, tautness is easily shown to be invariant under *Möbius transformations*, that is, conformal transformations of S^n onto itself, since such transformations transformations map hyperspheres to hyperspheres in S^n.

These considerations lead to a fundamental result on the bound on the codimension of a taut embedding. Recall that an immersion $f : M \to \mathbb{R}^n$ is said to be *substantial* if the image $f(M)$ does not lie in any affine hyperplane in $\mathbb{R}^n$.

In one of the most remarkable results in the theory of tight immersions, Kuiper [1962] showed that if $f : M \to \mathbb{R}^n$ is a substantial TPP immersion of a compact manifold M of dimension k, then $n \leq k(k+3)/2$, and if equality holds, then f must be a Veronese embedding of the real projective space P^k. (The case where equality holds is due to Little and Pohl [1971] for $k > 2$.) Since tautness implies tightness, Banchoff [1970] observed that the following basic theorem holds for compact manifolds; see [Cecil and Ryan 1985, p. 124] for a proof. Carter and West [1972] extended the result to noncompact manifolds.

THEOREM 1.1. *Let $f : M \to \mathbb{R}^n$ be a substantial taut embedding of a k-dimensional manifold M.*

(a) *If M is compact, then $n \leq k(k+3)/2$. If $n = k(k+3)/2$, then f is a spherical Veronese embedding of a real projective space P^k.*

(b) *If M is noncompact, then $n < k(k+3)/2$. If $n = k(k+3)/2 - 1$, then $f(M)$ is the image under stereographic projection of a Veronese manifold, where the pole of the projection is on the Veronese manifold.*

Actually, in the compact case only the STPP is required, and in the noncompact case all that is required is that every nondegenerate distance function L_p have exactly one local minimum and no local maxima on M.

Suppose that f is a taut embedding of a compact $(n-1)$-dimensional manifold M into S^n. Then $f(M)$ is orientable, so let ξ be a field of unit normal vectors to $f(M)$ in S^n. The *parallel hypersurface* to f at signed distance t is given by the map $f_t : M \to S^n$ with equation

$$f_t(x) = \cos t \, f(x) + \sin t \, \xi(x). \tag{1.3}$$

Thus, $f_t(x)$ is obtained by travelling a signed distance t along the normal geodesic to $f(M)$ through $f(x)$. For sufficiently small values of t, f_t is also an embedding of M, and f is taut if and only if f_t is taut, since each linear height function l_p has the same critical points on $f_t(M)$ as on $f(M)$. This type of consideration is also valid for submanifolds of codimension greater than one in S^n. If $\phi : V \to S^n$ is a compact submanifold of codimension greater than one, we consider the tube ϕ_t of radius t around $\phi(V)$ in S^n. For sufficiently small t, the map ϕ_t is an embedding of the unit normal bundle B^{n-1} of $\phi(V)$ into S^n. Furthermore, ϕ is a taut embedding of V if and only if ϕ_t is a taut embedding of B^{n-1}. To see this, one first computes that every nondegenerate height function l_p has twice as many critical points on the tube ϕ_t as it has on ϕ. Since the sum of the $\mathbb{Z}_2$-Betti numbers of B^{n-1} is twice the sum of the $\mathbb{Z}_2$-Betti numbers of V [Pinkall 1986], tautness is preserved.

Thus, in a certain sense, tautness is preserved by the group of Lie sphere transformations, since this group is generated by Möbius transformations and parallel transformations (those that map a hypersurface to a parallel hypersurface). In fact, one can show that any Lie sphere transformation T is of the form $T = \phi P_t \psi$, where ϕ and ψ are Möbius transformations, and P_t is parallel transformation with respect to either the spherical metric on S^n, a Euclidean metric on $S^n - \{p\}$, or a hyperbolic metric on an open hemisphere in S^n [Cecil and Chern 1987; Cecil 1992, p. 63]. As with the spherical metric, if $f(M)$ is taut, a parallel hypersurface in the Euclidean or hyperbolic metric is also taut if it is an immersed hypersurface. Thus, as long as $f(M)$ lies in the appropriate space and the image $Tf(M)$ is an immersed hypersurface, tautness is preserved. However, for certain values of t, a parallel hypersurface f_t contains focal points of the original hypersurface f, corresponding to singularities of the map f_t. In this case, a Lie invariant notion of tautness has yet to be established.

As noted earlier, the concept of tautness can be defined for maps that are not immersions. However, when the parallel map f_t is not an immersion, it is not necessarily true that f_t is a taut map of M into S^n, even though the original immersion $f : M \to S^n$ is taut. An example of this phenomenon is most easily described in Euclidean space rather than in the sphere.

Let M be a two-dimensional torus T^2 and let f be an embedding of T^2 as the torus of revolution obtained by revolving the circle with center $(2,0)$ and radius 1 in the xy-plane about the y-axis in $\mathbb{R}^3$. Then f is a taut embedding (see [Banchoff 1970] or Section 2). Let ξ be the field of unit outer normals on the torus of revolution. For $0 < t < 1$, the parallel hypersurface f_t is also a torus of revolution that is tautly embedded in $\mathbb{R}^3$. However, for $t \geq 1$, the parallel hypersurface f_t has singularities. If $t > 1$, two latitude circles of the original torus of revolution are mapped by f_t to single points where the profile circle intersects the axis of revolution. In this case, it is easy to see that the map f_t no longer has the STPP. Specifically, if p is a point on the positive x-axis that is outside the image of f_t, then L_p has two local maxima and two local minima at the points where the x-axis intersects the surface $f_t(T^2)$. Thus, the map f_t does not have the STPP, so it is not a taut map of T^2 into $\mathbb{R}^3$. Classically, these surfaces $f_t(T^2)$ with $t > 1$ were known as *spindle tori* (see [Cecil and Ryan 1985, pp. 151–165] for more detail).

There is a natural way to extend the notion of tautness to the Lie sphere geometric setting in the case where f is not an immersion by using the concept of minimal total absolute curvature, as we will describe below. This formulation of tautness is invariant under parallel transformation, but it has not been proved to be invariant under Möbius transformations, unlike the definition of a taut map given above. When the map f is an immersion, the two definitions give the same results.

Suppose that $f : M \to S^n \subset \mathbb{R}^{n+1}$ is an embedding of a compact $(n-1)$-dimensional manifold M and that ξ is a field of unit normals to $f(M)$ in S^n. We can consider ξ as a map from M into S^n by parallel translating each vector $\xi(x)$ to the origin in $\mathbb{R}^{n+1}$. We have the normal exponential map $F : M \times \mathbb{R} \to S^n$ defined by

$$F(x,t) = f_t(x) = \cos t \, f(x) + \sin t \, \xi(x). \tag{1.4}$$

If $p \in S^n$ is not a focal point of $f(M)$, then the linear height function l_p has a finite number $\mu(p)$ of critical points on M. In fact, $\mu(p)$ is the number of points $x \in M$ such that $p = F(x, t)$ for some $t \in [0, 2\pi)$. That is, $\mu(p)$ is the number of geodesics normal to $f(M)$ that go through the point p. The focal set C of the embedding f has measure zero in S^n. The *total absolute curvature* τ of f (see [Cecil and Ryan 1985, pp. 12–13], for example) is defined by the expression

$$\tau = \frac{1}{c_n} \int_{S^n - C} \mu(p) \, da, \tag{1.5}$$

where c_n is the volume of the unit sphere S^n. Of course, in the usual development of the theory, the embedding $f : M \to S^n \subset \mathbb{R}^{n+1}$ is tight (and hence taut, since f is spherical) if and only if $\tau = \beta(M)$, the sum of the $\mathbb{Z}_2$-Betti numbers of M.

Now consider briefly how the notion of a hypersurface is generalized in the setting of Lie sphere geometry (see Section 5, [Pinkall 1985a] or [Cecil 1992, pp. 65–78] for more detail). We consider $T_1 S^n$, the bundle of unit tangent vectors to S^n, as the $(2n - 1)$-dimensional submanifold of $S^n \times S^n \subset \mathbb{R}^{n+1} \times \mathbb{R}^{n+1}$ given by

$$T_1 S^n = \{(x, \xi) : |x| = 1, |\xi| = 1, \langle x, \xi \rangle = 0\}.$$

The manifold $T_1 S^n$ has a *contact structure*, that is, a globally defined one-form ω such that $\omega \wedge d\omega^{n-1}$ never vanishes on $T_1 S^n$. An immersion $\lambda : M \to T_1 S^n$ of an $(n - 1)$-dimensional manifold M into $T_1 S^n$ is called a *Legendre submanifold* if $\lambda^* \omega = 0$ on M. A Legendre submanifold is determined by two maps f and ξ from M into S^n that satisfy the following three conditions:

(L1) $\langle f(x), f(x) \rangle = 1$, $\langle \xi(x), \xi(x) \rangle = 1$, $\langle f(x), \xi(x) \rangle = 0$, for all $x \in M$.

(L2) There is no nonzero $X \in T_x M$, for any $x \in M$, such that $f_*(X)$ and $\xi_*(X)$ are both zero.

(L3) $\langle f_*(X), \xi(x) \rangle = 0$ for all $X \in T_x M$, for all $x \in M$.

Condition (L1) is precisely what is necessary for $\lambda = (f, \xi)$ to be a map into $T_1 S^n$. Condition (L2) is what is needed for λ to be an immersion, and (L3) is equivalent to $\lambda^* \omega = 0$ on M.

If $f : M \to S^n$ is an immersed hypersurface with field of unit normals ξ, it is easy to check that $\lambda = (f, \xi)$ is a Legendre submanifold. However, for a general Legendre submanifold λ, the map f into the first factor S^n may not be an immersion.

Suppose now that $\lambda = (f, \xi)$ is a Legendre submanifold with maps f and ξ satisfying the conditions (L1)–(L3), and that M is compact. For each $x \in M$, there is a well-defined "normal geodesic" $\gamma(t) = f_t(x)$, where f_t is given by 1.3. We can define the normal exponential map $F : M \times \mathbb{R} \to S^n$ by the same formula 1.4 used in the case where the map f is assumed to be an immersion. As before, we define the focal points of λ to be the critical values of F, and, again by Sard's Theorem, the set C of focal points has measure zero in S^n. If $p \in S^n$ is not a focal point, we define $\mu(p)$ to be the number of points $x \in M$ such that

$p = F(x,t)$ for some $t \in [0, 2\pi)$, that is, the number of normal geodesics that pass through p. Then we can define the total absolute curvature τ of λ by the same formula 1.5 used in the case where f is an immersion. We then define the Legendre submanifold λ to be taut if $\tau(\lambda) = \beta(M)$.

If $\lambda = (f, \xi)$ is a Legendre submanifold, the parallel hypersurface λ_t at signed distance t is defined by $\lambda_t = (f_t, \xi_t)$, where f_t is given by 1.3 and

$$\xi_t(x) = -\sin t\, f(x) + \cos t\, \xi(x).$$

One can show [Pinkall 1985a] that λ and λ_t have precisely the same focal set in S^n, and that, if p is not a focal point, then $\mu(p)$ is the same for λ_t as for λ, since λ_t and λ determine exactly the same family of normal geodesics. Thus, the total absolute curvature τ is invariant under parallel transformation. Although the usual proof of invariance under Möbius transformations does not work for this formulation of tautness when f is not an immersion, Möbius invariance may in fact hold. If so, that would be a satisfactory resolution of the question of the Lie invariance of tautness for Legendre submanifolds.

We close this section by noting that the torus of revolution in the example above was obtained from the taut circle by embedding the xy-plane into $\mathbb{R}^3$ and then forming a surface of revolution. This construction can be generalized to higher dimensions. Let M be a taut compact hypersurface in $\mathbb{R}^{k+1}$ that is disjoint from a hyperplane $\mathbb{R}^k$ in $\mathbb{R}^{k+1}$. Now embed $\mathbb{R}^{k+1}$ into $\mathbb{R}^{n+1}$ as a totally geodesic subspace. Then the hypersurface W obtained by revolving M about the axis $\mathbb{R}^k$ is a taut embedding of $M \times S^{n-k}$ into $\mathbb{R}^{n+1}$. See [Cecil and Ryan 1985, p. 187] for more detail.

2. Isoparametric Submanifolds

An important class of examples of taut submanifolds are the isoparametric submanifolds in $\mathbb{R}^n$ or in S^n. We begin with a discussion of isoparametric hypersurfaces and later treat the case of codimension greater than one.

A hypersurface $f : M \to \mathbb{R}^n$ (or S^n) is said to be *isoparametric* if it has constant principal curvatures. An isoparametric hypersurface in $\mathbb{R}^n$ must be an open subset of a hyperplane, hypersphere, or spherical cylinder $S^k \times \mathbb{R}^{n-k-1}$. This was shown by Levi-Civita [1937] for $n = 3$ and by B. Segre [1938] for arbitrary n. As E. Cartan demonstrated in a remarkable series of papers [Cartan 1938; 1939a; 1939b; 1940], the situation is much more interesting for isoparametric hypersurfaces in S^n. Despite the depth and beauty of Cartan's work, however, this topic was largely ignored until it was revived in the 1970's by Nomizu [1973; 1975] and Münzner [1980; 1981].

Among other things, Cartan showed that isoparametric hypersurfaces come as a parallel family of hypersurfaces; that is, if $f : M \to S^n$ is an isoparametric hypersurface, then so is any parallel hypersurface f_t. Of course, f_t is not an immersion if $\mu = \cot t$ is a principal curvature of M. Then, however, f_t factors through an immersion of the space of leaves M/T_μ of the principal foliation T_μ. Thus, f_t is a submanifold of codimension $m + 1$ in S^n, where m is the multiplicity of μ. Much more can be said about this parallel family, however. Münzner [1980; 1981] showed that a parallel family of isoparametric hypersurfaces in S^n always consists of the level sets in S^n of a homogeneous polynomial defined on $\mathbb{R}^{n+1}$. This implies that any local piece of an isoparametric hypersurface can be extended to a unique compact isoparametric hypersurface. Further, he showed that regardless of the number of distinct principal curvatures of M, there are only two distinct focal submanifolds in a parallel family of isoparametric hypersurfaces, and these are minimal submanifolds of the sphere. This minimality of the focal submanifolds was established independently by Nomizu [1973; 1975]. Münzner uses this information and a difficult topological argument to prove the following important theorem. The key fact here is that any isoparametric hypersurface divides the sphere into two ball bundles over the two focal submanifolds.

THEOREM 2.1. *The number g of distinct principal curvatures of an isoparametric hypersurface in S^n must be 1, 2, 3, 4, or 6.*

Cartan classified isoparametric hypersurfaces with $g \le 3$ principal curvatures. Of course, if $g = 1$, then M must is umbilic and it must be a great or small sphere. If $g = 2$, then M must be a standard product of two spheres

$$S^k(r) \times S^{n-k-1}(s) \subset S^n, \quad \text{with } r^2 + s^2 = 1.$$

In the case $g = 3$, Cartan [1939a] showed that all the principal curvatures must have the same multiplicity $m \in \{1, 2, 4, 8\}$, and the isoparametric hypersurface must be a tube of constant radius over a standard Veronese embedding of a projective plane FP^2 into S^{3m+1}, where F is the division algebra $\mathbb{R}$, $\mathbb{C}$, $\mathbb{H}$ (quaternions), $\mathbb{O}$ (Cayley numbers) for $m = 1, 2, 4, 8$, respectively. Thus, up to congruence, there is only one such family for each value of m. This was a remarkable result with a difficult proof. These isoparametric hypersurfaces with three principal curvatures are often referred to as *Cartan hypersurfaces*.

Isoparametric hypersurfaces with $g = 4$ or 6 principal curvatures have yet to be classified. In the case $g = 4$, Ferus, Karcher and Münzner [Ferus et al. 1981] use representations of Clifford algebras to construct for any positive integer m_1 an infinite series of isoparametric hypersurfaces with four principal curvatures having respective multiplicities (m_1, m_2, m_1, m_2), where m_2 is nondecreasing and unbounded in each series. In each case, one of the focal submanifolds is a Clifford–Stiefel manifold (see also [Pinkall and Thorbergsson 1989a; Wang 1988]). This class contains all known examples with $g = 4$ with the exception of two homogeneous examples.

Other notable facts in the case $g = 4$ are that the four principal curvatures can only have two distinct multiplicities m_1 and m_2, and several restrictions on these multiplicities have been obtained [Münzner 1980; 1981; Abresch 1983; Tang 1991; Fang 1995a; 1996]. Also, many of the isoparametric families with $g = 4$

are inhomogeneous [Ozeki and Takeuchi 1975; 1976; Ferus et al. 1981]. Wang [1988] found many results concerning the topology of the Clifford examples, and Wu [1994] showed that there are only finitely many diffeomorphism classes of compact isoparametric hypersurfaces with four distinct principal curvatures.

In the case $g = 6$, Münzner showed that all of the principal curvatures must have the same multiplicity m, and Abresch [1983] showed that m must be 1 or 2. There is only one homogeneous family in each case, and in the case $m = 1$, Dorfmeister and Neher [1985] showed that an isoparametric hypersurface must be homogeneous. However, for $m = 2$, it is still unknown whether or not M must be homogeneous. In the case $m = 1$, Miyaoka [1993a] has shown that a homogeneous isoparametric hypersurface M^6 in S^7 can be obtained as the inverse image under the Hopf fibration $h : S^7 \to S^4$ of an isoparametric hypersurface with three principal curvatures of multiplicity 1 in S^4. She also shows that the two focal submanifolds of M^6 are not congruent, even though they are lifts under h^{-1} of congruent Veronese surfaces in S^4. Thus, these focal submanifolds are two noncongruent minimal taut homogeneous embeddings of $\mathbb{RP}^2 \times S^3$ in S^7. Peng and Hou [1989] gave explicit forms for the isoparametric polynomials of degree six for the homogeneous isoparametric hypersurfaces with $g = 6$. Recently, Fang [1995b] has obtained results concerning the topology of isoparametric and compact proper Dupin hypersurfaces with six principal curvatures.

In a series of papers, Dorfmeister and Neher [1983a; 1983b; 1983c; 1983d; 1985; 1990] gave an algebraic approach to the study of isoparametric hypersurfaces and isoparametric triple systems.

All isoparametric hypersurfaces in S^n are taut. This was established in [Cecil and Ryan 1981] using the results of Münzner. In particular, any normal geodesic to an isoparametric hypersurface is also normal to each parallel hypersurface and to the focal submanifolds. Using Münzner's results, one can show that every nonfocal point $p \in S^n$ lies on exactly $2g$ normal geodesics from points in M; that is, the height function l_p has exactly $2g$ critical points on M. Since Münzner showed that the sum of the $\mathbb{Z}_2$-Betti of an isoparametric hypersurface with g principal curvatures is $2g$, the hypersurface M is taut. A similar argument shows that the focal submanifolds are taut.

In the early 1980's, a theory of isoparametric submanifolds of codimension greater than 1 was introduced independently by several mathematicians [Carter and West 1985b; West 1989; Harle 1982; Strübing 1986; Terng 1985]. An immersed submanifold $\phi : V \to \mathbb{R}^n$ (or S^n) is called *isoparametric* if its normal bundle $N(V)$ is flat and if, for any locally defined normal field ξ that is parallel with respect to the normal connection $\nabla^\perp$, the eigenvalues of the shape operator A_ξ are constant. Recall that $N(V)$ is flat if and only if for every $x \in V$ all the shape operators A_η, where $\eta \in N_x(V)$, are simultaneously diagonalized. In the decade after this definition of isoparametric submanifolds was formulated, intense research by several mathematicians produced a remarkable theory; see [Palais and Terng 1988; Terng 1993] for more detail. Among other things, Terng

[1985] showed that a compact isoparametric submanifold in Euclidean space must lie in a standard hypersphere. Palais and Terng [1987] showed that the only homogeneous isoparametric submanifolds in Euclidean spaces are the principal orbits of the isotropy representations of symmetric spaces, which had been studied extensively by Bott and Samelson [1958], who showed that the orbits are taut. The orbits of the isotropy representations of symmetric spaces, also called *R-spaces*, were studied independently by Takeuchi and Kobayashi [1968], who also showed that they were taut. The class of *R*-spaces contains many special subclasses of homogeneous submanifolds that were shown to be taut by various special arguments. (See, for example, [Kobayashi 1967; Tai 1968; Wilson 1969; Kuiper 1970; 1980; Ferus 1982; Kühnel 1994].) Later, Hsiang, Palais, and Terng [1988] obtained many facts about the geometry and topology of isoparametric submanifolds, including the fact that they and their focal submanifolds are taut. Thorbergsson [1991] then used the extensive results that had been obtained for isoparametric submanifolds along with the theory of Tits buildings to prove that all isoparametric submanifolds of codimension greater than one in the sphere are homogeneous. (See [Olmos 1993] for an alternate proof of this result.)

In a series of papers, Carter and West [1978; 1981; 1982; 1990] studied the relationship between isoparametric and totally focal submanifolds. Recall that a submanifold $\phi : V \to \mathbb{R}^n$ is said to be *totally focal* if the critical points of every Euclidean distance function L_p are either all nondegenerate or all degenerate. An isoparametric submanifold is totally focal, and the main result of [Carter and West 1990] is that a totally focal submanifold must be isoparametric. However, Terng and Thorbergsson [Terng and Thorbergsson 1997] have noted that there is a gap in the proof of this assertion, specifically in the proof of Theorem 5.1 of [Carter and West 1990].

A slight variation of the notion of isoparametric submanifolds is the following. A submanifold $\phi : V \to \mathbb{R}^n$ (or S^n) is said to have *constant principal curvatures* if, for any smooth curve γ on V and any parallel normal vector field $\xi(t)$ along γ, the shape operator $A_{\xi(t)}$ has constant eigenvalues along γ. If the normal bundle $N(M)$ is flat, then having constant principal curvatures is equivalent to being isoparametric. Heintze, Olmos, and Thorbergsson [Heintze et al. 1991] showed that a submanifold with constant principal curvatures is either isoparametric or a focal submanifold of an isoparametric submanifold. In a related work, Olmos [Olmos 1994] defines the *rank* of a submanifold in Euclidean space to be the maximal number of linearly independent (locally defined) parallel normal vector fields. He then shows that a compact homogeneous irreducible submanifold M substantially embedded in Euclidean space with rank greater than one must be an orbit of the isotropy representation of a simple symmetric space.

In closing this section, we note that many results on real hypersurfaces with constant principal curvatures in complex space forms are surveyed in the article by Niebergall and Ryan [1997] in this volume. There is also a theory of isoparametric hypersurfaces in pseudo-Riemannian space forms [Nomizu 1981; Hahn

1984; 1988; Magid 1985; Kashani 1993b; 1993a; 1992]. Wu [1992] extended the theory of isoparametric submanifolds of arbitrary codimension to submanifolds of hyperbolic space (see also [Zhao 1993]), and Verhóczki [1992] developed a theory of isoparametric submanifolds for Riemannian manifolds that do not have constant curvature. West [1993] and Mullen [1994] have formulated a theory of isoparametric systems on symmetric spaces, while Terng and Thorbergsson [1995] have generalized the notion of isoparametric to submanifolds of symmetric spaces using the concept of equifocal submanifolds. In a different direction, Carter and Şentürk [1994] have considered the space of immersions parallel to a given immersion whose normal bundle has trivial holonomy group. In yet another direction, Niebergall and Ryan [1993; 1994a; 1994b; 1996] have generalized the notions of isoparametric and Dupin hypersurfaces to the context of affine differential geometry.

More generally, Q.-M. Wang [1987; 1988; 1986] has extended Cartan's theory of isoparametric functions to arbitrary Riemannian manifolds. A smooth function $\phi : \tilde{M} \to \mathbb{R}$ on a Riemannian manifold $\tilde{M}$ is said to be *transnormal* if there is a smooth function b such that $|d\phi|^2 = b(\phi)$. The function ϕ is said to be *isoparametric* if it is transnormal and if there exists a smooth function a such that the Laplacian $\Delta\phi = a(\phi)$. This agrees with the definition of an isoparametric function used by Cartan in his work in the case where $\tilde{M}$ is a real space form. Wang shows that if $\tilde{M}$ is complete and ϕ is a transnormal function on $\tilde{M}$, the focal varieties of ϕ (of which there are at most two) are smooth submanifolds of $\tilde{M}$, and each regular level set of ϕ is a tube over either of the focal varieties. Moreover, if $\tilde{M}$ is S^n or $\mathbb{R}^n$, a transnormal function must, in fact, be isoparametric. However, this is not true if $\tilde{M}$ is a hyperbolic space H^n.

Finally, Solomon [1990a; 1990b; 1992] has studied the spectrum of the Laplacian of isoparametric hypersurfaces in S^n with three or four principal curvatures, while Eschenburg and Schroeder [Eschenburg and Schroeder 1991] have studied the behavior of the Tits metric on isoparametric hypersurfaces.

3. Dupin Submanifolds

In this section, we introduce the notion of Dupin submanifolds, beginning with hypersurfaces. Let $f : M \to \mathbb{R}^n$ be an immersed hypersurface. Let ξ be a locally defined field of unit normals to $f(M)$. A *curvature surface* of M is a smooth submanifold S such that, for each point $x \in S$, the tangent space T_xS is equal to a principal space of the shape operator A of M at x. This generalizes the classical notion of a line of curvature on a surface in $\mathbb{R}^3$. The hypersurface M is said to be *Dupin* if

(a) along each curvature surface, the corresponding principal curvature is constant.

The hypersurface M is called *proper Dupin* if, in addition,

(b) the number g of distinct principal curvatures is constant on M.

Several remarks about these definitions are in order. The proofs can be found in [Cecil and Ryan 1985, Section 2.4]. First, if the dimension of a curvature surface S is greater than one, the corresponding principal curvature is automatically constant on S. This is proved using the Codazzi equation. Second, Condition (b) is equivalent to requiring that each continuous principal curvature have constant multiplicity on M.

There is an open dense subset of M on which the multiplicities of the continuous principal curvatures of M are locally constant. (See [Singley 1975], for example.) Suppose now that a continuous principal curvature μ has constant multiplicity m on an open subset $U \subset M$. Then μ and its principal distribution T_μ are smooth on U. Furthermore, again using the Codazzi equation, one can show that T_μ is integrable, and thus it is called the *principal foliation* corresponding to μ. The leaves of this principal foliation are the curvature surfaces corresponding to μ on U. The principal curvature μ is constant along each of its curvature surfaces in U if and only if these curvature surfaces are open subsets of m-dimensional Euclidean spheres or planes. The *focal map* f_μ corresponding to μ is the map that maps $x \in M$ to the focal point $f_\mu(x)$ corresponding to μ, i.e.,

$$f_\mu(x) = f(x) + \frac{1}{\mu(x)}\xi(x).$$

A direct calculation shows that μ is constant along each of its curvature surfaces in U if and only if the focal map f_μ factors through an immersion of the $(n - 1 - m)$-dimensional space of leaves M/T_μ into $\mathbb{R}^n$.

In summary, on an open subset U on which the number of distinct principal curvatures is constant, Condition (a) is equivalent to requiring that each curvature surface in each principal foliation be an open subset of a Euclidean sphere or plane with dimension equal to the multiplicity of the corresponding principal curvature. On U, Condition (a) is also equivalent to the condition that each focal map be a submanifold of codimension greater than one in $\mathbb{R}^n$.

Like tautness, both the Dupin and proper Dupin conditions are invariant under Möbius transformations and under stereographic projection from S^n to $\mathbb{R}^n$ (see [Cecil and Ryan 1985, pp. 147–148]). These conditions are also invariant under parallel transformations and thus under the group of Lie sphere transformations. A proof of these claims must be formulated in the setting of Lie sphere geometry in order to handle the fact that a parallel hypersurface f_t to $f(M)$ may not be an immersion on all of M [Pinkall 1985a]. See also [Cecil 1992, p. 87].

An important class of compact proper Dupin hypersurfaces in $\mathbb{R}^n$ is obtained by taking the images under stereographic projection of isoparametric hypersurfaces in S^n. Of course, for these examples, the number g of distinct principal curvatures must be 1, 2, 3, 4, or 6. In fact, Thorbergsson [1983a] has shown that this restriction on g holds for any compact proper Dupin hypersurface embedded in $\mathbb{R}^n$ (see Section 4). Thorbergsson first showed that a compact proper Dupin hypersurface M embedded in S^n must be taut and then used this fact to show that M divides the sphere into two ball bundles over the first focal submanifolds on either side of M. One can then invoke Münzner's theorem to get the restriction on g. All of the restrictions on the multiplicities of the principal curvatures that follow from this topological situation also apply in this case as well. For some time, it was conjectured [Cecil and Ryan 1985, p. 184] that every compact proper Dupin hypersurface embedded in S^n was Lie equivalent to an isoparametric hypersurface. This is not true, however, as was shown independently by Pinkall and Thorbergsson [1989a] and by Miyaoka and Ozawa [1989]. Their constructions of counterexamples to this conjecture will be discussed in Section 6.

In contrast with Thorbergsson's result, Pinkall [1985a] showed how to construct a proper Dupin hypersurface with an arbitrary number of distinct principal curvatures having any prescribed multiplicities. (See also [Cecil and Ryan 1985, p. 179].) This is done using the following basic constructions. Start with a Dupin hypersurface W^{n-1} in $\mathbb{R}^n$ and then consider $\mathbb{R}^n$ as the linear subspace $\mathbb{R}^n \times \{0\}$ in $\mathbb{R}^{n+1}$. The following constructions yield a Dupin hypersurface M^n in $\mathbb{R}^{n+1}$.

(1) Let M^n be the cylinder $W^{n-1} \times \mathbb{R}$ in $\mathbb{R}^{n+1}$.
(2) Let M^n be the hypersurface in $\mathbb{R}^{n+1}$ obtained by rotating W^{n-1} around an axis $\mathbb{R}^{n-1} \subset \mathbb{R}^n$.
(3) Project W^{n-1} stereographically onto a hypersurface $V^{n-1} \subset S^n \subset \mathbb{R}^{n+1}$. Let M^n be the cone over V^{n-1} in $\mathbb{R}^{n+1}$.
(4) Let M^n be a tube in $\mathbb{R}^{n+1}$ around W^{n-1}.

These constructions introduce a new principal curvature of multiplicity one that is easily seen to be constant along its lines of curvature. The other principal curvatures are determined by the principal curvatures of W^{n-1}, and the Dupin property is preserved for these principal curvatures. These constructions can easily be generalized to produce a new principal curvature of multiplicity m by considering $\mathbb{R}^n$ as a subset of $\mathbb{R}^n \times \mathbb{R}^m$ rather than $\mathbb{R}^n \times \mathbb{R}$. By repeated use of these constructions, Pinkall [1985a] proved the following basic existence theorem, and we repeat his proof here.

THEOREM 3.1. *Given positive integers $m_1, \ldots, m_g$ with $m_1 + \cdots + m_g = n - 1$, there exists a proper Dupin hypersurface M^{n-1} in $\mathbb{R}^n$ with g distinct principal curvatures having respective multiplicities $m_1, \ldots, m_g$.*

PROOF. The proof is by an inductive construction that will be clear once the first few cases are handled. First we construct a proper Dupin hypersurface M^3 in $\mathbb{R}^4$ with three principal curvatures of multiplicity one. Begin with an open subset U of a torus of revolution in $\mathbb{R}^3$ on which neither principal curvature vanishes. Take M^3 to be the cylinder $U \times \mathbb{R}$ in $\mathbb{R}^3 \times \mathbb{R} = \mathbb{R}^4$. Then M^3 has three distinct principal curvatures at each point, one of which is identically zero. These curvatures are clearly constant along their corresponding lines of curvature. Next, to construct a proper Dupin hypersurface in $\mathbb{R}^5$ with three distinct principal curvatures having respective multiplicities 1, 1, 2, take a cylinder $U \times \mathbb{R}^2$ in $\mathbb{R}^3 \times \mathbb{R}^2$. Finally, to get a proper Dupin hypersurface V^4 in $\mathbb{R}^5$ with four principal curvatures, first invert the hypersurface M^3 above in a three-sphere in $\mathbb{R}^4$ chosen so that the image of M^3 contains an open set W^3 on which no principal curvature vanishes. Now take V^4 to be the cylinder $W^3 \times \mathbb{R}$. □

These constructions only yield a compact proper Dupin hypersurface if the original manifold W^{n-1} is itself a sphere [Cecil 1989]. Otherwise, the number of distinct principal curvatures is not constant on a compact manifold M^n obtained in this way, because there are points where the new principal curvature is equal to one of the original principal curvatures. For example, in the cylinder construction, the new principal curvature is identically zero, while the other principal curvatures of M^n are equal to those of W^{n-1}. Thus, if one of the principal curvatures of W^{n-1} is zero at some points but not identically zero, the number of distinct principal curvatures is not constant on M^n. For a tube of radius ε over W^{n-1}, there are always only two distinct principal curvatures at the points on the set $W^{n-1} \times \{\pm\varepsilon\}$ in M^n, regardless of the number of distinct principal curvatures on W^{n-1}. In the surface of revolution construction, the new principal curvature is equal to one of the original principal curvatures if the focal point corresponding to the original principal curvature lies on the axis of revolution.

A second problem is that the constructions may not yield an immersed hypersurface in $\mathbb{R}^{n+1}$. For example, in the tube construction, a singularity occurs if the radius of the tube is the reciprocal of one of the principal curvatures of W^{n-1} at some point, that is, if the tube contains a focal point of W^{n-1}. In the surface of revolution construction, a singularity occurs if W^{n-1} intersects the axis of revolution. These problems are resolved by working in the context of Lie sphere geometry. See [Cecil 1989; Cecil 1992, Section 4.2] for more detail.

A proper Dupin hypersurface that is locally Lie equivalent to a hypersurface M^n obtained by one of these constructions is said to be *reducible*, and a proper Dupin hypersurface that does not contain any reducible open subset is said to be *locally irreducible*. These are useful concepts in attempting to obtain local classifications of proper Dupin hypersurfaces (see Section 7).

In order to extend the notion of Dupin to submanifolds of codimension greater than one, we first need the definition, due to Reckziegel [1979], of a curvature surface in that case. Suppose that $\phi : V \to \mathbb{R}^n$ is a submanifold of codimension greater than one, and let B^{n-1} denote the unit normal bundle of $\phi(V)$. Then a *curvature surface* is a connected submanifold $S \subset V$ for which there is a parallel section $\eta : S \to B^{n-1}$ such that, for each $x \in S$, the tangent space $T_x S$ is equal to some smooth eigenspace of the shape operator A_η. We then define $\phi(V)$ to be *Dupin* if, along each curvature surface, the corresponding principal curvature of A_η is constant. The Dupin submanifold $\phi(V)$ is *proper Dupin* if the number of distinct principal curvatures of A_ξ is constant on the unit normal bundle B^{n-1}. From the definitions, it is clear that an isoparametric submanifold is always Dupin, but it may not be proper Dupin. (See [Terng 1993, pp. 464–469] for more discussion.) Dupin submanifolds of codimension greater than one are handled quite naturally in the setting of Lie sphere geometry (see Section 5).

Pinkall [1985a, p. 439] showed that every extrinsically symmetric submanifold of a real space form is Dupin. Takeuchi [Takeuchi 1991] then determined which of these are proper Dupin.

4. Relationship Between the Taut and Dupin Conditions

All of the results in this section apply to submanifolds of S^n as well as to those in $\mathbb{R}^n$.

Thorbergsson [1983a] showed that if M is a complete proper Dupin hypersurface embedded in $\mathbb{R}^n$, then M is tautly embedded. Pinkall [1986] then showed the following.

THEOREM 4.1. *Every taut submanifold in $\mathbb{R}^n$ is Dupin.*

This result was also obtained independently by Miyaoka [1984b]. (See also [Cecil and Ryan 1985, p. 195] for a proof.) Note that a taut submanifold need not be proper Dupin. For example, a tube M^3 of sufficiently small radius ε over a torus of revolution $T^2 \subset \mathbb{R}^3 \subset \mathbb{R}^4$ is taut but not proper Dupin, since there are only two distinct principal curvatures on the set $T^2 \times \pm\{\varepsilon\}$ but three distinct principal curvatures elsewhere on M. Many isoparametric submanifolds of codimension greater than one in S^n are also not proper Dupin; that is, the number of distinct principal curvatures is not constant on the unit normal bundle.

Pinkall also showed that, if M is an embedded submanifold of $\mathbb{R}^n$ and M_ε is a tube of sufficiently small radius ε as to be embedded in $\mathbb{R}^n$, then M is taut with respect to $\mathbb{Z}_2$-coefficients if and only if M_ε is taut with respect to $\mathbb{Z}_2$-coefficients. This can be combined with Thorbergsson's result to yield the following.

THEOREM 4.2. *Let M be a complete proper Dupin submanifold of $\mathbb{R}^n$. Then M is taut with respect to $\mathbb{Z}_2$-coefficients.*

Using tautness, Thorbergsson then showed that a compact proper Dupin hypersurface embedded in S^n divides the sphere into two ball bundles over the first focal submanifolds on either side of M in S^n. He could then invoke Münzner's [1980; 1981] results to obtain the following.

THEOREM 4.3. *Let M be a compact proper Dupin hypersurface in $\mathbb{R}^n$. Then:*

(1) *The number g of distinct principal curvatures of M must be 1, 2, 3, 4, or 6.*
(2) *The sum of the $\mathbb{Z}_2$-Betti numbers of M is $2g$.*

Later, Grove and Halperin [1987] derived many other results about the topology of compact proper Dupin hypersurfaces.

One consequence of Theorem 4.3 is that any compact proper Dupin hypersurface M with $g \geq 3$ principal curvatures must be irreducible [Cecil 1989, p. 297; Cecil 1992, p. 148]. To see this, suppose that M is reducible to a compact proper Dupin hypersurface V in a lower-dimensional Euclidean space. If $g \geq 3$, then M must be obtained from V by the surface of revolution construction, since a compact proper Dupin hypersurface obtained via the other constructions always has some points where the number of distinct principal curvatures is two. In that case, M is diffeomorphic to $V \times S^m$ for some positive integer m, and the sum $\beta(M)$ of the $\mathbb{Z}_2$-Betti numbers of M is

$$\beta(M) = \beta(V \times S^m) = 2\beta(V). \qquad (4.1)$$

An analysis of the surface of revolution construction shows that the number k of distinct principal curvatures on V must be $g - 1$ or g. Since $k = 2\beta(V)$ by Theorem 4.3, and $\beta(V) \neq \beta(M)$, it is impossible for V and M to have the same number of distinct principal curvatures. So $k = g - 1$, and 4.1 implies that $2g = 4k = 4g - 4$. This implies that $g = 2$, a contradiction, which shows that M is not reducible.

In the case where M is compact, we can use a theorem of Ozawa [1986] to prove something slightly stronger than Theorem 4.1. As we noted in Section 3, if a curvature surface has dimension greater than one, the corresponding principal curvature is always constant along it, even without the assumption of tautness. Thus, Pinkall's proof of Theorem 4.1 consisted in showing that tautness implies that each principal curvature is constant along each of its curvature surfaces of dimension one (lines of curvature). Note that we are using Pinkall's definition of Dupin, which does not insist that given any principal space T_μ at any point $x \in M$ there exists a curvature surface S through x whose tangent space at x is T_μ. However, using the next theorem, due to Ozawa [1986], we will show as a corollary that tautness implies the existence of such a curvature surface at each $x \in M$. (See [Terng and Thorbergsson 1997] in this volume for a generalization of Ozawa's theorem for taut immersions into arbitrary complete Riemannian manifolds.)

THEOREM 4.4. *Let M be a taut compact submanifold of $\mathbb{R}^n$, and let L_p be a Euclidean distance function on M. Let $x \in M$ be a critical point of L_p and let S be the connected component of the critical set of L_p that contains x. Then S is*

(1) *a smooth compact manifold of dimension equal to the nullity of the Hessian of L_p at the critical point x,*

(2) *nondegenerate as a critical manifold, and*

(3) *taut in $\mathbb{R}^n$.*

Using this result, we can prove the following theorem.

THEOREM 4.5. *Let M be a taut compact submanifold of $\mathbb{R}^n$. Then:*

(1) *M is a Dupin submanifold.*
(2) *Given any principal space T_μ of any shape operator A_ξ at any point $x \in M$, there exists a curvature surface S through x whose tangent space at x is equal to T_μ.*

PROOF. Let $f : M \to \mathbb{R}^n$ be a taut embedding. Let ξ be any unit normal vector at any given point $x \in M$, and let μ be a principal curvature of A_ξ. Let $p = f(x) + (1/\mu)\xi$ be the focal point of (M, x) determined by the principal curvature μ of A_ξ. Then the distance function L_p has a degenerate critical point at x, and the nullity of the Hessian of L_p at x is equal to the multiplicity m of μ as a principal curvature of A_ξ [Milnor 1963, p. 36]. By Theorem 4.4, the connected component S of the critical set of L_p containing x is a smooth submanifold of dimension m. We will now show that S is the desired curvature surface and that the corresponding principal curvature is constant along S, i.e., that M is Dupin.

The function L_p has a constant value, which must be $1/\mu^2$, on the critical submanifold S. Thus, for every point $y \in S$, the vector $p - f(y)$ is normal to $f(M)$ at $f(y)$, and it has length $1/\mu$. So we can extend the normal vector ξ to a unit normal vector field to $f(M)$ along S, which we also denote by ξ, by setting $\xi(y) = \mu(p - f(y))$. Note that p is a focal point of (M, y) for every point $y \in S$, and Theorem 4.4 implies that the number μ is a principal curvature of $A_{\xi(y)}$ of multiplicity $m = \dim S$ for every point $y \in S$. Thus, the principal curvature μ is constant along S. We now complete the proof of Theorem 4.5 by showing that $T_y S$ equals the principal space $T_\mu(y)$ at each point $y \in S$ and that the normal field ξ is parallel along S with respect to the normal connection. Consider the focal map

$$f_\mu(y) = f(y) + \frac{1}{\mu}\xi(y),$$

for $y \in S$. Then $f_\mu(y) = p$ for all $y \in S$. Let X be any tangent vector to S at any point $y \in S$. Then $(f_\mu)_* X = 0$, since f_μ is constant on S. On the other hand,

$$(f_\mu)_* X = f_* X + \frac{1}{\mu}\xi_* X,$$

and $\xi_* X = D_X \xi = f_*(-A_\xi X) + \nabla^\perp_X \xi$. Therefore,

$$(f_\mu)_* X = f_*\left(X - \frac{1}{\mu}A_\xi X\right) + \frac{1}{\mu}\nabla^\perp_X \xi.$$

Since $(f_\mu)_* X = 0$, we see that $A_\xi X = \mu X$ and $\nabla^\perp_X \xi = 0$. Thus, ξ is parallel along S and $T_y S \subset T_\mu(y)$. But since $T_y S$ and $T_\mu(y)$ have the same dimension, they must be equal. So S is the curvature surface through y corresponding to μ. $\qquad \square$

Certainly, a major open problem in this area is whether the converse to Theorem 4.5 is true. That is, suppose M is a compact Dupin submanifold embedded in $\mathbb{R}^n$ with the property that, given any principal space T_μ of any shape operator A_ξ at any point $x \in M$, there exists a curvature surface S through x whose tangent space at x is equal to T_μ. Must M be taut? Thorbergsson's proof in the case where M is proper Dupin relies on the fact that all of the curvature surfaces are spheres, and this is not true if M is not proper Dupin.

Tautness has been established for *Dupin submanifolds with constant multiplicities* by Terng [1987; 1993, p. 467]. These are Dupin submanifolds M such that the multiplicities of the principal curvatures of any parallel normal field $\xi(t)$ along any piecewise smooth curve on M are constant.

5. Submanifolds in Lie Sphere Geometry

In this section, we give a brief description of the method for studying submanifolds of Euclidean space and the sphere using Lie sphere geometry (see [Cecil 1992; Cecil and Chern 1987; Chern 1991; Pinkall 1985a] for more detail). As we noted earlier, the Dupin property is invariant under stereographic projection from $\mathbb{R}^n$ to S^n (see [Cecil and Ryan 1985, pp. 147–148]). At times, it is simpler to work in S^n, and we will give our description in those terms here. The formulation given here in terms of projective geometry has some advantages over the formulation given in Section 1 in terms of the unit tangent bundle to S^n. In practice, it is helpful to keep both models in mind.

Let $\mathbb{R}_2^{n+3}$ be a real vector space of dimension $n + 3$ endowed with a metric of signature $(n+1, 2)$,

$$\langle x, y \rangle = -x_1 y_1 + x_2 y_2 + \cdots + x_{n+2} y_{n+2} - x_{n+3} y_{n+3}. \qquad (5.1)$$

Let $e_1, \ldots, e_{n+3}$ denote the standard orthonormal basis with respect to this metric, with e_1 and e_{n+3} timelike. Let P^{n+2} be the real projective space of lines through the origin in $\mathbb{R}_2^{n+3}$, and let Q^{n+1} be the quadric hypersurface determined by the equation $\langle x, x \rangle = 0$. This hypersurface is called the *Lie quadric*. We consider S^n to be the unit sphere in the Euclidean space $\mathbb{R}^{n+1}$ spanned by the vectors $e_2, \ldots, e_{n+2}$.

The points in Q^{n+1} are in bijective correspondence with the set of all oriented hyperspheres and point spheres in S^n. Specifically, the oriented hypersphere with center $p \in S^n$ and signed radius ρ corresponds to the point $[(\cos \rho, p, \sin \rho)]$ in Q^{n+1}, where the square brackets denote the point in P^{n+2} given by the homogeneous coordinates within the parentheses. The point spheres in S^n correspond to those points with $\rho = 0$.

The Lie quadric contains projective lines but no linear subspaces of P^{n+2} of higher dimension. The line $[x, y]$ determined by two points $[x]$ and $[y]$ of Q^{n+1} lies on the quadric if and only if $\langle x, y \rangle = 0$. In terms of the geometry of S^n, this means that the two hyperspheres corresponding to $[x]$ and $[y]$ are in oriented contact. The points on a line on the quadric correspond to the parabolic pencil of oriented hyperspheres in S^n in oriented contact at a point (p, ξ) in the unit tangent bundle T_1S^n to S^n, where ξ is a unit tangent vector to S^n at the point p. This leads to a natural diffeomorphism from T_1S^n to the manifold Λ^{2n-1} of projective lines on Q^{n+1} given by $(p, \xi) \to [k_1, k_2]$, where $k_1 = (1, p, 0)$ and $k_2 = (0, \xi, 1)$. In terms of the geometry of spheres in S^n, k_1 corresponds to the point sphere in the parabolic pencil and k_2 corresponds to the great sphere in the pencil. We will refer to the elements of T_1S^n as *contact elements*.

A *Lie sphere transformation* is a projective transformation of P^{n+2} that maps Q^{n+1} to itself. In terms of the geometry of S^n, a Lie sphere transformation maps oriented hyperspheres to oriented hyperspheres. Furthermore, a Lie sphere transformation preserves oriented contact of spheres, since it takes lines on Q^{n+1} to lines on Q^{n+1}. The group of Lie sphere transformations is isomorphic to $O(n+1, 2)/\{\pm I\}$, where $O(n+1, 2)$ is the orthogonal group for the metric in 5.1. A *Möbius transformation* is a Lie sphere transformation that takes point spheres to point spheres. As a transformation on S^n itself, a Möbius transformation is conformal. The Lie sphere group is generated by Möbius transformations and parallel transformations P_t, which fix the center of each sphere but add t to its signed radius.

The manifold Λ^{2n-1} has a *contact structure*, that is, a globally defined one-form ω such that $\omega \wedge d\omega^{n-1}$ never vanishes on Λ^{2n-1}. The condition $\omega = 0$ defines a codimension-one distribution D on Λ^{2n-1} that has integral submanifolds of dimension $n-1$ but none of higher dimension. A *Legendre submanifold* is one of these integral submanifolds of maximal dimension, that is, an immersion $\lambda : M^{n-1} \to \Lambda^{2n-1}$ such that $\lambda^*\omega = 0$.

A Legendre submanifold is determined by two functions k_1, k_2 from an $(n-1)$-dimensional manifold M to $\mathbb{R}_2^{n+3}$ satisfying these conditions:

(L1) For all $x \in M$, the vectors $k_1(x)$ and $k_2(x)$ are linearly independent and $\langle k_i(x), k_j(x) \rangle = 0$, for $i, j = 1, 2$.

(L2) There is no nonzero $X \in T_xM$, for any $x \in M$, such that $dk_1(X)$ and $dk_2(X)$ are both in $\mathrm{Span}\{k_1(x), k_2(x)\}$.

(L3) $\langle dk_1(X), k_2(x) \rangle = 0$, for all $X \in T_xM$, for all $x \in M$.

The Legendre submanifold is then defined by $\lambda(x) = [k_1(x), k_2(x)]$. Conditions (L1)–(L3) are preserved if one reparametrizes by taking $\tilde{k}_1 = \alpha k_1 + \beta k_2$ and $\tilde{k}_2 = \gamma k_1 + \delta k_2$, where $\alpha, \beta, \gamma, \delta$ are smooth real-valued functions on M such that $\alpha\delta - \beta\gamma$ never vanishes.

Condition (L1) means that k_1 and k_2 determine a line on the quadric for each $x \in M$. Condition (L2) means that λ is an immersion, and (L3) means that $\lambda^*\omega = 0$. These conditions correspond precisely to the conditions (L1)–(L3) given in Section 1, where we used T_1S^n rather than Λ^{2n-1} as our contact manifold.

An immersion $f : M^{n-1} \to S^n$ with field of unit normals $\xi : M^{n-1} \to S^n$ naturally induces a Legendre submanifold $\lambda = [k_1, k_2]$, where

$$k_1 = (1, f, 0), \quad k_2 = (0, \xi, 1). \tag{5.2}$$

For each $x \in M^{n-1}, [k_1(x)]$ is the point sphere in the pencil of spheres in S^n corresponding to $\lambda(x)$, and $[k_2(x)]$ is the great sphere in the pencil.

An immersed submanifold $\phi : V \to S^n$ of codimension greater than one also induces a Legendre submanifold whose domain is the bundle B^{n-1} of unit normal vectors to $\phi(V)$. For a unit normal ξ to $\phi(V)$ at a point $\phi(v)$, we define $\lambda(v, \xi)$ to be the line on Q^{n+1} corresponding to the contact element $(\phi(v), \xi)$. In this case, the point sphere map $k_1(v, \xi) = (1, \phi(v), 0)$ has constant rank equal to the dimension of V. For a general Legendre submanifold λ, the point sphere map does not have constant rank.

A Lie sphere transformation β maps lines on Q^{n+1} to lines on Q^{n+1}, so it naturally induces a map $\tilde{\beta}$ from Λ^{2n-1} to itself. If λ is a Legendre submanifold, then $\tilde{\beta}\lambda$ is also a Legendre submanifold, which is denoted $\beta\lambda$ for short. These two Legendre submanifolds are said to be *Lie equivalent*. If β is a Möbius transformation, then the two Legendre submanifolds are said to be *Möbius equivalent*. Finally, if β is the parallel transformation P_t and λ is the Legendre submanifold induced by an immersed hypersurface $f : M \to S^n$, then $P_t\lambda$ is the Legendre submanifold induced by the parallel hypersurface f_{-t} (see [Cecil 1992, p. 88]).

Suppose that $\lambda = [k_1, k_2]$ is a Legendre submanifold. Let $x \in M$ and let r and s be real numbers at least one of which is nonzero. The sphere corresponding to the point

$$[K] = [rk_1(x) + sk_2(x)]$$

is called a *curvature sphere* of λ at x if there exists a nonzero $X \in T_xM$ such that

$$rdk_1(X) + sdk_2(X) \in \mathrm{Span}\{k_1(x), k_2(x)\}.$$

This definition is invariant under a reparametrization of λ by a pair $\{\tilde{k}_1, \tilde{k}_2\}$.

To see the relationship between curvature spheres and principal curvatures, suppose now that $\lambda = [k_1, k_2]$ as in 5.2. At a given $x \in M$, we can write the distinct curvature spheres in the form

$$[K_i] = [\mu_i k_1 + k_2], \quad \text{for } 1 \le i \le g. \tag{5.3}$$

When the map f in 5.2 is an immersion, these μ_i are the principal curvatures of f at x. In terms of the geometry of S^n, the curvature sphere at x corresponding to a principal curvature μ_i is the oriented hypersphere in oriented contact with $f(M)$ at $f(x)$ and centered at the focal point determined by the principal curvature μ_i.

We refer to the μ_i in 5.3 as the *principal curvatures* of λ. These principal curvatures are not Lie invariant, and they depend on the special parametrization 5.2 for λ. However, Miyaoka [1989a] pointed out that the cross-ratio of any four of these principal curvatures is Lie invariant. These cross-ratios are known as the *Lie curvatures* of λ. Such a cross-ratio is Lie invariant because it is equal to the cross-ratio of the corresponding four curvature spheres on the line $\lambda(x)$. Since a Lie sphere transformation β is a projective transformation and it maps the curvature spheres of λ to the curvature spheres of $\beta\lambda$, it preserves these cross-ratios.

A Legendre submanifold is said to be *Dupin* if, along each curvature surface, the corresponding curvature sphere map is constant, and it is *proper Dupin* if the number of distinct curvature spheres is constant. Pinkall [1985a] showed that both of these concepts are invariant under Lie sphere transformation. In the case where a Legendre submanifold is induced from a submanifold of S^n, these definitions agree with those given in Section 3.

Recall from Section 3 that a proper Dupin submanifold is reducible if it is Lie equivalent to a proper Dupin submanifold obtained as a result of one of Pinkall's four standard constructions. Pinkall [1985a] found a simple formulation for reducibility in terms of Lie sphere geometry as follows.

THEOREM 5.1. *Let* $\lambda : M^{n-1} \to \Lambda^{2n-1}$ *be a proper Dupin submanifold with distinct curvature spheres* $K_1, \ldots K_g$. *Then* λ *is reducible if and only if, for some* i *with* $1 \le i \le g$, *the image of the curvature sphere map* K_i *is contained in an n-dimensional linear subspace of* P^{n+2}.

Another important question in classifying proper Dupin submanifolds is when is the submanifold Lie equivalent to an isoparametric hypersurface in S^n. This condition also has a natural formulation in Lie sphere geometry. Recall that a line in P^{n+2} is said to be *timelike* if it contains only timelike points. This means that that an orthonormal basis for the two-plane in $\mathbb{R}_2^{n+3}$, determined by the timelike line, consists of two timelike vectors. An example is the line $[e_1, e_{n+3}]$. The following theorem was obtained in [Cecil 1990].

THEOREM 5.2. *Let* $\lambda : M^{n-1} \to \Lambda^{2n-1}$ *be a Legendre submanifold with g distinct curvature spheres* $K_1, \ldots, K_g$ *at each point. Then* λ *is Lie equivalent to the Legendre submanifold induced by an isoparametric hypersurface in S^n if and only if there exist g points* $P_1, \ldots, P_g$ *on a timelike line in* P^{n+2} *such that* $\langle K_i, P_i \rangle = 0$ *for* $1 \le i \le g$.

For more general considerations of Lie contact structures on manifolds, see [Miyaoka 1991b; 1991a; 1993b].

6. Compact Proper Dupin Submanifolds

In this section, we consider compact proper Dupin submanifolds embedded in the sphere S^n. Since a tube of sufficiently small radius ε over a compact proper Dupin submanifold of codimension greater than one is a compact proper Dupin hypersurface, we will for the most part restrict our attention to the codimension-one case. Let M be a compact proper Dupin hypersurface embedded in S^n with

g distinct principal curvatures. As noted in Theorem 4.3, the number g must be 1, 2, 3, 4, or 6. Of course, in the case $g = 1$, the hypersurface M is totally umbilic and must be a great or small hypersphere in S^n. The case $g = 2$ was handled in [Cecil and Ryan 1978]:

THEOREM 6.1. *A compact, connected proper Dupin hypersurface embedded in S^n with two distinct principal curvatures is Möbius equivalent to an isoparametric hypersurface, that is, a standard product of two spheres.*

Next Miyaoka [1984a] handled the case $g = 3$, where the full Lie sphere group was needed to get equivalence with an isoparametric hypersurface:

THEOREM 6.2. *A compact, connected proper Dupin hypersurface embedded in S^n with three distinct principal curvatures is Lie equivalent to an isoparametric hypersurface.*

For some time after that, it was widely thought that every compact, connected proper Dupin hypersurface embedded in S^n is Lie equivalent to an isoparametric hypersurface [Cecil and Ryan 1985, p. 184]. Further evidence for this conjecture was provided by Grove and Halperin [1987], who found many restrictions on the topology of a compact proper Dupin hypersurface and on the multiplicities of its principal curvatures. These restrictions included almost all of the known restrictions for isoparametric hypersurfaces due to Münzner [1980; 1981] and Abresch [1983]. However, in 1989, the conjecture was shown to be false by Pinkall and Thorbergsson [1989a] and Miyaoka and Ozawa [1989], who independently gave different methods for constructing counterexamples to the conjecture in the case $g = 4$. The method of Miyaoka and Ozawa also works in the case $g = 6$.

These constructions are also described in detail in [Cecil 1992]. The construction of Pinkall and Thorbergsson begins with the isoparametric hypersurfaces constructed using representations of Clifford algebras by Ferus, Karcher, and Münzner [Ferus et al. 1981]. For each of these isoparametric hypersurfaces, one of the focal submanifolds is a so-called Clifford–Stiefel manifold. Here we will only describe the simplest case where the Clifford algebra is $\mathbb{R}$.

Let $\mathbb{R}^{2n+2} = \mathbb{R}^{n+1} \times \mathbb{R}^{n+1}$ and let S^{2n+1} denote the unit sphere in $\mathbb{R}^{2n+2}$. The Stiefel manifold V of orthogonal two-frames in $\mathbb{R}^{n+1}$ of length $1/\sqrt{2}$ is given by

$$V = \{(u, v) \in \mathbb{R}^{2n+2} : u \cdot v = 0, |u| = |v| = 1/\sqrt{2}\}.$$

The submanifold V lies in S^{2n+1} with codimension two, so V has dimension $2n - 1$. Note that

$$V \subset F^{-1}(0) \cap G^{-1}(0),$$

where F and G are the real-valued functions defined on S^{2n+1} by

$$F(u, v) = (v \cdot v - u \cdot u)/2, \quad G(u, v) = -u \cdot v.$$

The gradients $\xi = (-u, v)$ and $\eta = (-v, -u)$ of F and G are two orthogonal fields of unit normals to V in S^{2n+1}. One can show by a direct calculation

[Cecil 1992, pp. 115–117; Pinkall and Thorbergsson 1989a] that, at every point of V, the shape operators A_ξ and A_η have three distinct principal curvatures $-1, 0, 1$, with respective multiplicities $n-1, 1, n-1$, although A_ξ and A_η are not simultaneously diagonalizable. From this one can calculate that if ζ is any unit normal at any point of V, then A_ζ has these same eigenvalues and multiplicities. Thus, the principal curvatures

$$-1 = \cot(3\pi/4), \quad 0 = \cot(\pi/2), \quad 1 = \cot(\pi/4)$$

are constant on the unit normal bundle $B(V)$. Therefore, a tube M_t^{2n} of radius t around V in S^{2n+1} has four constant principal curvatures (see [Cecil and Ryan 1985, pp. 131–132])

$$\cot(3\pi/4 - t), \ \cot(\pi/2 - t), \ \cot(\pi/4 - t), \ \cot(-t), \qquad (6.1)$$

with respective multiplicities $n-1, 1, n-1, 1$. This is the family of isoparametric hypersurfaces with focal submanifold V. For a general Clifford algebra, one considers the Clifford–Stiefel manifold of "Clifford orthogonal" pairs of vectors in defining the focal submanifold [Pinkall and Thorbergsson 1989a].

For a proper Dupin submanifold with $g = 4$, we can order the principal curvatures,

$$\mu_1 < \mu_2 < \mu_3 < \mu_4, \qquad (6.2)$$

and thereby determine a unique Lie curvature Ψ by the cross-ratio

$$\Psi = \frac{(\mu_4 - \mu_3)(\mu_1 - \mu_2)}{(\mu_4 - \mu_2)(\mu_1 - \mu_3)}. \qquad (6.3)$$

Note that the ordering 6.2 implies that $0 < \Psi < 1$. Münzner [1980] showed that the principal curvatures of an isoparametric hypersurface with $g = 4$ must always take the values in 6.1 for an appropriate value of t, and one can directly compute that Ψ always takes the constant value $\frac{1}{2}$ on an isoparametric hypersurface with four principal curvatures. In terms of projective geometry, this means that the four curvature spheres along each line $\lambda(x)$ form a harmonic set.

The construction of Pinkall and Thorbergsson then proceeds as follows. Let α and β be positive real numbers satisfying $\alpha^2 + \beta^2 = 1$, and let $T_{\alpha,\beta}$ be the linear transformation of $\mathbb{R}^{2n+2}$ defined by

$$T_{\alpha,\beta}(u, v) = \sqrt{2}(\alpha u, \beta v).$$

The image $W^{\alpha,\beta} = T_{\alpha,\beta}V$ is contained in S^{2n+1}, and it is proper Dupin. To see this, one first notes that $W^{\alpha,\beta}$ has codimension two in S^{2n+1}, just as V does. This means that the point sphere map k_1 in 5.2 is a curvature sphere in both cases, and so by 5.3 they both have a constant principal curvature $\kappa_4 = \infty$ of multiplicity one (see [Cecil 1992, p. 97] for more detail on this point). For the other principal curvatures, note that a hypersphere Σ that is tangent to V along a curvature surface S lies in a hyperplane π in $\mathbb{R}^{2n+2}$ that is tangent to V along S. The image $T_{\alpha,\beta}(\pi)$ cuts S^{2n+1} in a hypersphere $\tilde{\Sigma}$ that is tangent to $W^{\alpha,\beta}$ along

$T(S)$. Thus, $\tilde{\Sigma}$ is a curvature sphere of $W^{\alpha,\beta}$ with corresponding curvature surface $T(S)$, and $\tilde{\Sigma}$ is constant along $T(S)$. Therefore, we have established a bijective correspondence between the curvature surfaces of V and those of $W^{\alpha,\beta}$ and have shown that the Dupin condition is satisfied on $W^{\alpha,\beta}$. Hence, $W^{\alpha,\beta}$ is a proper Dupin submanifold with four distinct principal curvatures, including $\kappa_4 = \infty$.

The other principal curvatures of $W^{\alpha,\beta}$ can be computed in the same way as for V. Then one can find a certain unit normal ζ to $W^{\alpha,\beta}$ such that the shape operator A_ζ has principal curvatures

$$\kappa_1 = -\alpha/\beta, \quad \kappa_2 = 0, \quad \kappa_3 = \beta/\alpha,$$

with respective multiplicities $n - 1$, 1, $n - 1$. When these principal curvatures are taken along with $\kappa_4 = \infty$, the cross-ratio Ψ equals α^2. Thus, if $\alpha \neq 1/\sqrt{2}$, that is, if $T_{\alpha,\beta} \neq I$, then the Lie curvature Ψ is not $\frac{1}{2}$ at this point ζ, and $W^{\alpha,\beta}$ is not Lie equivalent to an isoparametric hypersurface. In fact, one can show that the Lie curvature is not constant on $W^{\alpha,\beta}$ if $\alpha \neq 1/\sqrt{2}$.

The construction of Miyaoka and Ozawa [Miyaoka and Ozawa 1989] (see also [Cecil 1992, pp. 120–128]) uses the *Hopf fibration* of S^7 over S^4. Let $\mathbb{R}^8 = \mathbb{H} \times \mathbb{H}$, where $\mathbb{H}$ is the division ring of quaternions. The Hopf fibering of the unit sphere S^7 in $\mathbb{R}^8$ over the unit sphere S^4 in $\mathbb{R}^5 = \mathbb{H} \times \mathbb{R}$ is given by

$$h(u, v) = (2u\bar{v}, |u|^2 - |v|^2), \quad \text{for } u, v \in \mathbb{H}.$$

Miyaoka and Ozawa begin by showing that, if M is a taut compact submanifold of S^4, then $h^{-1}M$ is taut in S^7. They use this to show that if M is proper Dupin in S^4 with g distinct principal curvatures, then $h^{-1}M$ is proper Dupin in S^7 with $2g$ principal curvatures. Finally, they show that if M is proper Dupin but not isoparametric, then the Lie curvatures of $h^{-1}M$ are not constant, and so $h^{-1}M$ is not Lie equivalent to an isoparametric hypersurface with $2g$ principal curvatures in S^7. Taking $g = 2$ or 3, respectively, yields a compact proper Dupin hypersurface with 4 or 6 principal curvatures in S^7 that is not Lie equivalent to an isoparametric hypersurface.

As noted in Section 2, Miyaoka [1993a] has recently shown that if M is an isoparametric hypersurface with three principal curvatures in S^4, then $h^{-1}M$ is an isoparametric hypersurface with six principal curvatures in S^7. According to Dorfmeister and Neher [Dorfmeister and Neher 1985], this family of isoparametric hypersurfaces with six principal curvatures in S^7 is unique up to congruence.

A problem for further study is to determine the strength of the assumption that the Lie curvatures are constant on a proper Dupin submanifold. Miyaoka [Miyaoka 1989a; Miyaoka 1989b] proved that this assumption on a compact proper Dupin hypersurface M in S^n with $g = 4$ or 6 principal curvatures, together with an additional assumption regarding the intersections of curvature surfaces from different principal foliations, implies that M is Lie equivalent to an isoparametric hypersurface. It is not known whether this additional assumption on the intersections of the curvature surfaces can be dropped. However, the compactness assumption of Miyaoka is definitely needed because there exist noncompact proper Dupin hypersurfaces with $g = 4$ on which $\Psi = c$, for every value $0 < c < 1$ [Cecil 1990; 1992, pp. 106–108]. These examples are all obtained as open subsets of a tube in S^n over an isoparametric hypersurface with three principal curvatures $V^{k-1} \subset S^k \subset S^n$, and they cannot be completed to be compact proper Dupin hypersurfaces. They are also reducible as Dupin hypersurfaces.

7. Local Results on Dupin Submanifolds

Most local classifications of proper Dupin submanifolds have been obtained in the context of Lie sphere geometry. We will state these results for hypersurfaces of S^n, since we can always arrange that the point sphere map of a Legendre submanifold is locally an immersion by taking a parallel submanifold if necessary.

The known results depend on the number g of distinct principal curvatures, and they are progressively harder to prove as g increases. In fact, results have only been obtained up to the case $g = 4$, and much remains to be done in that case. Of course, a connected proper Dupin hypersurface in S^n with one distinct principal curvature must be an open subset of a hypersphere. In the case $g = 2$, Pinkall [1985a] has obtained a complete classification, which we now describe. A proper Dupin hypersurface in S^n (or $\mathbb{R}^n$) with two distinct principal curvatures of respective multiplicities p and q is called a *cyclide of Dupin of characteristic* (p,q). The compact cyclides embedded in $\mathbb{R}^3$ can all be obtained through stereographic projection from a standard product of two circles in the unit sphere $S^3 \subset \mathbb{R}^4$. This construction obviously can be generalized to higher dimensions. Cecil and Ryan [1978] showed that a connected, compact cyclide M^{n-1} of characteristic (p,q) embedded in S^n must be Möbius equivalent to a standard product of spheres

$$S^p \times S^q \subset S^n(1) \subset \mathbb{R}^{p+1} \times \mathbb{R}^{q+1} = \mathbb{R}^{n+1}, \quad \text{for } r^2 + s^2 = 1, \ n = p + q + 1.$$

Varying the value of r in this product produces a family of parallel hypersurfaces. These are Lie equivalent by parallel transformation, but they are not Möbius equivalent for different values of r.

The proof of Cecil and Ryan uses the compactness assumption in an essential way, whereas the classification of Dupin surfaces in $\mathbb{R}^3$ obtained in the nineteenth century does not need such an assumption (see [Cecil and Ryan 1985, pp. 151–166], for example). Using Lie sphere geometry, Pinkall obtained the following classification in arbitrary dimensions, which does not need the assumption of compactness.

THEOREM 7.1. (a) *Every connected cyclide of Dupin is contained in a unique compact, connected cyclide.*

(b) *Any two cyclides of the same characteristic are locally Lie equivalent.*

A consequence of part (a) of the theorem is that any connected piece of a Dupin cyclide of characteristic (p,q) immersed in S^n determines a unique compact Legendre submanifold with domain $S^p \times S^q$.

Pinkall's result can be used to obtain the following Möbius geometric characterization of the cyclides in $\mathbb{R}^n$ [Cecil 1991; 1992, p. 154].

THEOREM 7.2. (a) *Every connected cyclide of Dupin M^{n-1} of characteristic (p,q) embedded in $\mathbb{R}^n$ is Möbius equivalent to an open subset of a surface of revolution obtained by revolving a q-sphere $S^q \subset \mathbb{R}^{q+1} \subset \mathbb{R}^n$ about an axis of revolution $\mathbb{R}^q \subset \mathbb{R}^{q+1}$ or a p-sphere $S^p \subset \mathbb{R}^{p+1} \subset \mathbb{R}^n$ about an axis $\mathbb{R}^p \subset \mathbb{R}^{p+1}$.*

(b) *Two such surfaces are Möbius equivalent if and only if they have the same value of $\rho = |r|/a$, where r is the signed radius of the profile sphere S^q and $a > 0$ is the distance from the center of S^q to the axis of revolution.*

Note that the profile sphere is allowed to intersect the axis of revolution, thereby resulting in singularities. However, in the context of Lie sphere geometry, the corresponding Legendre map is an immersion.

The classical cyclides of Dupin in $\mathbb{R}^3$ are the only surfaces for which all lines of curvature are circles (or straight lines). Using exterior differential systems, Ivey [1995] showed that any surface in $\mathbb{R}^3$ containing two orthogonal families of circles is a cyclide of Dupin.

Finally, we note that recently the cyclides of Dupin have been used in computer-aided geometric design of surfaces. See, for example, [Degen 1994; Pratt 1990; 1995; [Srinivas and Dutta 1994a]; 1994b; 1995a; 1995b].

The case $g = 3$ has proved to be much more difficult than the case of two principal curvatures. In his dissertation, Pinkall [1981] did obtain a complete local classification up to Lie sphere transformation for Dupin hypersurfaces with three principal curvatures in $\mathbb{R}^4$, but it is quite involved. (See also [Pinkall 1985b; Cecil and Chern 1989; Cecil 1992, pp. 171–190].) Recall from Section 3 that a proper Dupin hypersurface is *reducible* if it locally Lie equivalent to a proper Dupin hypersurface obtained from one of Pinkall's standard constructions. It is *locally irreducible* if it does not contain any reducible open subset. Pinkall found that any two irreducible proper Dupin hypersurfaces in $\mathbb{R}^4$ are locally Lie equivalent, each being Lie equivalent to an open subset of an isoparametric hypersurface in S^4. However, he found a one-parameter family of Lie equivalence classes among the reducible proper Dupin hypersurfaces with $g = 3$ in $\mathbb{R}^4$. In higher dimensions, the focus has been on classifying the locally irreducible Dupin hypersurfaces and little has been done in attempting to classify the reducible ones up to Lie equivalence, although this appears to be a problem where some progress could be made.

The first result in higher dimensions is due to Niebergall [1991], who showed that every proper Dupin hypersurface in $\mathbb{R}^5$ with three principal curvatures is reducible. Recently, Cecil and Jensen [1997] have generalized the results of Pinkall and Niebergall as follows.

THEOREM 7.3. *Let* $f : M \to S^n$ *be a proper Dupin hypersurface with three distinct principal curvatures of multiplicities* m_1, m_2, *and* m_3. *If the hypersurface is locally irreducible, then* $m_1 = m_2 = m_3$, *and* M *is Lie equivalent to an isoparametric hypersurface in a sphere.*

We now give a brief outline of the proof of this theorem. We work in the context of Lie sphere geometry and consider a proper Dupin submanifold $\lambda : M^{n-1} \to \Lambda^{2n-1}$ with three curvature spheres. As in Section 5, we can parametrize the Dupin submanifold as $\lambda = [k_1, k_2]$, where $[k_1]$ and $[k_2]$ are two curvature sphere maps. We can also arrange that the third curvature sphere map have the form $[k_3] = [k_1 + k_2]$. We first compute the derivatives of the $[k_i]$ using the method of moving frames. In the case where M has dimension three, Pinkall found one function c with the property that all of the terms arising in the exterior differentiation of the frame fields could eventually be expressed in terms of c and its derivatives. If c is identically zero, then λ is reducible. If c is never zero on M, then one can arrange that $c = 1$ with an appropriate choice of frame, and all such hypersurfaces are locally Lie equivalent to Cartan's isoparametric hypersurface in S^4.

In the general case where the three curvature spheres have respective multiplicities m_1, m_2, and m_3, there are $m_1 \cdot m_2 \cdot m_3$ functions F_{ap}^α, where

$$1 \le a \le m_1, \quad m_1 + 1 \le p \le m_1 + m_2, \quad m_1 + m_2 + 1 \le \alpha \le n - 1,$$

which are defined in a similar way to the one function c of Pinkall. They can be arranged in vector form, $v_{p\alpha} = (F_{ap}^\alpha)$, for $1 \le a \le m_1$, with $v_{a\alpha}$ and v_{ap} defined in a similar way. One first shows that if a column or row of any of the arrays $[v_{p\alpha}], [v_{a\alpha}], [v_{ap}]$ is identically zero on an open subset $V \subset M$, then the restriction of λ to V is reducible. This result is applied to show that unless all of the multiplicities are equal, λ must contain a reducible open subset, i.e., it is not locally irreducible. In the case where all the multiplicities equal m, one next shows that all of the vectors in all of the arrays have the same length ρ. This one function ρ actually plays the same role that c did in the case $m = 1$. If λ is locally irreducible, then ρ is nonzero on M, and it can be made locally constant with an appropriate choice of frame. The proof is completed by showing that there exist three points P_1, P_2, P_3 on a certain timelike line in P^{n+2} such that $\langle k_i, P_i \rangle = 0$, for $1 \le i \le 3$, where the $[k_i]$ are the curvature sphere maps of λ. By Theorem 5.2, this implies that λ is Lie equivalent to an isoparametric hypersurface.

The next case $g = 4$ is still more complicated, but many aspects of the approach outlined above apply. As in Section 6, one can order the principal curvatures as in 6.2 and determine a unique Lie curvature Ψ by 6.3 satisfying $0 < \Psi < 1$. As in the $g = 3$ case, one can reparametrize the Dupin submanifold as $\lambda = [k_1, k_2]$, where $[k_1]$ and $[k_2]$ are two of the curvature sphere maps, and then arrange that a third curvature sphere map satisfies $[k_3] = [k_1 + k_2]$. Then the fourth curvature sphere $[k_4]$ is determined by the Lie curvature Ψ. Niebergall [1992] used this framework to find some sufficient conditions for a proper Dupin hypersurface with $g = 4$ in S^5 to be Lie equivalent to an isoparametric hypersurface.

For $g = 4$ or 6, it is not true that every locally irreducible Dupin hypersurface is Lie equivalent to an isoparametric hypersurface. The examples of Pinkall and Thorbergsson [1989a] and Miyaoka and Ozawa [1989] discussed in Section 6 are locally irreducible and are not Lie equivalent to an isoparametric hypersurface. On the other hand, Miyaoka [1989a; 1989b] has shown that a compact proper Dupin hypersurface embedded in $\mathbb{R}^n$ is Lie equivalent to an isoparametric hypersurface if it has constant Lie curvatures and it satisfies certain global conditions regarding the intersections of leaves of its various principal foliations. As in the $g = 3$ case, in trying to obtain local results, one can replace the global conditions of Miyaoka with the assumption of local irreducibility. This yields the following question. In the cases $g = 4$ or 6, is every locally irreducible proper Dupin hypersurface with constant Lie curvatures Lie equivalent to an isoparametric hypersurface?

Note that the hypothesis of local irreducibility is definitely necessary here because of the noncompact reducible proper Dupin hypersurfaces with constant Lie curvature of [Cecil 1990] mentioned in Section 6.

A more general problem is to attempt to identify key local Lie invariants of Dupin submanifolds within the context of moving Lie frames. Niebergall [1992] made some progress in this direction in the case of a proper Dupin hypersurface M^4 in S^5 with four principal curvatures. He assumed that the Lie curvature is constant and then found four other invariants, analogous to the F_{ap}^α in the $g = 3$ case, that when suitably prescribed yield the conclusion that M^4 is Lie equivalent to an isoparametric hypersurface. Moreover, the vanishing of any three of these invariants implies that M^4 is reducible. At this point, however, the geometric meaning of these invariants is not yet clear, nor is it obvious that these are the only invariants to be considered. This is clearly a complicated problem, but a systematic study may yield new results.

Another problem is to attempt to obtain a complete local classification of reducible Dupin hypersurfaces with three principal curvatures up to Lie equivalence. As mentioned above, Pinkall obtained such a classification in the case of $M^3 \subset \mathbb{R}^4$. In that case, while there is only one class of irreducible Dupin hypersurfaces, the reducible ones determine a one-parameter family of Lie equivalence classes. It may be possible to obtain a similar classification in the higher dimensional reducible case by using the framework established to prove Theorem 7.3.

The approach of Lie sphere geometry can also be used to obtain results in Möbius (conformal) geometry. As noted in Theorem 7.2, one can derive a local Möbius classification in the case $g = 2$ from Pinkall's Lie-geometric classification. Pinkall and Thorbergsson [1989a] introduced a Möbius invariant, called the *Möbius curvature*, that can distinguish among the Lie equivalent parallel hypersurfaces in a family of isoparametric hypersurfaces. Recently, C.-P. Wang used the method of moving frames to determine a complete set of Möbius invariants for surfaces in $\mathbb{R}^3$ without umbilic points [Wang 1992] and for hypersurfaces in $\mathbb{R}^4$ with three distinct principal curvatures at each point [Wang 1995]. He then applied this result to derive a local classification of Dupin hypersurfaces in $\mathbb{R}^4$ with three principal curvatures up to Möbius transformation. A natural problem is to try to extend Wang's result to proper Dupin hypersurfaces in $\mathbb{R}^n$, for $n > 4$, and to thoroughly investigate the geometric significance of his invariants, including the Möbius curvature, in $\mathbb{R}^4$.

Ferapontov [1995a; 1995b] has explored the relationship between Dupin and isoparametric hypersurfaces and Hamiltonian systems of hydrodynamic type. Ferapontov poses several research problems in that context.

8. Classifications of Taut Submanifolds

In this section, we survey the known classification results on taut submanifolds. To a reasonable extent, we have attempted to make this section self-contained, although some references to the previous sections are inevitable.

In the paper [Banchoff 1970] that introduced the STPP, it was shown that a taut embedding of S^1 into $\mathbb{R}^n$ must be a metric circle in a plane. In the same paper, Banchoff also obtained a complete classification of compact taut (two-dimensional) surfaces in Euclidean spaces. We now give a brief outline of his proof. As noted in Section 1, Banchoff observed that, because tautness is invariant under stereographic projection, there exists a substantial taut nonspherical embedding of a compact manifold M into $\mathbb{R}^n$ if and only if there exists a substantial taut spherical embedding of M into $S^n \subset \mathbb{R}^{n+1}$. As a consequence, one can invoke Kuiper's result on the bound on the codimension to obtain Theorem 1.1. In particular, if $f : M^2 \to \mathbb{R}^n$ is a substantial taut embedding, then $n \leq 5$, and if $n = 5$, then f is an embedding of P^2 as a Veronese surface in $S^4 \subset \mathbb{R}^5$. Next, if $f : M^2 \to \mathbb{R}^4$ is a taut nonspherical embedding, then $f(M^2)$ must be the image under stereographic projection of a spherical Veronese surface in $\mathbb{R}^5$. Thus, the problem is reduced to finding all compact taut surfaces in S^3. Again by stereographic projection, this is equivalent to finding all compact taut surfaces in $\mathbb{R}^3$.

A key step in Banchoff's classification of compact taut surfaces in $\mathbb{R}^3$ is showing that if $f : M^2 \to \mathbb{R}^3$ is a taut embedding of a compact surface M^2, then $f(M^2)$ lies in between the two spheres S_1 and S_2 tangent to $f(M^2)$ at $f(x)$ and centered at the focal points $p_i = f(x) + (1/\mu_i)\xi(x)$, for $i = 1, 2$, respectively, where ξ is a field of unit normals to $f(M^2)$, and the μ_i are the principal curvatures of f. Thus, if $f(M^2)$ has one umbilic point, it must be a metric sphere, because it lies between two identical spheres S_1 and S_2 at the umbilic point. This implies that $f(M^2)$ must be either a metric sphere or smooth torus, because any embedding of a surface of higher genus would necessarily have an umbilic point. Suppose now that $f(M^2)$ is a taut torus with no umbilic points. Then Banchoff shows that the principal curvatures must be constant along their corresponding lines of curvature, i.e., $f(M^2)$ is Dupin, and so it is a cyclide of Dupin in $\mathbb{R}^3$.

Cecil [1976] then generalized Banchoff's argument to the noncompact case to again show that a taut $f(M^2) \subset \mathbb{R}^3$ must be Dupin. This implies that $f(M^2)$ must a plane, circular cylinder, or parabolic ring cyclide. The latter surface is obtained by by inverting a torus of revolution in a sphere centered at a point on the torus; its name comes from the fact that its focal set consists of a pair of parabolas [Cecil and Ryan 1985, pp. 151–166]. These results are combined in the following theorem, which is a complete classification of taut surfaces in $\mathbb{R}^n$.

THEOREM 8.1. *Let $f : M^2 \to \mathbb{R}^n$ be a substantial taut embedding of a surface M^2.*

(a) *If M^2 is compact, then $f(M^2)$ is a metric sphere or a cyclide of Dupin in $\mathbb{R}^3$, a spherical Veronese surface in $S^4 \subset \mathbb{R}^5$, or a surface in $\mathbb{R}^4$ related to one of these by stereographic projection.*

(b) *If M^2 is noncompact, then it is a plane, circular cylinder or parabolic ring cyclide in $\mathbb{R}^3$, or it is the image in $\mathbb{R}^4$ of a punctured spherical Veronese surface under stereographic projection from S^4.*

The first results after Banchoff's paper dealt with taut embeddings of submanifolds with relatively simple topology. Nomizu and Rodríguez [1972] proved the following.

THEOREM 8.2. *Let M be a complete Riemannian manifold of dimension $k \geq 2$, isometrically immersed in $\mathbb{R}^n$. If every nondegenerate Euclidean distance function L_p has index 0 or k at any of its critical points, then M is embedded as a k-plane or a metric k-sphere $S^k \subset \mathbb{R}^{k+1} \subset \mathbb{R}^n$.*

The proof is accomplished by showing that M is a totally umbilic submanifold. This is a consequence of the following elementary but important argument. Let $f : M \to \mathbb{R}^n$ be the isometric immersion. Let ξ be any unit normal to $f(M)$ at any point $f(x)$. We want to show that the shape operator A_ξ is a scalar multiple of the identity. If $A_\xi = 0$, we are done. If not, then by replacing ξ by $-\xi$, if necessary, we can assume that A_ξ has a positive eigenvalue. Let λ be the largest eigenvalue of A_ξ, and let t satisfy $1/\lambda < t < 1/\mu$, where μ is the next largest positive eigenvalue of A_ξ (just consider $t > 1/\lambda$ if there are no other positive eigenvalues). If $q = f(x) + t\xi$, then L_q has a nondegenerate critical point at x with index equal to the multiplicity m of λ. It may be that L_q is not a Morse function. If so, there exists a Morse function L_p, with p near q, such that L_p has a critical point near x of the same index m. Now since $m > 0$, the hypotheses imply that $m = k$, and thus $A_\xi = \lambda I$, as desired.

An immediate consequence is the following result, obtained independently by
Carter and West [1972]. The result is true in the case $k = 1$ by the work of
Banchoff mentioned earlier. See [Cecil 1974] for a similar characterization of
metric spheres in hyperbolic space.

THEOREM 8.3. *Let* $f : S^k \to \mathbb{R}^n$ *be a taut embedding. Then* $f(S^k)$ *is a metric
sphere in a* $(k+1)$-*dimensional affine subspace* $\mathbb{R}^{k+1} \subset \mathbb{R}^n$.

Approaching the problem from a different point of view, Hebda [1981] asked
which ambient spaces admit taut embeddings of hyperspheres. He found that a
complete simply connected n-dimensional manifold that admits a taut embed-
ding of S^{n-1} is either homeomorphic to S^n, diffeomorphic to $\mathbb{R}^n$ or diffeomorphic
to $S^{n-1} \times \mathbb{R}$.

Next Carter and West [1972] obtained the following characterization of spherical cylinders.

THEOREM 8.4. *Let* $f : M^{n-1} \to \mathbb{R}^n$ *be a taut embedding of a noncompact
manifold such that* $H_k(M^{n-1}; \mathbb{Z}_2) = \mathbb{Z}_2$ *for some* k *with* $0 < k < n-1$, *and such
that* $H_i(M^{n-1}; \mathbb{Z}_2) = 0$ *for* $i \neq 0, k$. *Then* M^{n-1} *is diffeomorphic to* $S^k \times \mathbb{R}^{n-k-1}$,
and f *is a standard product embedding.*

Thorbergsson [1983b; 1986] then obtained the following characterization of highly
connected taut submanifolds of arbitrary codimension. (See [Jorge and Mercuri
1984] for a related result involving minimal submanifolds.)

THEOREM 8.5. *Let* M *be a compact* $2k$-*dimensional taut submanifold of* $\mathbb{R}^n$ *that
is* $(k-1)$-*connected but not* k-*connected, and that does not lie in any totally
umbilic hypersurface of* $\mathbb{R}^n$. *Then either*

(a) $n = 2k + 1$ *and* M *is a cyclide of Dupin diffeomorphic to* $S^k \times S^k$, *or*
(b) $n = 3k + 1$ *and* M *is a spherical Veronese embedding of the projective plane*
    FP^2, *where* F *is the division algebra* $\mathbb{R}, \mathbb{C}, \mathbb{H}, \mathbb{O}$ *for* $k = 1, 2, 4, 8$, *respec-
    tively.*

In a related paper, Hebda [1984] constructs tight smooth embeddings of arbitrarily many copies of $S^k \times S^k$ into $\mathbb{R}^{2k+1}$. No taut embeddings of these manifolds
exist by Theorem 8.5.

The next case in terms of the homology is when M has the same homology
as $S^k \times S^m$, where $k \neq m$. In that case, Cecil and Ryan [1978; 1985, p. 202]
obtained the following.

THEOREM 8.6. *A taut hypersurface* $M \subset \mathbb{R}^n$ *with the same* $\mathbb{Z}_2$-*homology as
$S^k \times S^{n-k-1}$ is a cyclide of Dupin.*

To prove this, one first uses the Index Theorem for L_p functions to prove that,
at each point of M, the number of distinct principal curvatures must be either
2 or 3. The most difficult part of the proof is to then show that the number of
distinct principal curvatures must be constant on M. Then, since taut implies

Dupin, M is a proper Dupin hypersurface with $g = 2$ or 3 principal curvatures. Then it is fairly easy given the homology of M to show that $g = 2$, and so M must be a cyclide of Dupin.

Ozawa [1986] generalized this result by showing that, if an embedding of $S^k \times S^m$ into S^n, where $k < m$, is taut and substantial, then the codimension of the embedding is either 1 or $m - k + 1$. He also showed that the r-times connected sum of $S^k \times S^m$, for $k < m$, cannot be tautly embedded into any Euclidean space if $r > 1$.

There is a related result that takes into account the intrinsic geometry of M. Recall that a Riemannian manifold (M, g) is *conformally flat* if every point has a neighborhood conformal to an open subset in Euclidean space. Schouten [1921] showed that a hypersurface M of dimension $n \geq 4$ and immersed in $\mathbb{R}^{n+1}$ is conformally flat in the induced metric if and only if at least $n - 1$ of the principal curvatures coincide at each point. (This characterization fails when $n = 3$ [Lancaster 1973].) Using Schouten's result, Theorem 8.6, and some basic results on tautness, Cecil and Ryan [1980] proved the following.

THEOREM 8.7. *Let M be a taut hypersurface of dimension $n \geq 4$ in $\mathbb{R}^{n+1}$. Then M is conformally flat in the induced metric if and only if it is one of the following:*

(a) *a hyperplane or metric sphere;*
(b) *a cylinder over a circle or over an $(n-1)$-sphere;*
(c) *a cyclide of Dupin (diffeomorphic to $S^1 \times S^{n-1}$);*
(d) *a parabolic ring cyclide (diffeomorphic to $S^1 \times S^{n-1} - \{p\}$).*

Concerning taut embeddings of three-manifolds, Pinkall and Thorbergsson have proved the following [1989b].

THEOREM 8.8. *A compact taut three-dimensional submanifold in Euclidean space is diffeomorphic to one of the following seven manifolds: S^3, $\mathbb{RP}^3$, the quaternion space $S^3/\{\pm 1, \pm i, \pm j, \pm k\}$, the three-torus T^3, $S^1 \times S^2$, $S^1 \times \mathbb{RP}^2$, or $S^1 \times_h S^2$, where h denotes an orientation-reversing diffeomorphism of S^2. All of these manifolds admit taut embeddings.*

Pinkall and Thorbergsson actually determine much more about the geometric structure of these taut embeddings. Because of the invariance of tautness under stereographic projection, it makes sense to attempt to classify *spherically substantial* embeddings, that is, those that do not lie in any hypersphere. In the following description of the results of Pinkall and Thorbergsson, the codimension means the spherically substantial codimension.

A taut embedding of S^3 must be a metric hypersphere. The projective space $\mathbb{RP}^3$ can be tautly embedded with codimension 2 as the Stiefel manifold $V_{2,3} \subset S^5 \subset \mathbb{R}^6$ and with codimension 5 as SO(3) in the unit sphere in the space of 3×3 matrices. It is not known whether the codimensions 3 and 4 are possible. The quaternion space is realized as Cartan's isoparametric hypersurface in S^4, where it is unique up to Lie equivalence, and no other codimensions are possible. The three-torus can be tautly embedded with codimension one as a tube in $\mathbb{R}^4$ around a torus of revolution $T^2 \subset \mathbb{R}^3 \subset \mathbb{R}^4$, and with codimension 2 as $T^2 \times S^1 \subset \mathbb{R}^5$. One can tautly embed $S^1 \times S^2$ with codimension 1 as a cyclide of Dupin (see Theorem 8.6), and no other codimension is possible. The manifold $S^1 \times \mathbb{RP}^2$ can be tautly embedded with codimension 3 as the product of a metric circle and a Veronese surface. It can be tautly embedded with codimension 2 as a rotational submanifold with profile submanifold $\mathbb{RP}^2$, and the only codimensions possible are 2 and 3. Finally, $S^1 \times_h S^2$ can be tautly embedded with codimension 2 as the "complexified unit sphere"

$$\{e^{i\theta}x : \theta \in \mathbb{R}, \, x \in S^2 \subset \mathbb{R}^3\} \subset S^5 \subset \mathbb{C}^3.$$

This is one of the focal submanifolds of a homogeneous family of isoparametric hypersurfaces with four principal curvatures in S^5, the other being a Stiefel manifold $V_{2,3}$ (see [Cecil and Ryan 1985, pp. 299–304], for example). No other codimensions are possible for a taut embedding of $S^1 \times_h S^2$.

As we have seen, many examples of taut embeddings, such as the R-spaces, are homogeneous spaces. However, the question of which homogeneous spaces admit taut embeddings remains open. Thorbergsson [1988] found some necessary topological conditions for the existence of a taut embedding, which allowed him to prove that certain homogeneous spaces do not admit taut embeddings. In the same vein, Hebda [1988] found certain necessary cohomological conditions for the existence of a taut embedding, and he used these results to give examples of manifolds that cannot be tautly embedded. In the case where M is a compact homogeneous submanifold substantially embedded in Euclidean space with flat normal bundle, Olmos [1994] has shown that the following statements are equivalent: (i) M is taut; (ii) M is Dupin; (iii) M has constant principal curvatures; (iv) M is an orbit of the isotropy representation of a symmetric space; (v) the first normal space of M coincides with the normal space.

There have been various generalizations of tautness in terms of the distance functions used and the ambient space considered. Carter, Mansour, and West [Carter et al. 1982] introduced a notion of k-cylindrical taut immersion $f : M \to \mathbb{R}^n$ by using distance functions from k-planes in $\mathbb{R}^n$ (see also [Carter and Şentürk 1994; Carter and West 1985a]). For $k = 0$, this is equivalent to tautness, and for $k = n - 1$ it is equivalent to tightness. This theory turns out to closely related to the theory of convex sets and many of the results concern embeddings of spheres. (See also [Wegner 1984] for more on cylindrical distance functions.)

There is also a theory of tight and taut immersions into hyperbolic space H^n [Cecil and Ryan 1979a; 1979b]. This involves consideration of three types of distance functions on $\mathbb{H}^n$ whose level sets are respectively spheres, horospheres, and hypersurfaces equidistant from a hyperplane. Buyske [1989] introduced a notion of tautness in semi-Riemannian space forms and treated the issue of tightness of focal sets of certain taut submanifolds obtained by Lie sphere transformations.

Beltagy [1986] extended the TPP and STPP to subsets of a general Riemannian manifold without focal points. Finally, in a paper published in this volume, Terng and Thorbergsson [1997] have extended the notion of tautness to submanifolds of complete Riemannian manifolds.

Acknowledgements

I thank Peter Breuer, Wolfgang Kühnel, Chuu-Lian Terng, and Gudlaugur Thorbergsson for their comments on earlier versions of this paper.

References

[Abresch 1983] U. Abresch, "Isoparametric hypersurfaces with four or six distinct principal curvatures. Necessary conditions on the multiplicities", *Math. Ann.* **264**:3 (1983), 283–302.

[Banchoff 1970] T. F. Banchoff, "The spherical two-piece property and tight surfaces in spheres", *J. Differential Geometry* **4** (1970), 193–205.

[Banchoff and Kühnel 1997] T. F. Banchoff and W. Kühnel, "Tight submanifolds, smooth and polyhedral", pp. 51–118 in *Tight and Taut Submanifolds*, edited by T. E. Cecil and S.-s. Chern, Cambridge U. Press, 1997.

[Beltagy 1986] M. A. Beltagy, "Two-piece property in manifolds without focal points", *Indian J. Pure Appl. Math.* **17**:7 (1986), 883–889.

[Bott and Samelson 1958] R. Bott and H. Samelson, "Applications of the theory of Morse to symmetric spaces", *Amer. J. Math.* **80** (1958), 964–1029. Corrections in **83** (1961), pp. 207–208.

[Buyske 1989] S. G. Buyske, "Lie sphere transformations and the focal sets of certain taut immersions", *Trans. Amer. Math. Soc.* **311**:1 (1989), 117–133.

[Cartan 1938] E. Cartan, "Familles de surfaces isoparamétriques dans les espaces à courbure constante", *Annali di Mat.* **17** (1938), 177–191.

[Cartan 1939a] E. Cartan, "Sur des familles remarquables d'hypersurfaces isoparamétriques dans les espaces sphériques", *Math. Z.* **45** (1939), 335–367.

[Cartan 1939b] E. Cartan, "Sur quelques familles remarquables d'hypersurfaces", pp. 30–41 in *Comptes rendus du congrès des sciences mathématiques de Liège*, Georges Thone, Liège, 1939.

[Cartan 1940] E. Cartan, "Sur des familles d'hypersurfaces isoparamétriques des espaces sphériques à 5 et à 9 dimensions", *Revista Univ. Tucumán Ser. A* **1** (1940), 5–22.

[Carter and Şentürk 1994] S. Carter and Z. Şentürk, "The space of immersions parallel to a given immersion", *J. London Math. Soc.* (2) **50**:2 (1994), 404–416.

[Carter and West 1972] S. Carter and A. West, "Tight and taut immersions", *Proc. London Math. Soc.* (3) **25** (1972), 701–720.

[Carter and West 1978] S. Carter and A. West, "Totally focal embeddings", *J. Diff. Geom.* **13** (1978), 251–261.

[Carter and West 1981] S. Carter and A. West, "Totally focal embeddings: special cases", *J. Diff. Geom.* **16** (1981), 685–697.

[Carter and West 1982] S. Carter and A. West, "A characterisation of isoparametric hypersurfaces in spheres", *J. London Math. Soc.* (2) **26** (1982), 183–192.

[Carter and West 1985a] S. Carter and A. West, "Convexity and cylindrical two-piece properties", *Illinois J. Math.* **29**:1 (1985), 39–50.

[Carter and West 1985b] S. Carter and A. West, "Isoparametric systems and transnormality", *Proc. London Math. Soc.* (3) **51**:3 (1985), 520–542.

[Carter and West 1990] S. Carter and A. West, "Isoparametric and totally focal submanifolds", *Proc. London Math. Soc.* (3) **60**:3 (1990), 609–624.

[Carter et al. 1982] S. Carter, N. G. Mansour, and A. West, "Cylindrically taut immersions", *Math. Ann.* **261** (1982), 133–139.

[Cayley 1873] A. Cayley, "On the cyclide", *Quart. J. of Pure and Appl. Math.* **12** (1873), 148–165. Reprinted as pp. 64–78 in *Collected mathematical papers*, vol. 9, Cambridge U. Press, 1896.

[Cecil 1974] T. E. Cecil, "A characterization of metric spheres in hyperbolic space by Morse theory", *Tôhoku Math. J.* (2) **26** (1974), 341–351.

[Cecil 1976] T. E. Cecil, "Taut immersions of noncompact surfaces into a Euclidean 3-space", *J. Differential Geometry* **11**:3 (1976), 451–459.

[Cecil 1989] T. E. Cecil, "Reducible Dupin submanifolds", *Geom. Dedicata* **32**:3 (1989), 281–300.

[Cecil 1990] T. E. Cecil, "On the Lie curvature of Dupin hypersurfaces", *Kodai Math. J.* **13**:1 (1990), 143–153.

[Cecil 1991] T. E. Cecil, "Lie sphere geometry and Dupin submanifolds", pp. 90–107 in *Geometry and topology of submanifolds, III* (Leeds, 1990), edited by L. Verstraelen and A. West, World Sci. Publishing, River Edge, NJ, 1991.

[Cecil 1992] T. E. Cecil, *Lie sphere geometry, with applications to submanifolds*, Universitext, Springer, New York, 1992.

[Cecil and Chern 1987] T. E. Cecil and S.-s. Chern, "Tautness and Lie sphere geometry", *Math. Ann.* **278**:1-4 (1987), 381–399.

[Cecil and Chern 1989] T. E. Cecil and S.-s. Chern, "Dupin submanifolds in Lie sphere geometry", pp. 1–48 in *Differential geometry and topology* (Tianjin, 1986–87), edited by B. Jiang et al., Lecture Notes in Math. **1369**, Springer, Berlin, 1989.

[Cecil and Jensen 1997] T. E. Cecil and G. R. Jensen, "Dupin hypersurfaces with three principal curvatures", 1997. To appear in *Invent. Math.*

[Cecil and Ryan 1978] T. E. Cecil and P. J. Ryan, "Focal sets, taut embeddings and the cyclides of Dupin", *Math. Ann.* **236** (1978), 177–190.

[Cecil and Ryan 1979a] T. E. Cecil and P. J. Ryan, "Distance functions and umbilic submanifolds of hyperbolic space", *Nagoya Math. J.* **74** (1979), 67–75.

[Cecil and Ryan 1979b] T. E. Cecil and P. J. Ryan, "Tight and taut immersions into hyperbolic space", *J. London Math. Soc.* (2) **19** (1979), 561–572.

[Cecil and Ryan 1980] T. E. Cecil and P. J. Ryan, "Conformal geometry and the cyclides of Dupin", *Can. J. Math.* **32** (1980), 767–782.

[Cecil and Ryan 1981] T. E. Cecil and P. J. Ryan, "Tight spherical embeddings", pp. 94–104 in *Global Differential Geometry and Global Analysis* (Berlin, 1979), edited by D. Ferus et al., Lecture Notes in Math. **838**, Springer, Berlin, 1981.

[Cecil and Ryan 1985] T. E. Cecil and P. J. Ryan, *Tight and taut immersions of manifolds*, Research Notes in Mathematics **107**, Pitman, Boston, 1985.

[Chern 1991] S.-s. Chern, "An introduction to Dupin submanifolds", pp. 95–102 in *Differential geometry: A symposium in honour of Manfredo do Carmo* (Rio de Janeiro, 1988), edited by H. B. Lawson and K. Tenenblat, Pitman Monographs Surveys Pure Appl. Math. **52**, Longman Sci. Tech., Harlow, 1991.

[Curtin 1991] E. Curtin, "Manifolds with intermediate tautness", *Geom. Dedicata* **38**:3 (1991), 245–255.

[Curtin 1994] E. Curtin, "Tautness for manifolds with boundary", *Houston J. Math.* **20**:3 (1994), 409–424.

[Degen 1994] W. Degen, "Generalized cyclides for use in CAGD", pp. 349–363 in *Computer-aided surface geometry and design: the mathematics of surfaces IV* (Bath, 1990), edited by A. Bowyer, Inst. Math. Appl. Conf. (New) Series **48**, Oxford U. Press, New York, 1994.

[Dorfmeister and Neher 1983a] J. Dorfmeister and E. Neher, "An algebraic approach to isoparametric hypersurfaces I, II", *Tôhoku Math. J.* **35** (1983), 187–224, 225–247.

[Dorfmeister and Neher 1983b] J. Dorfmeister and E. Neher, "Isoparametric triple systems of algebra type", *Osaka J. Math.* **20**:1 (1983), 145–175.

[Dorfmeister and Neher 1983c] J. Dorfmeister and E. Neher, "Isoparametric triple systems of FKM-type I", *Abh. Math. Sem. Univ. Hamburg* **53** (1983), 191–216.

[Dorfmeister and Neher 1983d] J. Dorfmeister and E. Neher, "Isoparametric triple systems of FKM-type II", *Manuscripta Math.* **43**:1 (1983), 13–44.

[Dorfmeister and Neher 1985] J. Dorfmeister and E. Neher, "Isoparametric hypersurfaces, case $g = 6$, $m = 1$", *Comm. Algebra* **13**:11 (1985), 2299–2368.

[Dorfmeister and Neher 1990] J. Dorfmeister and E. Neher, "Isoparametric triple systems with special **Z**-structure", *Algebras Groups Geom.* **7**:1 (1990), 21–94.

[Dupin 1822] P. C. F. Dupin, *Applications de géométrie et de méchanique: à la marine, aux ponts et chaussées, etc.*, Bachelier, Paris, 1822.

[Eschenburg and Schroeder 1991] J.-H. Eschenburg and V. Schroeder, "Tits distance of Hadamard manifolds and isoparametric hypersurfaces", *Geom. Dedicata* **40**:1 (1991), 97–101.

[Fang 1995a] F. Fang, "On the topology of isoparametric hypersurfaces with four distinct principal curvatures", preprint, Nankai Institute of Mathematics, 1995.

[Fang 1995b] F. Fang, "Topology of Dupin hypersurfaces with six principal curvatures", preprint, University of Bielefeld, 1995.

[Fang 1996] F. Fang, "Multiplicities of principal curvatures of isoparametric hypersurfaces", preprint, Max-Planck–Institut für Mathematik, Bonn, 1996.

[Ferapontov 1995a] E. V. Ferapontov, "Dupin hypersurfaces and integrable Hamiltonian systems of hydrodynamic type, which do not possess Riemann invariants", *Differential Geom. Appl.* **5**:2 (1995), 121–152.

[Ferapontov 1995b] E. V. Ferapontov, "Isoparametric hypersurfaces in spheres, integrable nondiagonalizable systems of hydrodynamic type, and N-wave systems", *Differential Geom. Appl.* **5**:4 (1995), 335–369.

[Ferus 1982] D. Ferus, "The tightness of extrinsic symmetric submanifolds", *Math. Z.* **181** (1982), 563–565.

[Ferus et al. 1981] D. Ferus, H. Karcher, and H. F. Münzner, "Cliffordalgebren und neue isoparametrische Hyperflächen", *Math. Z.* **177** (1981), 479–502.

[Grove and Halperin 1987] K. Grove and S. Halperin, "Dupin hypersurfaces, group actions and the double mapping cylinder", *J. Differential Geom.* **26**:3 (1987), 429–459.

[Hahn 1984] J. Hahn, "Isoparametric hypersurfaces in the pseudo-Riemannian space forms", *Math. Z.* **187**:2 (1984), 195–208.

[Hahn 1988] J. Hahn, "Isotropy representations of semisimple symmetric spaces and homogeneous hypersurfaces", *J. Math. Soc. Japan* **40**:2 (1988), 271–288.

[Harle 1982] C. E. Harle, "Isoparametric families of submanifolds", *Bol. Soc. Brasil. Mat.* **13**:2 (1982), 35–48.

[Hebda 1981] J. J. Hebda, "Manifolds admitting taut hyperspheres", *Pacific J. Math.* **97** (1981), 119–124.

[Hebda 1984] J. J. Hebda, "Some new tight embeddings which cannot be made taut", *Geom. Dedicata* **17**:1 (1984), 49–60.

[Hebda 1988] J. J. Hebda, "The possible cohomology of certain types of taut submanifolds", *Nagoya Math. J.* **111** (1988), 85–97.

[Heintze et al. 1991] E. Heintze, C. Olmos, and G. Thorbergsson, "Submanifolds with constant principal curvatures and normal holonomy groups", *Internat. J. Math.* **2**:2 (1991), 167–175.

[Hsiang et al. 1988] W.-Y. Hsiang, R. S. Palais, and C.-L. Terng, "The topology of isoparametric submanifolds", *J. Differential Geom.* **27**:3 (1988), 423–460.

[Ivey 1995] T. Ivey, "Surfaces with orthogonal families of circles", *Proc. Amer. Math. Soc.* **123**:3 (1995), 865–872.

[Jorge and Mercuri 1984] L. P. Jorge and F. Mercuri, "Minimal immersions into space forms with two principal curvatures", *Math. Z.* **187**:3 (1984), 325–333.

[Kashani 1992] S. M. B. Kashani, "On quadratic isoparametric submanifolds", *Bull. Iranian Math. Soc.* **18**:2 (1992), 31–39.

[Kashani 1993a] S. M. B. Kashani, "Isoparametric functions and submanifolds", *Glasgow Math. J.* **35**:2 (1993), 145–152.

[Kashani 1993b] S. M. B. Kashani, "Quadratic isoparametric systems in $\mathbb{R}_p^{n+m}$", *Glasgow Math. J.* **35**:2 (1993), 135–143.

[Kobayashi 1967] S. Kobayashi, "Imbeddings of homogeneous spaces with minimum total curvature", *Tôhoku Math. J.* (2) **19** (1967), 63–70.

[Kühnel 1994] W. Kühnel, "Tensor products of spheres", pp. 106–109 in *Geometry and topology of submanifolds, VI* (Leuven/Brussels, 1993), edited by F. Dillen et al., World Scientific, River Edge, NJ, 1994.

[Kuiper 1962] N. H. Kuiper, "On convex maps", *Nieuw Arch. Wisk.* (3) **10** (1962), 147–164.

[Kuiper 1970] N. H. Kuiper, "Minimal total absolute curvature for immersions", *Invent. Math.* **10** (1970), 209–238.

[Kuiper 1980] N. H. Kuiper, "Tight embeddings and maps: Submanifolds of geometrical class three in E^{nn}", pp. 97–145 in *The Chern Symposium* (Berkeley, 1979), edited by W.-Y. Hsiang et al., Springer, New York, 1980.

[Kuiper 1984] N. H. Kuiper, "Taut sets in three space are very special", *Topology* **23**:3 (1984), 323–336.

[Lancaster 1973] G. M. Lancaster, "Canonical metrics for certain conformally Euclidean spaces of dimension three and codimension one", *Duke Math. J.* **40** (1973), 1–8.

[Levi-Civita 1937] T. Levi-Civita, "Famiglie di superficie isoparametrische nell'ordinario spacio euclideo", *Atti Accad. Naz. Lincei Rend. Cl. Sci. Fis. Mat. Natur.* **26** (1937), 355–362.

[Liouville 1847] J. Liouville, "Note au sujet de l'article précedént", *J. de Math. Pure et Appl.* (1) **12** (1847), 265–290.

[Little and Pohl 1971] J. A. Little and W. F. Pohl, "On tight immersions of maximal codimension", *Invent. Math.* **13** (1971), 179–204.

[Magid 1985] M. A. Magid, "Lorentzian isoparametric hypersurfaces", *Pacific J. Math.* **118**:1 (1985), 165–197.

[Maxwell 1867] J. C. Maxwell, "On the cyclide", *Quart. J. of Pure and Appl. Math.* **34** (1867). Reprinted as pp. 144–159 in *Scientific papers*, vol. 2, Cambridge U. Press, 1890.

[Milnor 1963] J. W. Milnor, *Morse Theory*, Ann. Math. Stud. **51**, Princeton U. Press, 1963.

[Miyaoka 1984a] R. Miyaoka, "Compact Dupin hypersurfaces with three principal curvatures", *Math. Z.* **187**:4 (1984), 433–452.

[Miyaoka 1984b] R. Miyaoka, "Taut embeddings and Dupin hypersurfaces", pp. 15–23 in *Differential geometry of submanifolds* (Kyoto, 1984), edited by K. Kenmotsu, Lecture Notes in Math. **1090**, Springer, Berlin, 1984.

[Miyaoka 1989a] R. Miyaoka, "Dupin hypersurfaces and a Lie invariant", *Kodai Math. J.* **12**:2 (1989), 228–256.

[Miyaoka 1989b] R. Miyaoka, "Dupin hypersurfaces with six principal curvatures", *Kodai Math. J.* **12**:3 (1989), 308–315.

[Miyaoka 1991a] R. Miyaoka, "Lie contact structures and conformal structures", *Kodai Math. J.* **14**:1 (1991), 42–71.

[Miyaoka 1991b] R. Miyaoka, "Lie contact structures and normal Cartan connections", *Kodai Math. J.* **14**:1 (1991), 13–41.

[Miyaoka 1993a] R. Miyaoka, "The linear isotropy group of $G_2/SO(4)$, the Hopf fibering and isoparametric hypersurfaces", *Osaka J. Math.* **30**:2 (1993), 179–202.

[Miyaoka 1993b] R. Miyaoka, "A note on Lie contact manifolds", pp. 169–187 in *Progress in differential geometry*, Adv. Stud. Pure Math., 22, Math. Soc. Japan, Tokyo, 1993.

[Miyaoka and Ozawa 1989]   R. Miyaoka and T. Ozawa, "Construction of taut embeddings and Cecil–Ryan conjecture", pp. 181–189 in *Geometry of manifolds* (Matsumoto, 1988), edited by K. Shiohama, Perspect. Math. **8**, Academic Press, Boston, MA, 1989.

[Morse and Cairns 1969]   M. Morse and S. S. Cairns, *Critical point theory in global analysis and differential topology: An introduction*, Pure and Applied Mathematics **33**, Academic Press, New York, 1969.

[Mullen 1994]   S. Mullen, "Isoparametric systems on symmetric spaces", pp. 152–154 in *Geometry and topology of submanifolds, VI* (Leuven/Brussels, 1993), edited by F. Dillen et al., World Sci. Publishing, River Edge, NJ, 1994.

[Münzner 1980]   H. F. Münzner, "Isoparametrische Hyperflächen in Sphären", *Math. Ann.* **251** (1980), 57–71.

[Münzner 1981]   H. F. Münzner, "Isoparametrische Hyperflächen in Sphären II: Über die Zerlegung der Sphäre in Ballbündel", *Math. Ann.* **256** (1981), 215–232.

[Niebergall 1991]   R. Niebergall, "Dupin hypersurfaces in R^5 I", *Geom. Dedicata* **40**:1 (1991), 1–22.

[Niebergall 1992]   R. Niebergall, "Dupin hypersurfaces in $\mathbb{R}^5$ II", *Geom. Dedicata* **41**:1 (1992), 5–38.

[Niebergall and Ryan 1993]   R. Niebergall and P. J. Ryan, "Isoparametric hypersurfaces—the affine case", pp. 201–214 in *Geometry and topology of submanifolds, V* (Leuven/Brussels, 1992), edited by F. Dillen et al., World Sci. Publishing, River Edge, NJ, 1993.

[Niebergall and Ryan 1994a]   R. Niebergall and P. J. Ryan, "Affine isoparametric hypersurfaces", *Math. Z.* **217**:3 (1994), 479–485.

[Niebergall and Ryan 1994b]   R. Niebergall and P. J. Ryan, "Focal sets in affine geometry", pp. 155–164 in *Geometry and topology of submanifolds, VI* (Leuven/Brussels, 1993), edited by I. V. d. W. Franki Dillen and L. Verstraelen, World Sci. Publishing, River Edge, NJ, 1994.

[Niebergall and Ryan 1996]   R. Niebergall and P. J. Ryan, "Affine Dupin surfaces", *Trans. Amer. Math. Soc.* **348**:3 (1996), 1093–1115.

[Niebergall and Ryan 1997]   R. Niebergall and P. J. Ryan, "Real hypersurfaces in complex space forms", pp. 233–305 in *Tight and Taut Submanifolds*, edited by T. E. Cecil and S.-s. Chern, Cambridge U. Press, 1997.

[Nomizu 1973]   K. Nomizu, "Some results in E. Cartan's theory of isoparametric families of hypersurfaces", *Bull. Amer. Math. Soc.* **79** (1973), 1184–1188.

[Nomizu 1975]   K. Nomizu, "Élie Cartan's work on isoparametric families of hypersurfaces", pp. 191–200 in *Differential geometry* (Stanford, 1973), vol. 1, edited by S.-s. Chern and R. Osserman, Proc. Sympos. Pure Math. **27**, Amer. Math. Soc., Providence, 1975.

[Nomizu 1981]   K. Nomizu, "On isoparametric hypersurfaces in Lorentzian space forms", *Jap. J. Math.* (N.S.) **7** (1981), 217–226.

[Nomizu and Rodríguez 1972]   K. Nomizu and L. L. Rodríguez, "Umbilical submanifolds and Morse functions", *Nagoya Math. J.* **48** (1972), 197–201.

[Olmos 1993]   C. Olmos, "Isoparametric submanifolds and their homogeneous structures", *J. Differential Geom.* **38**:2 (1993), 225–234.

[Olmos 1994] C. Olmos, "Homogeneous submanifolds of higher rank and parallel mean curvature", *J. Differential Geom.* **39**:3 (1994), 605–627.

[Ozawa 1986] T. Ozawa, "On critical sets of distance functions to a taut submanifold", *Math. Ann.* **276**:1 (1986), 91–96.

[Ozeki and Takeuchi 1975] H. Ozeki and M. Takeuchi, "On some types of isoparametric hypersurfaces in spheres I", *Tôhoku Math. J.* (2) **27**:4 (1975), 515–559.

[Ozeki and Takeuchi 1976] H. Ozeki and M. Takeuchi, "On some types of isoparametric hypersurfaces in spheres II", *Tôhoku Math. J.* (2) **28**:1 (1976), 7–55.

[Palais and Terng 1987] R. S. Palais and C.-L. Terng, "A general theory of canonical forms", *Trans. Amer. Math. Soc.* **300**:2 (1987), 771–789.

[Palais and Terng 1988] R. S. Palais and C.-L. Terng, *Critical point theory and submanifold geometry*, Lecture Notes in Math. **1353**, Springer, Berlin, 1988.

[Peng and Hou 1989] C. K. Peng and Z. X. Hou, "A remark on the isoparametric polynomials of degree 6", pp. 222–224 in *Differential geometry and topology* (Tianjin, 1986–87), edited by B. Jiang et al., Lecture Notes in Math. **1369**, Springer, Berlin, 1989.

[Pinkall 1981] U. Pinkall, *Dupinsche Hyperflächen*, Ph.D. thesis, Univ. Freiburg, 1981.

[Pinkall 1985a] U. Pinkall, "Dupin hypersurfaces", *Math. Ann.* **270**:3 (1985), 427–440.

[Pinkall 1985b] U. Pinkall, "Dupinsche Hyperflächen in E^4", *Manuscripta Math.* **51**:1-3 (1985), 89–119.

[Pinkall 1986] U. Pinkall, "Curvature properties of taut submanifolds", *Geom. Dedicata* **20**:1 (1986), 79–83.

[Pinkall and Thorbergsson 1989a] U. Pinkall and G. Thorbergsson, "Deformations of Dupin hypersurfaces", *Proc. Amer. Math. Soc.* **107**:4 (1989), 1037–1043.

[Pinkall and Thorbergsson 1989b] U. Pinkall and G. Thorbergsson, "Taut 3-manifolds", *Topology* **28**:4 (1989), 389–401.

[Pratt 1990] M. J. Pratt, "Cyclides in computer aided geometric design", *Computer Aided Geometric Design* **7** (1990), 221–242.

[Pratt 1995] M. J. Pratt, "Cyclides in computer aided geometric design II", *Comput. Aided Geom. Design* **12** (1995), 131–152.

[Reckziegel 1979] H. Reckziegel, "On the eigenvalues of the shape operator of an isometric immersion into a space of constant curvature", *Math. Ann.* **243** (1979), 71–82.

[Schouten 1921] J. A. Schouten, "Über die konforme Abbildung n-dimensionaler Mannigfaltigkeiten mit quadratischer Maßbestimmung auf eine Mannigfaltigkeit mit euklidischer Maßbestimmung", *Math. Z.* **11** (1921), 58–88.

[Segre 1938] B. Segre, "Famiglie di ipersuperficie isoparametrische negli spazi euclidei ad un qualunque numero di dimensioni", *Atti Accad. Naz. Lincei Rend. Cl. Sci. Fis. Mat. Natur.* **27** (1938), 203–207.

[Singley 1975] D. H. Singley, "Smoothness theorems for the principal curvatures and principal vectors of a hypersurface", *Rocky Mountain J. Math.* **5** (1975), 135–144.

[Solomon 1990a] B. Solomon, "The harmonic analysis of cubic isoparametric minimal hypersurfaces, I: Dimensions 3 and 6", *Amer. J. Math.* **112**:2 (1990), 157–203.

[Solomon 1990b] B. Solomon, "The harmonic analysis of cubic isoparametric minimal hypersurfaces, II: Dimensions 12 and 24", *Amer. J. Math.* **112**:2 (1990), 205–241.

[Solomon 1992] B. Solomon, "Quartic isoparametric hypersurfaces and quadratic forms", *Math. Ann.* **293**:3 (1992), 387–398.

[Srinivas and Dutta 1994a] Y. L. Srinivas and D. Dutta, "Blending and joining using cyclides", *ASME Trans. Journal of Mechanical Design* **116** (1994), 1034–1041.

[Srinivas and Dutta 1994b] Y. L. Srinivas and D. Dutta, "An intuitive procedure for constructing complex objects using cyclides", *Computer-Aided Design* **26** (1994), 327–335.

[Srinivas and Dutta 1995a] Y. L. Srinivas and D. Dutta, "Cyclides in geometric modeling: Computational tools for an algorithmic infrastructure", *ASME Trans. Journal of Mechanical Design* **117** (1995), 363–373.

[Srinivas and Dutta 1995b] Y. L. Srinivas and D. Dutta, "Rational parametric representation of parabolic cyclide: formulation and applications", *Comput. Aided Geom. Design* **12**:6 (1995), 551–566.

[Strübing 1986] W. Strübing, "Isoparametric submanifolds", *Geom. Dedicata* **20**:3 (1986), 367–387.

[Tai 1968] S.-S. Tai, "Minimum imbeddings of compact symmetric spaces of rank one", *J. Diff. Geom.* **2** (1968), 55–66.

[Takeuchi 1991] M. Takeuchi, "Proper Dupin hypersurfaces generated by symmetric submanifolds", *Osaka J. Math.* **28**:1 (1991), 153–161.

[Takeuchi and Kobayashi 1968] M. Takeuchi and S. Kobayashi, "Minimal imbeddings of R-spaces", *J. Differential Geometry* **2** (1968), 203–215.

[Tang 1991] Z. Z. Tang, "Isoparametric hypersurfaces with four distinct principal curvatures", *Chinese Sci. Bull.* **36**:15 (1991), 1237–1240.

[Terng 1985] C.-L. Terng, "Isoparametric submanifolds and their Coxeter groups", *J. Differential Geom.* **21**:1 (1985), 79–107.

[Terng 1987] C.-L. Terng, "Submanifolds with flat normal bundle", *Math. Ann.* **277**:1 (1987), 95–111.

[Terng 1993] C.-L. Terng, "Recent progress in submanifold geometry", pp. 439–484 in *Differential geometry: partial differential equations on manifolds* (Los Angeles, 1990), edited by R. Greene and S. T. Yau, Proc. Sympos. Pure Math. **54** (part 1), Amer. Math. Soc., Providence, RI, 1993.

[Terng and Thorbergsson 1995] C.-L. Terng and G. Thorbergsson, "Submanifold geometry in symmetric spacs", *J. Differential Geom.* **42**:3 (1995), 665–718.

[Terng and Thorbergsson 1997] C.-L. Terng and G. Thorbergsson, "Taut immersions into complete Riemannian manifolds", pp. 181–228 in *Tight and Taut Submanifolds*, edited by T. E. Cecil and S.-s. Chern, Cambridge U. Press, 1997.

[Thorbergsson 1983a] G. Thorbergsson, "Dupin hypersurfaces", *Bull. London Math. Soc.* (2) **15**:5 (1983), 493–498.

[Thorbergsson 1983b] G. Thorbergsson, "Highly connected taut submanifolds", *Math. Ann.* **265**:3 (1983), 399–405.

[Thorbergsson 1986] G. Thorbergsson, *Geometrie hochzusammenhängender Unterman-nigfaltigkeiten*, Bonner Mathematische Schriften **170**, Universität Bonn Math. Insti-tut, Bonn, 1986. Habilitationsschrift, Friedrich-Wilhelms-Universität, Bonn, 1985.

[Thorbergsson 1988] G. Thorbergsson, "Homogeneous spaces without taut embed-dings", *Duke Math. J.* **57**:1 (1988), 347–355.

[Thorbergsson 1991] G. Thorbergsson, "Isoparametric foliations and their buildings", *Ann. of Math.* (2) **133**:2 (1991), 429–446.

[Verhóczki 1992] L. Verhóczki, "Isoparametric submanifolds of general Riemannian manifolds", pp. 691–705 in *Differential geometry and its applications* (Eger, 1989), edited by J. Szenthe and L. Tamássy, Colloq. Math. Soc. János Bolyai **56**, North-Holland, Amsterdam, 1992.

[Wang 1986] Q. M. Wang, "Isoparametric maps of Riemannian manifolds and their applications", pp. 79–103 in *Advances in science of China. Mathematics*, vol. 2, edited by C. H. Gu and Y. Wang, Wiley, New York, 1986.

[Wang 1987] Q. M. Wang, "Isoparametric functions on Riemannian manifolds I", *Math. Ann.* **277**:4 (1987), 639–646.

[Wang 1988] Q. M. Wang, "On the topology of Clifford isoparametric hypersurfaces", *J. Differential Geom.* **27**:1 (1988), 55–66.

[Wang 1992] C. P. Wang, "Surfaces in Möbius geometry", *Nagoya Math. J.* **125** (1992), 53–72.

[Wang 1995] C. P. Wang, "Möbius geometry for hypersurfaces in S^4", *Nagoya Math. J.* **139** (1995), 1–20.

[Wegner 1984] B. Wegner, "Morse theory for distance functions to affine subspaces of Euclidean spaces", pp. 165–168 in *Proceedings of the conference on differential geometry and its applications. Part 1, Differential Geometry* (Nové Město na Moravě, 1983), edited by O. Kowalski, Charles Univ., Prague, 1984.

[West 1989] A. West, "Isoparametric systems", pp. 222–230 in *Geometry and topology of submanifolds* (Marseille, 1987), edited by J.-M. Morvan and L. Verstraelen, World Sci. Publishing, Teaneck, NJ, 1989.

[West 1993] A. West, "Isoparametric systems on symmetric spaces", pp. 281–287 in *Geometry and topology of submanifolds, V* (Leuven/Brussels, 1992), edited by F. Dillen et al., World Sci. Publishing, River Edge, NJ, 1993.

[Wilson 1969] J. P. Wilson, "Some minimal imbeddings of homogenous spaces", *J. London Math. Soc.* (2) **1** (1969), 335–340.

[Wu 1992] B.-L. Wu, "Isoparametric submanifolds of hyperbolic spaces", *Trans. Amer. Math. Soc.* **331**:2 (1992), 609–626.

[Wu 1994] B.-L. Wu, "A finiteness theorem for isoparametric hypersurfaces", *Geom. Dedicata* **50**:3 (1994), 247–250.

[Zhao 1993] Q. Zhao, "Isoparametric submanifolds of hyperbolic space", *Chinese J. Contemp. Math.* **14**:4 (1993), 339–346.

THOMAS E. CECIL
DEPARTMENT OF MATHEMATICS
COLLEGE OF THE HOLY CROSS
WORCESTER, MASSACHUSETTS 01610
    cecil@math.holycross.edu

Tight and Taut Submanifolds
MSRI Publications
Volume **32**, 1997

Taut Immersions into Complete Riemannian Manifolds

CHUU-LIAN TERNG AND GUDLAUGUR THORBERGSSON

ABSTRACT. The main purpose of this paper is to propose a natural generalization of the notion of a taut immersion into a complete Riemannian manifold. We explain the motivation behind our definition, give many examples, note some interesting topological and geometric properties of such immersions, and remark on many intriguing relations to other well-known topics in geometry such as transformation groups, Morse theory, Blaschke manifolds and tori without conjugate points.

1. Introduction

Chern and Lashof started the study of tight immersions of compact manifolds into $\mathbb{R}^n$ in the 1950's. Recall that the *total absolute curvature* $\tau(M,\phi)$ of an immersion $\phi : M \to \mathbb{R}^n$ is the volume of the normal bundle map ξ from the unit normal bundle $\nu^1(M)$ to S^{n-1} (defined by $\xi(v) = v$). Since $-d\xi_v$ is equal to the shape operator A_v of M in the direction of v, we can write

$$\tau(M,\phi) = \frac{1}{c_{n-1}} \int_{\nu^1(M)} |\det(A_v)| \, dv,$$

where dv is the natural volume element on $\nu^1(M)$ and c_{n-1} is the volume of the unit sphere S^{n-1}. Chern and Lashof [1957; 1958] proved that

$$\tau(M,\phi) \geq \sum_i b_i(M),$$

where $b_i(M)$ is the i-th Betti number of M with respect to $\mathbb{Z}_2$. The number

$$\tau(M) = \inf_\phi \tau(M,\phi)$$

is clearly a differential invariant of M. An immersion $\phi : M \to \mathbb{R}^n$ is called *an immersion with minimal total absolute curvature* if $\tau(M,\phi) = \tau(M)$. The main problem studied by Chern and Lashof was to characterize such immersions. For

---

Terng's research is supported in part by NSF grant DMS-9304285.

example, they proved that an immersion of S^{n-1} into $\mathbb{R}^n$ with minimal total absolute curvature must be a convex hypersurface.

Kuiper [1959] reformulated the study of immersions with minimal total absolute curvature in terms of the Morse theory of height functions. First recall that the Morse number $\mu(M)$ of M is defined to be the minimum number of critical points a Morse function on M can have. It follows from the Morse inequalities that $\mu(M) \geq \sum_i b_i(M)$. Chern and Lashof's proof in fact gives $\tau(M) \geq \mu(M)$. Sharpe [1988] proved that $\tau(M) = \mu(M)$ under mild assumptions on the dimensions.

If $\mu(M) = \sum_i b_i(M)$, then for an immersion $\phi : M \to \mathbb{R}^n$ the following statements are equivalent:

(i) ϕ has minimal total absolute curvature.
(ii) $\tau(M, \phi) = \sum_i b_i(M)$.
(iii) For generic $a \in \mathbb{R}^n$ the height function $h_a : M \to \mathbb{R}$ defined by $h_a(x) = \phi(x) \cdot a$ is a perfect Morse function.

(Recall that a Morse function on M is called perfect if the number of index k critical points is equal to the k-th Betti number of M for all k. Here, as always in this paper, homology groups have coefficients in $\mathbb{Z}_2$).

An immersion satisfying conditions (i)–(iii) is called *tight*. If M is a tight submanifold in $\mathbb{R}^n$ and M lies in S^{n-1}, the restriction of the distance squared function $f_a(x) = \|x - a\|^2$ to M is also a perfect Morse function for generic $a \in \mathbb{R}^n$. Banchoff [1970] proved that tight surfaces contained in a round sphere satisfy the so-called spherical two-piece property. This motivated Carter and West [1972] to define a notion of tautness, as follows. A submanifold M of $\mathbb{R}^n$ is *taut* if, for generic $a \in \mathbb{R}^n$, the distance squared function $f_a : M \to \mathbb{R}$ defined by $f_a(x) = \|x - a\|^2$ is perfect, and a submanifold M in S^n is taut if M is taut in $\mathbb{R}^{n+1}$. This is equivalent to saying that M in S^n is taut if a squared spherical distance function f_a of M is a perfect Morse function for a generic a.

There have been much progress and many beautiful results in the study of tight and taut immersions in space forms; see for example [Cecil and Ryan 1985] and the articles in this volume. But there has not been a notion of tautness for submanifolds in arbitrary Riemannian manifolds. One reason is that the function $f_p : N \to \mathbb{R}$, defined by setting $f_p(x)$ equal to the square of the distance from p to x, is not differentiable if the cut locus of p is not empty [Wolter 1979]. This is not a problem if N is S^n, since a submanifold M does not meet the cut locus of p for generic p. But this is not the case in general, and there is no simple and direct way to generalize the notion of tautness.

To explain our definition of a taut immersion into a complete Riemannian manifold, we first review the definitions of focal points, the energy functional on path spaces, and the relations between them.

Let (N, g) be a Riemannian manifold, $\phi : M \to N$ an immersion, and $\nu(M)$ the normal bundle of M. We will assume throughout the paper that all immersions are proper. The *endpoint map* $\eta : \nu(M) \to N$ of M is by definition the restriction of the exponential map $\exp$ to $\nu(M)$. If $v \in \nu(M)_x$ is a singular point of η and the dimension of the kernel of $d\eta_v$ is m, then v is called a *focal normal of multiplicity* m and $\exp(v)$ is called a *focal point of multiplicity* m of M with respect to x in N. When $M = \{p\}$ is a single point, a focal point of multiplicity k of M is called a *conjugate point of order* k. The *focal data*, $\Gamma(M)$, is defined to be the set of all pairs (v, m) such that v is a focal normal of multiplicity m of M. The *focal variety* $\mathcal{V}(M)$ is the set of all pairs $(\eta(v), m)$ with $(v, m) \in \Gamma(M)$.

For $B \subset N \times N$, let $P(N, B)$ denote the set of all H^1-paths γ in N such that $(\gamma(0), \gamma(1)) \in B$. (A path is H^1 if it is absolutely continuous and the norm of its derivative is square integrable.) For a fixed $p \in N$, let $\pi : P(N, N \times p) \to N$ be the fibration defined by $\pi(\gamma) = \gamma(0)$, and let $P(N, \phi \times p)$ denote $\phi^*(P(N, N \times p))$, that is, the space of pairs (q, γ) such that $q \in M$ and γ a H^1-path $\gamma : [0, 1] \to N$ such that $(\gamma(0), \gamma(1)) = (\phi(q), p)$. The space $P(N, \phi \times p)$ is a Hilbert manifold [Palais 1963]. If $M \subset N$ and ϕ is the inclusion map, then $P(N, \phi \times p)$ is diffeomorphic to the space $P(N, M \times p)$. Let

$$E_p : P(N, \phi \times p) \to \mathbb{R}, \qquad E_p(q, \gamma) = \int_0^1 \|\gamma'(t)\|^2 \, dt$$

be the energy functional. Then it is well-known that $(q, \gamma) \in P(N, \phi \times p)$ is a critical point of E_p if and only if γ is a geodesic normal to $\phi(U)$ at $\phi(q) = \gamma(0)$ parametrized proportional to arc length, where U is a neighborhood of q on which ϕ is injective. It is also well-known that E_p is a Morse function if and only if p is not a focal point of M. Notice that we do not require in this paper that the levels of critical points of a Morse function are different, only that all critical points are nondegenerate. The Morse index theorem says that the index of E_p at a critical point (q, γ) is the sum of the integers m such that $\gamma(t)$ is a focal point of multiplicity m of M with respect to q with $0 < t < 1$.

Let μ_k denote the number of critical points of index k of E_p in $P(N, \phi \times p)$, and let b_k denote the k-th Betti number of $P(N, \phi \times p)$. It is known that E_p is bounded below and satisfies the Palais–Smale condition [Palais and Smale 1964]. So μ_k is finite for all k, and the weak Morse inequalities say that $\mu_k \geq b_k$ for all k. The function E_p is called *perfect* if $\mu_k = b_k$ for all k.

We will prove in Section 2 that, if $N = \mathbb{R}^n$ or S^n, a submanifold $\phi : M \to N$ is taut if and only if the energy functional $E_p : P(N, \phi \times p) \to \mathbb{R}$ is perfect for generic $p \in N$. This leads to a natural generalization of the notion of a taut immersion into any complete Riemannian manifold N, namely:

DEFINITION 1.1. An immersion $\phi : M \to N$ of (N, g) is called *taut* if the energy functional $E_p : P(N, \phi \times p) \to \mathbb{R}$ is perfect for every p in N that is not a focal point of M. In particular, a point $q \in N$ is called a *taut point* if $\{q\}$ is a taut submanifold of N, that is, if $E_p : P(N, q \times p) \to \mathbb{R}$ is perfect for every $p \in N$ that is not conjugate to q along some geodesic.

Taut immersions of M in S^n can also be defined in terms of the lower bound of volumes of certain images under the endpoint map. In fact, if M is an immersed submanifold of S^n then

$$\mathrm{vol}(\nu_k(M)) \geq b_k(M)\,\mathrm{vol}(S^n)$$

for all k, and the equalities hold if and only if M is taut in S^n, where $\nu_k(M)$ is the set of $v \in \nu(M)$ such that the foot point of v is a nondegenerate critical point of $f_{\exp(v)}$ with index k and $\nu_k(M)$ is equipped with the metric induced from N via η. A similar statement is true for an immersion $\phi : M \to N$ if N is a Riemannian manifold with finite volume: $\mathrm{vol}(\nu_k(M)) \geq b_k\,\mathrm{vol}(N)$, and the equalities hold if and only if M is taut. Here b_k is the k-th Betti number of $P(N, \phi \times p)$ and $\nu_k(M)$ is the set of $v \in \nu(M)$ such that $\gamma(t) = \exp(tv)$ is a nondegenerate critical point of $E_{\exp(v)}$ with index k.

Before we explain our results concerning taut immersions into arbitrary complete Riemannian manifolds, we review some fundamental properties of taut immersions in $\mathbb{R}^n$:

(1) A taut immersion is an embedding.

(2) Ozawa's theorem: If M is a taut submanifold in $\mathbb{R}^n$ and $a \in \mathbb{R}^n$ is a focal point of M then $f_a : M \to \mathbb{R}$ is a perfect Morse–Bott function. By a *Morse–Bott function* we mean a function with the property that every connected component of the set of critical points is a nondegenerate critical submanifold.

(3) A distance sphere in $\mathbb{R}^n$ is taut, and conversely a taut immersion of S^{n-1} into $\mathbb{R}^n$ must be a distance sphere.

(4) Orbits of the isotropy representation of a symmetric space (s-representation) are taut in Euclidean space. Modeled on these orbits, isoparametric and weakly isoparametric submanifolds of $\mathbb{R}^n$ are introduced and shown to be taut [Terng 1985; 1987; Hsiang et al. 1988]. Recall that a submanifold of $\mathbb{R}^n$ is *isoparametric* [Harle 1982; Carter and West 1985; Terng 1985] if its normal bundle is flat and the principal curvatures along any parallel normal field are constant. A submanifold of $\mathbb{R}^n$ is called *weakly isoparametric* [Terng 1987] if its normal bundle is flat, the principal curvatures along any parallel normal field have constant multiplicities, and the lines of curvatures are standard circles. In fact, principal orbits of s-representations are isoparametric, and isoparametric submanifolds are weakly isoparametric.

The goal of this paper is to investigate whether a taut immersion of M into an arbitrary Riemannian manifold has the above properties. We will explain our results item by item:

Property (1): It will be proved in section 2 that if N is simply connected and $\phi : M \to N$ is a taut immersion, then ϕ is an embedding. If N is not simply connected, ϕ may have self-intersections. For example, let N be a flat n-dimensional torus and $r > 0$ a number slightly bigger than the injectivity radius.

Then the restriction of the exponential map at a point $p \in T^n$ to the sphere of radius r gives a taut immersion of S^{n-1} into T^n with self-intersections.

Property (2): We prove an analogue of Ozawa's theorem in Section 2: If $\phi : M \to N$ is taut then for any focal point p of M in N the energy functional E_p is a perfect Morse–Bott function on $P(N, \phi \times p)$. This is a very useful tool for the study of taut immersions because it allows us to use Kuiper's top set technique and Bott's technique of using the critical submanifolds of E_p to obtain geometric and topological properties of M and N.

Property (3): Assume that N is an n-dimensional simply connected, complete Riemannian manifold that is not a rational homology sphere. We prove in Section 6 that if S^{n-1} is a taut hypersurface in N and S^{n-1} is null-homotopic, then S^{n-1} is a distance sphere. In particular, a null-homotopic taut S^{n-1} in a simply connected symmetric space N^n is a distance sphere.

We also prove that the distance sphere of radius r centered at p is taut in N if and only if p is a taut point of N. So the question whether a distance sphere centered at p in N is taut is equivalent to the question whether p is a taut point in N.

It is obvious that for $\mathbb{R}^n$ and S^n all points are taut. The questions we study in Section 6 are: Is every point of N taut? Is there some taut point in N? The answers are definitely no for both questions for an arbitrary Riemannian manifold N, because if p is a taut point of N then E_p is perfect on the loop space $\Omega(N) = P(N, p \times p)$. This gives a lot of restrictions on the structure of conjugate points of p. But there are many Riemannian manifolds for which all points are taut. For example, we will prove that all points in symmetric spaces and Blaschke manifolds are taut. Roughly speaking, a simply connected manifold N is Blaschke at a point p if the conjugate point data of geodesics starting in p is the same as that of a compact rank one symmetric space. In particular, if N is *Blaschke* at p then there exist $l > 0$ and an integer a such that the first conjugate point along any geodesic ray at p occurs at length l with multiplicity a [Besse 1978]. N is called a *Blaschke manifold* if it is Blaschke at every point. For example, a simply connected rank-one symmetric space is Blaschke. In fact, $a = n - 1$ for S^n, $a = 1$ for $\mathbb{CP}^m$ with $n = 2m$, $a = 3$ for $\mathbb{HP}^m$ with $n = 4m$ and $a = 7$ for $\mathrm{Ca}\mathbb{P}^2$ with $n = 16$. (Here $\mathbb{HP}^m$ is the quaternionic projective space and $\mathrm{Ca}\mathbb{P}$ is the Cayley plane.) Notice that if N is a compact, simply connected rank-one symmetric space then the first three nonzero Betti numbers of $\Omega(N) = P(N, p \times p)$ are b_0, b_a, b_{a+n-1}, and they are all equal to 1.

So the following questions arise naturally: What can one say about the geometry and topology of a Riemannian manifold with all points taut? We have some results concerning this question:

(i) If the first three nonzero Betti numbers of the loop space $\Omega(N)$ of the compact Riemannian manifold N are b_0, b_a, b_{a+n-1}, and they are equal to 1 for some

$1 \leq a \leq n - 1$ and N has a taut point p, then N is Blaschke at p. So it follows from a theorem of Warner [1967] that homologically N is a rank-one symmetric space.

(ii) If the loop space of N is as in (i) and all points of N are taut then N is Blaschke. By work of Sato [1984] and Yang [1990], it follows that N is homeomorphic to a compact rank-one symmetric space.

(iii) If g is a Riemannian metric on S^n such that all points are taut then g must be the standard metric.

(iv) If g is a Riemannian metric on the n-torus T^n such that all points are taut then (T^n, g) is flat.

Property (4): An isometric G-action is called *hyperpolar* if there exists a closed, flat submanifold Σ that meets every G-orbit and meets orthogonally at every intersection point. Using results from [Bott and Samelson 1958; Conlon 1971], it follows that orbits of a hyperpolar action are taut. In fact, the isotropy representations of symmetric spaces are essentially all the hyperpolar actions on Euclidean spaces [Dadok 1985]. There are many hyperpolar actions on symmetric spaces. Hence there exist many homogeneous taut submanifolds in symmetric spaces.

A principal orbit of a hyperpolar action on a symmetric space N has flat normal bundle and its focal data is invariant under the parallel translation with respect to the induced normal connection. Motivated by these orbit examples, we introduced equifocal and weakly equifocal submanifolds in compact symmetric spaces, and proved that they are again taut [Terng and Thorbergsson 1995]. Recall that a submanifold M of a compact symmetric space N is called *equifocal* if the normal bundle is flat and abelian and the focal data is invariant under normal parallel translations. M is called *weakly equifocal* if the normal bundle is flat and abelian, the multiplicity of the k-th focal point along the normal geodesic ray $\exp(tv(x))$ is independent of $x \in M$ for a parallel normal field v, and the focal radius on a focal leaf of multiplicity one is constant. It was proved in [Terng and Thorbergsson 1995] that principal orbits of a hyperpolar action are equifocal, and equifocal submanifolds share many of the geometric and topological properties of the principal orbits of hyperpolar actions. It follows from the definitions that equifocal submanifolds are weakly equifocal.

A hypersurface M in $\mathbb{R}^n$ is *Dupin* if all its lines of curvatures are standard circles, and is *proper Dupin* if it is Dupin and all its principal curvatures have constant multiplicities. Pinkall [1986] proved that taut hypersurfaces are Dupin, and Thorbergsson [1983] proved that proper Dupin hypersurfaces are taut. In this paper, we also define a notion of Dupin submanifold in an arbitrary complete Riemannian manifold. We generalize the above results to symmetric spaces.

This is the beginning of our project on taut immersions into a complete Riemannian manifold. We have obtained some basic results, but many interesting questions remain open.

Our paper is organized as follows: In Section 2, we give general results concerning taut immersions into a complete Riemannian manifold N obtained from Morse theory. In Section 3, we review results on taut submanifolds that are orbits of some isometric actions. In Section 4, we review results on equifocal and weakly equifocal submanifolds in symmetric spaces, and their relation to tautness. In Section 5, we study Dupin submanifolds in complete Riemannian manifolds and submanifolds in Hilbert spaces. In Section 6, we study manifolds with taut points and taut spheres. In the Appendix, we use infinite-dimensional Morse theory to prove some of the results in Section 2.

2. Taut Immersions

In this section, we will prove that our new Definition 1.1 is equivalent to the original definition of tautness if the ambient space is $\mathbb{R}^n$ and S^n. If $\phi : M \to N$ is taut and the ambient space N is simply connected, or more generally if the path space $P(N, \phi \times p)$ is connected, then we will show that ϕ is injective. We will also prove an analogue of Ozawa's theorem for taut immersions into arbitrary Riemannian manifolds. It will again turn out in our more general situation that this result is one of the most important tools in dealing with taut submanifolds, see in particular Section 6.

We start with two propositions that prove that the two definitions of tautness are equivalent when the ambient space is $\mathbb{R}^n$ or S^n. We may assume that the submanifolds are embedded because a taut immersion in $\mathbb{R}^n$ or S^n is an embedding under either definition (see Theorem 2.5 for the case of the new definition).

PROPOSITION 2.1. *Let M be a submanifold of $\mathbb{R}^n$ and let $a \in \mathbb{R}^n$. Then*

(i) *$P(\mathbb{R}^n, M \times a)$ is homotopy equivalent to M, and*

(ii) *M is taut in $\mathbb{R}^n$ with respect to the original definition if and only if M is taut in $\mathbb{R}^n$ with respect to the new definition.*

PROOF. Given $b \in \mathbb{R}^n$, let $l_b(t) = (1 - t)b + ta$ denote the line segment joining b to a. Then $\phi(\gamma) = l_{\gamma(0)}$ defines a deformation retract of $P(\mathbb{R}^n, M \times a)$ to $M^* = \{l_x : x \in M\}$, which is diffeomorphic to M. This proves (i). To prove (ii), we note that $E_a(l_x) = \|x - a\|^2 = f_a(x)$ and γ is a critical point of E_a of index k if and only if $\gamma(0)$ is an index k critical point of f_a. $\qquad \square$

PROPOSITION 2.2. *Let M^m be a submanifold of S^n, set $\tilde{M} = P(S^n, M \times p)$, and let b_k be the k-th Betti number of M. Then*

(i) *the Poincaré polynomial of $\tilde{M}$ is*

$$\left(\sum_{j=0}^{m} b_j t^j \right) \left(\sum_{k=0}^{\infty} t^{k(n-1)} \right);$$

(ii) *M is taut in S^n with respect to the old definition if and only if M is taut in S^n with respect to the new definition.*

PROOF. Let $\pi : P(S^n, S^n \times p) \to S^n$ denote the projection $\pi(\gamma) = \gamma(0)$, and take $p_0 \in S^n \setminus M$. Then $S^n \setminus \{p_0\}$ is contractible, which implies that π is trivial over $S^n \setminus \{p_0\}$. So

$$\tilde{M} = \pi^{-1}(M) \simeq M \times \pi^{-1}(p_0) = M \times P(S^n, p_0 \times p_0).$$

Now (i) follows since the Poincaré polynomial for $P(S^n, p_0 \times p_0)$ is $\sum_{k=0}^{\infty} t^{k(n-1)}$ [Milnor 1963].

It is known that $q \in M$ is a critical point of f_p for $p \in S^n$, where f_p is a squared spherical distance function, if and only if there exists $v \in \nu(M)$ such that $\|v\| < \pi$ and $p = \exp_q(v)$. Given a critical point q of f_p, of index k, and a natural number j, let $\gamma_{q,j}$ denote the geodesic starting from q by going j times around the normal circle in the direction of v and then continuing to p. Then $\gamma_{q,j}$ is a critical point of E_p with index $k + j(n-1)$. Moreover, all critical points of E_p arise this way. By (i), this proves that the number of index r critical points of E_p is equal to $b_r(\tilde{M})$, which proves (ii). $\qquad \square$

Hence there is no ambiguity in the definition of taut immersions, and from now on we will always use the new definition. The following proposition follows immediately from the definition.

PROPOSITION 2.3. *Suppose $\phi : M \to N$ is a taut immersion, and that $f : N \to N$ is an isometry. Then $f \circ \phi : M \to N$ is also a taut immersion.*

Note that the set of focal points of M in N has measure zero and E_p has at least $b_k(P(N, \phi \times p))$ critical points of index k if p is not a focal point of M in N. Hence the following proposition is an easy consequence of the definition.

PROPOSITION 2.4. *Let $\phi : M \to N$ be an immersion, $\eta : \nu(M) \to N$ the endpoint map, and let $\nu_k(M)$ denote the set of $v \in \nu(M)$ such that $\gamma(t) = \exp(tv)$ is a nondegenerate critical point of index k of $E_{\exp(v)}$. If N has finite volume, then*

$$\mathrm{vol}(\nu_k(M)) \geq b_k \, \mathrm{vol}(N),$$

where b_k is the k-th Betti number of $P(N, \phi \times p)$ and $\nu_k(M)$ is equipped with the metric induced from N via η. Moreover, equality holds for all k if and only if M is taut.

THEOREM 2.5. *Suppose $\phi : M \to N$ is a taut immersion. If $P(N, \phi \times p)$ is connected, then ϕ is injective. In particular, if N is simply connected and M connected, then ϕ is injective.*

PROOF. Assume that $P(N, \phi \times p)$ is connected and that ϕ is not injective. Then there is a ball $B_\varepsilon(p)$ in N such that $\phi^{-1}(B_\varepsilon(p))$ is disconnected. We can choose p such that E_p is a Morse function. It follows that E_p has at least two critical points of index zero. This implies that E_p is not perfect, since $P(N, \phi \times p)$ being connected implies that the Betti number b_0 is equal to one. This is a contradiction, so ϕ is injective. The homotopy sequence of the fibration $P(N, \phi \times p) \to M$;

$(q, \gamma) \to q$ implies that $P(N, \phi \times p)$ is connected if N is simply connected and M connected. $\qquad \square$

REMARK 2.6. We will show in Section 6 that if N is a complete Riemannian manifold without conjugate points then $\exp : S_r(0) \to N$ is taut for every $r > 0$, where $S_r(0)$ is the sphere in TM_p centered at 0 with radius r. If N is not simply connected we will thus have noninjective taut immersions of spheres into N. Notice that $P(N, \phi \times p)$ can be connected, although N is not simply connected. An example is given by letting M be a projective subspace of $N = RP^n$ and ϕ the inclusion.

To simplify the notation we will often let $\mathcal{M}_p$ denote the path space $P(N, \phi \times p)$, where $\phi : M \to N$ is an immersion and $p \in N$. We also set

$$\mathcal{M}_p^r = \{(q, \gamma) \in \mathcal{M}_p : E_p(q, \gamma) \leq r\},$$
$$\mathcal{M}_p^{r-} = \{(q, \gamma) \in \mathcal{M}_p : E_p(q, \gamma) < r\}.$$

We will denote the set of regular values of E_p by $R(E_p)$.

Before we state the next proposition we review some well-known facts from Morse theory. Let us assume that $p \in N$ is not a focal point of the immersion $\phi : M \to N$. Then E_p is a Morse function. Let κ be a critical value of E_p and $\varepsilon > 0$ so small that $(\kappa - \varepsilon, \kappa) \subset R(E_p)$. Then we have

$$H_k(\mathcal{M}_p^\kappa, \mathcal{M}_p^{\kappa - \varepsilon}) = \mathbb{Z}_2^i,$$

where i is the number of critical points of E_p with index k and value κ. The nonzero homology classes of $H_k(\mathcal{M}_p^\kappa, \mathcal{M}_p^{\kappa - \varepsilon})$ can be represented by the local unstable manifolds of the critical points of E_p with index k and value κ. We will think of the local unstable manifolds as maps of k-dimensional closed disks with boundaries below the level $\kappa - \varepsilon$.

Now assume that $\phi : M \to N$ is a taut immersion. Then E_p is a perfect Morse function and the maps

$$H_k(\mathcal{M}_p^{\kappa - \varepsilon}) \to H_k(\mathcal{M}_p^\kappa) \qquad \text{and} \qquad H_k(\mathcal{M}_p^{\kappa -}) \to H_k(\mathcal{M}_p^\kappa)$$

are injective for all k. It follows from the homology sequence that

$$H_k(\mathcal{M}_p^\kappa) \to H_k(\mathcal{M}_p^\kappa, \mathcal{M}_p^{\kappa - \varepsilon})$$

is surjective for all k. Hence there is for every unstable manifold U of a critical point of E_p with index k and value κ a k-cycle z in $\mathcal{M}_p^\kappa$ that is homologous to U in $\mathcal{M}_p^\kappa$ modulo $\mathcal{M}_p^{\kappa - \varepsilon}$. By choosing ε smaller if necessary we can deform z into U in a Morse coordinate chart. We have thus seen that the local unstable manifold U can be completed to a cycle z in $\mathcal{M}_p^\kappa$. It is clear that such a z cannot be homologous within $\mathcal{M}_p^\kappa$ to a cycle in $\mathcal{M}_p^{\kappa -}$ since it maps onto a nontrivial cycle U in $H_k(\mathcal{M}_p^\kappa, \mathcal{M}_p^{\kappa -}) = H_k(\mathcal{M}_p^\kappa, \mathcal{M}_p^{\kappa - \varepsilon})$. Notice also that z cannot be homologous within $\mathcal{M}_p$ to a cycle in $\mathcal{M}_p^{\kappa -}$ since the maps $H_k(\mathcal{M}_p^{\kappa -}) \to H_k(\mathcal{M}_p^\kappa) \to H_k(\mathcal{M}_p)$ are injective and the class of z does not lie in the image of the first map.

In the next proposition we prove that these remarks hold true at a nondegenerate critical point even if E_p is not a Morse function.

PROPOSITION 2.7. *Let N be a Riemannian manifold and $\phi : M \to N$ a taut immersion. Let p be a point in N and $(q, \gamma) \in \mathcal{M}_p$ a nondegenerate critical point of E_p. Then the local unstable manifold at (q, γ) can be completed to a cycle z in $\mathcal{M}_p^\kappa$, where $\kappa = E_p(q, \gamma)$. Furthermore, if V is a Morse chart around (q, γ), the cycle z is not homologous within $\mathcal{M}_p^\kappa \cup V$ to any cycle in $\mathcal{M}_p^{\kappa-}$.*

PROOF. Let k denote the index of (q, γ). Since (q, γ) is a nondegenerate critical point there is a closed neighborhood U of p in N and differentiable functions $U \to M$ taking r to q_r and $U \to TM$ taking r to v_r such that $(q_p, \gamma_p) = (q, \gamma)$ and (q_r, γ_r) lies in $\mathcal{M}_r$ and is a nondegenerate critical point with index k of E_r, where $\gamma_r(t) = \exp t v_r$. There is a differentiable map $\Phi : U \times B^k \to P(N, N \times N)$ such that $U_r := \Phi(r, B^k)$ is a local unstable manifold in $\mathcal{M}_r$ of E_r at (q_r, γ_r) where B^k is a closed k-dimensional Euclidean ball.

By choosing U smaller if necessary, we can find an $\varepsilon > 0$ such that the boundaries of the unstable manifolds U_r of E_r at (q_r, γ_r) lie below the $\kappa - \varepsilon$ level of E_r for all $r \in U$. Set $\kappa(r) = E_r(q_r, \gamma_r)$. Again by choosing U smaller if necessary, we can assume that $\kappa(r) > \kappa - \varepsilon$.

Let $F_r : \mathcal{M}_r \to \mathcal{M}_p$ be the map that sends (s, f) to $(s, \tilde{f})$, where $\tilde{f}$ is the curve that one gets by adding the geodesic from r to p on f and then linearly reparametrizing so that it is still parametrized on $[0, 1]$. By choosing U smaller if necessary we can assume that $F_r(\mathcal{M}_r^{\kappa-\varepsilon}) \subset \mathcal{M}_p^{\kappa-\varepsilon/2}$. Again by choosing U smaller if necessary we can assume that $F_r(U_r)$ lies in $\mathcal{M}_p^\kappa \cup V$ for all $r \in U$ where V is some fixed Morse coordinate chart around (q, γ).

Now notice that there is a point s in every neighborhood of p such that E_s is a Morse function. Let s be such a point in U. Then by the observations before this proposition there is a cycle $\tilde{z}$ in $\mathcal{M}_s^{\kappa(s)}$ that agrees with the unstable manifold U_s above the E_s-level $\kappa - \varepsilon$. Then $F_r(\tilde{z})$ is a cycle in $\mathcal{M}_p^\kappa \cup V$ that can be deformed within the Morse chart V into a cycle z in $\mathcal{M}_p^\kappa$ that agrees with U_p above the E_p-level $\kappa - \varepsilon/2$. It follows that the homology class of z maps into the nontrivial homology class of U_p in $H_k(\mathcal{M}_p^\kappa, \mathcal{M}_p^{\kappa-})$ under the map induced by the inclusion. Notice that there is a deformation retraction of $\mathcal{M}_p^\kappa \cup V$ onto $\mathcal{M}_p^\kappa$. It follows that z cannot be homologous within $\mathcal{M}_p^\kappa \cup V$ to a class in $\mathcal{M}_p^{\kappa-}$. This finishes the proof of the proposition. $\square$

We now come to the generalization of the theorem of Ozawa [1986] on the distance functions of taut submanifolds in Euclidean spaces and in spheres. In the proof we will use the main idea of [Ozawa 1986].

THEOREM 2.8. *Let N be a Riemannian manifold and $\phi : M \to N$ a taut immersion. Then for every $p \in N$ the energy functional $E_p : P(N, \phi \times p) \to \mathbb{R}$ is a Morse–Bott function.*

PROOF. We will assume that M is embedded and work with $P(N, M \times q)$ instead of $P(N, \phi \times q)$ to simplify the notation. The general case does not require any new ideas.

We will work with finite-dimensional approximations of the path spaces $\mathcal{M}_p^{r-}$ as in [Milnor 1963]. Let $r > 0$ be some positive number. Let $i(N)$ denote the injectivity radius of N. Let n be a natural number greater than $r/i(N)^2$. We denote by $\mathcal{P}_p = \mathcal{P}_p(r, n)$ the space of continuous curves $\gamma : [0, 1] \to N$ that start in M and end in p and have the property that $\gamma \,|\, [(j-1)/n, j/n]$ is a geodesic of length less than $i(N)$ for all $j = 1, \ldots, n$. Then $E_p(\gamma) < r$. Notice that $\mathcal{P}_p$ is a finite-dimensional open manifold. We denote its dimension by d.

There is a deformation retraction of $\mathcal{M}_p^{r-}$ onto $\mathcal{P}_p$ [Milnor 1963]. The critical points of E_p in $\mathcal{M}_p^{r-}$ are of course contained in $\mathcal{P}_p$. Conversely, if γ is a critical point of the restriction of E_p to $\mathcal{P}_p$, then it is also a critical point of E_p on $\mathcal{M}_p^{r-}$. Their indices and nullity in $\mathcal{P}_p$ are the same as in $\mathcal{M}_p^{r-}$. A connected component of the set of critical points in $\mathcal{M}_p^{r-}$ is nondegenerate if and only if it is nondegenerate in $\mathcal{P}_p$.

Let γ be a geodesic in $\mathcal{P}_p$ whose index as a critical point of E_p we denote by i. We assume that the nullity of γ as a critical point is $n_0 > 0$. This means that p is a focal point of M along γ.

We now start following the arguments in [Ozawa 1986]. There are coordinates $x = (x_1, \ldots, x_d)$ around γ in $\mathcal{P}_p$ such that

$$E_p(x) = E_p(\gamma) - x_1^2 - \cdots - x_i^2 + x_{i+1}^2 + \cdots + x_{d-n_0}^2 + O(\|x\|^3)$$

in these coordinates and γ corresponds to 0.

For $c = (c_1, \ldots, c_{n_0})$ we set

$$A(c) = \{x_{d-n_0+j} = c_j \text{ for all } j = 1, \ldots, n_0\}.$$

Then the function $E_p \,|\, A(c)$ has a nondegenerate critical point of index i in $(0, c)$ with value $E_p(0, c) = E_p(\gamma) + O(\|(0, c)\|^3)$. We set $\kappa = E(\gamma)$ and $\kappa_c = E_p(0, c)$.

Our goal is to show that $\kappa_c = \kappa$ for c small. It then follows that $(0, c)$ is a critical point of E_p and hence that γ lies in a nondegenerate critical submanifold of dimension n_0. We do this in two steps. First we show that $\kappa_c \leq \kappa$ for c small. Then we prove that $\kappa_c \geq \kappa$ for c small.

We introduce the stable and unstable manifolds of $E_p \,|\, A(c)$ at $(0, c)$ before we start with the proof of the two steps explained above.

We can parametrize the family of local stable manifolds of $E_p \,|\, A(c)$ at $(0, c)$ by a differentiable map Φ into $\mathcal{P}_p$ depending on c and the elements of a closed ball $B_r(0)$ in $\mathbb{R}^{d-n_0-i}$. We set $S_c = \Phi(c, B_r(0))$. Then $S_c \in A(c)$ is the stable manifold of $E_p \,|\, A(c)$ at $(0, c)$.

Similarly we parametrize the family of local unstable manifolds of $E_p \,|\, A(c)$ at $(0, c)$ by a differentiable map Ψ depending on c and the elements of a closed ball $B_r(0)$ in $\mathbb{R}^i$. We set $U_c = \Psi(c, B_r(0))$.

We will use the notation $\mathcal{P}_p^s$ for the set of curves in $\mathcal{P}_p$ with E_p-value less than or equal to s and $\mathcal{P}_p^{s-}$ for the set of curves in $\mathcal{P}_p$ with E_p-value strictly less than s.

We choose $\varepsilon > 0$ and $\delta > 0$ so small that we have the following situation: $\|c\| \leq \delta$ implies that the local stable manifold S_c in $A(c)$ is a nontrivial cycle in $(\mathcal{P}_p - \mathcal{P}_p^{\kappa_c - \varepsilon}) \cap A(c) \mod (\mathcal{P}_p - \mathcal{P}_p^{\kappa_c + \varepsilon}) \cap A(c)$. This means that E_p is strictly greater than $\kappa_c + \varepsilon$ on ∂S_c. By choosing δ smaller if necessary we can assume that $\|c\| \leq \delta$ implies that E_p is strictly greater than $\kappa + \varepsilon$ on ∂S_c.

Furthermore, we assume ε and δ chosen such that the local unstable manifold U_c in $A(c)$ is a nontrivial cycle in $\mathcal{P}_p^{\kappa_c + \varepsilon} \cap A(c) \mod \mathcal{P}_p^{\kappa_c - \varepsilon} \cap A(c)$. This means that E_p is strictly smaller than $\kappa_c - \varepsilon$ on ∂U_c. After choosing δ smaller if necessary, we can assume that E_p is strictly smaller than $\kappa - \varepsilon$ on ∂U_c.

We can now begin the proof of the two cases.

(i) We continue the geodesic γ beyond p to a geodesic $\tilde{\gamma}$ defined on $[0, 1+t_0]$, where $t_0 > 0$ is so small that there is no focal point of M between p and $q :=$ $\tilde{\gamma}(1 + t_0)$ and the length of $\tilde{\gamma}$ between $(n-1)/n$ and $1 + t_0$ is less than $i(N)$, the injectivity radius of N. Denote by $\tilde{\mathcal{P}}_q = \tilde{\mathcal{P}}_q(r, n)$ the path space defined as above except that curves are parametrized between 0 and $1 + t_0$ instead of between 0 and 1 (but still with breaks at $1/n, \ldots, (n-1)/n$). The dimension of $\tilde{\mathcal{P}}_q$ is equal to that of $\mathcal{P}_p$, which was denoted by d. We assume that t_0 is so small that $\tilde{\gamma} \in \tilde{\mathcal{P}}_q$.

Notice that $\tilde{\gamma}$ is a nondegenerate critical point of E_q of index $i + n_0$. By tautness, the local unstable manifold of E_q at $\tilde{\gamma}$ can be completed to a cycle z in $\tilde{\mathcal{P}}_q$ under the $E_q(\tilde{\gamma})$-level; see Proposition 2.7.

Now we make a further restriction on t_0. We assume it is so small that $\tilde{f} \in \tilde{\mathcal{P}}_q$ for every $f \in S_c$ with $\|c\| \leq \delta$, where $\tilde{f}$ is the path we get by replacing the segment $f \mid [(n-1)/n, 1]$ of f by the geodesic segment between $f((n-1)/n)$ and q parametrized on the interval $[(n-1)/n, 1+t_0]$. This gives us a differentiable map of S_c into $\tilde{\mathcal{P}}_q$ that is also differentiable in the parameter c. We denote the images by $\tilde{S}_c$.

By choosing t_0 smaller if necessary we can arrange that the boundaries of $\tilde{S}_c$ for all $\|c\| \leq \delta$ have E_q-values strictly above $\kappa + \varepsilon$ and that $\kappa + \varepsilon > E_q(\tilde{\gamma})$. For t_0 small enough we have that $E_q \mid \tilde{S}_0$ has a nondegenerate absolute minimum in $\tilde{\gamma}$ and the absolute minimum is not reached in any other point. It follows that the intersection number of z and the relative cycle $\tilde{S}_0 \mod \tilde{\mathcal{P}}_q - \tilde{\mathcal{P}}_q^{\kappa + \varepsilon}$ is equal to one. (Notice that the dimensions of z and $\tilde{S}_0$ are complementary).

Now assume that there is a c_0 with $\|c_0\| \leq \delta$ such that $\kappa_{c_0} > \kappa$. Again by choosing t_0 smaller if necessary we can assume that E_q is strictly larger than $E_q(\tilde{\gamma})$ on $\tilde{S}_{c_0}$. Let $c(t)$ be a path between 0 and c_0 such that $\|c(t)\| \leq \delta$. Then $\tilde{S}_{c(t)}$ gives a homotopy between the cycles $\tilde{S}_0$ and $\tilde{S}_{c_0}$ keeping the boundaries above the E_q-level $\kappa + \varepsilon$. The intersection number of z and $\tilde{S}_{c_0}$ is equal to 0. This is a contradiction. It follows that $\kappa_c \leq \kappa$ for all $\|c\| < \delta$.

(ii) We now let $\tilde{\gamma}$ denote the restriction of the geodesic γ to the interval $[0, 1-t_0]$, where $t_0 > 0$ is so small that there is no focal point of M between $q := \gamma(1 - t_0)$ and p. Denote by $\tilde{\mathcal{P}}_q = \tilde{\mathcal{P}}(r, n)$ the path space defined as above except that curves are parametrized between 0 and $1 - t_0$ instead of between 0 and 1 (but still with breaks at $1/n$, ..., $(n - 1)/n$; that is, we also assume $t_0 < 1/n$). Notice that the dimension of $\tilde{\mathcal{P}}_q$ is again equal to d and that $\tilde{\gamma} \in \mathcal{P}_q$ and that $\tilde{\gamma}$ is a nondegenerate critical point of E_q with index i.

As above we make further restrictions on t_0. We assume it is so small that $\tilde{f} \in \tilde{\mathcal{P}}_q$ for every $f \in U_c$ with $\|c\| < \delta$, where $\tilde{f}$ is the path we get by replacing the segment $f \mid [(n-1)/n, 1]$ of f by the geodesic segment between $f((n - 1)/n)$ and q parametrized on the interval $[(n-1)/n, 1-t_0]$. This gives us a differentiable map of U_c into $\tilde{\mathcal{P}}_q$ that is also differentiable in the parameter c. We denote the images by $\tilde{U}_c$.

By choosing t_0 smaller if necessary we can arrange that the boundaries of $\tilde{U}_c$ for all $\|c\| \leq \delta$ have E_q-values strictly below $\kappa - \varepsilon$ and that $\kappa - \varepsilon < E_q(\tilde{\gamma})$. For t_0 small enough we have that $E_q \mid \tilde{U}_0$ has a nondegenerate absolute maximum in $\tilde{\gamma}$ and that $E_q \mid \tilde{U}_0$ does not have any further critical points. It follows that $\tilde{U}_0$ is homologous to the unstable manifold of E_q at $\tilde{\gamma}$ mod $\tilde{\mathcal{P}}_q^{\kappa-\varepsilon}$. Proposition 2.7 now implies that $\tilde{U}_0$ can be completed to a cycle z below the E_q level $\kappa - \varepsilon$ and that z is not homologous within $\tilde{\mathcal{P}}_q^{\tilde{\kappa}} \cup V$ to any cycle strictly below the level $\tilde{\kappa} := E_q(\tilde{\gamma})$, where V is some Morse chart around $\tilde{\gamma}$; see Proposition 2.7.

We now fix a Morse chart V around $\tilde{\gamma}$. By choosing δ smaller if necessary, we can assume that $\tilde{U}_c$ lies in $\tilde{\mathcal{P}}_q^{\tilde{\kappa}} \cup V$ for all $\|c\| \leq \delta$.

We now assume that there is a c_0 with $\|c_0\| \leq \delta$ such that $\kappa_{c_0} < \kappa$. By choosing t_0 smaller if necessary we can assume that E_q is strictly smaller than $E_q(\tilde{\gamma})$ on $\tilde{U}_{c_0}$. Let $c(t)$ be a path between 0 and c_0 such that $\|c(t)\| \leq \delta$. Then $\tilde{U}_{c(t)}$ gives a homotopy between the cycles $\tilde{U}_0$ and $\tilde{U}_{c_0}$ keeping the boundaries below the E_q level $\kappa - \varepsilon$. This induces a homotopy of z within $\tilde{\mathcal{P}}_q^{\tilde{\kappa}} \cup V$ that deforms z below the E_q-level $E_q(\tilde{\gamma})$. This contradicts Proposition 2.7. It follows that $\kappa_c = \kappa$ for all $\|c\| < \delta$, thus finishing the proof. $\qquad \square$

THEOREM 2.9. *Let* $\phi : M \to N$ *be a taut immersion. Then, given any* $p \in N$, *the map between the homology groups*

$$H_*(\mathcal{M}_p^r) \to H_*(\mathcal{M}_p)$$

induced by the inclusion of $\mathcal{M}_p^r$ *into* $\mathcal{M}_p$ *is injective for all* $r \geq 0$. *In particular, the energy function* E_p *is a perfect Morse–Bott function.*

PROOF. If E_p is a Morse function, it is perfect by the definition of tautness, and the claim of the theorem follows by standard Morse theory (see the remarks before Proposition 2.7). We therefore assume that E_p is not a Morse function. Then it is a Morse–Bott function by Theorem 2.8. We have to show that it is a perfect Morse–Bott function. Now assume that $H_*(\mathcal{M}_p^r) \to H_*(\mathcal{M}_p)$ is not injective. Let z be a nontrivial cycle in $\mathcal{M}_p^r$ that is homologous to zero in $\mathcal{M}_p$.

Let w be a chain in $\mathcal{M}_p$ such that $\partial w = z$. Let q be a point close to p such that E_q is a Morse function. Let $F_p : \mathcal{M}_p \to \mathcal{M}_q$ be the map that sends $(r, \gamma) \in \mathcal{M}_p$ to $(r, \tilde{\gamma})$, where $\tilde{\gamma}$ is the curve we get by adding onto γ the geodesic segment between p and q and then reparametrizing it between 0 and 1. We define $F_q : \mathcal{M}_q \to \mathcal{M}_p$ similarly. It follows, since E_p is a Morse–Bott function, that the critical levels of E_p are isolated. We assume that q is so close to p that there is an $\varepsilon > 0$ such that $(r, r+3\varepsilon) \subset R(E_p)$ and $F_p(\mathcal{M}_p^r) \subset \mathcal{M}_q^{r+\varepsilon}$ and $F_q(\mathcal{M}_q^{r+\varepsilon}) \subset \mathcal{M}_p^{r+2\varepsilon}$. There is an obvious continuous deformation of $F_q(F_p(\gamma))$ into γ since these curves only differ up to parametrization by paths that go back and forth between p and q. We assume that q is so close to p that the deformation of $F_q(F_p(\mathcal{M}_p^r))$ into $\mathcal{M}_p^r$ takes place within $\mathcal{M}_p^{r+3\varepsilon}$. Now let $\tilde{z} = F_p(z)$ and $\tilde{w} = F_p(w)$. Since $\tilde{z} = \partial \tilde{w}$, we see that $\tilde{z}$ is homologous to zero in $\mathcal{M}_p$. The injectivity of $H_*(\mathcal{M}_p^{r+\varepsilon}) \to H_*(\mathcal{M}_q)$ implies that there is a chain $\tilde{y}$ in $\mathcal{M}_q^{r+\varepsilon}$ such that $\tilde{z} = \partial \tilde{y}$. Set $y = F_q(\tilde{y})$. Notice that y lies in $\mathcal{M}_p^{r+2\varepsilon}$. Notice also that $F_q(\tilde{z})$ and z are homologous within $\mathcal{M}_p^{r+3\varepsilon}$. It follows that z is homologous to zero in $\mathcal{M}_p^{r+3\varepsilon}$ since $\partial y = F_q(\tilde{z})$, and z and $F_q(\tilde{z})$ are homologous in $\mathcal{M}_p^{r+3\varepsilon}$. This is a contradiction since $\mathcal{M}_p^r$ is a deformation retract of $\mathcal{M}_p^{r+3\varepsilon}$. This finishes the proof of the theorem. $\qquad\square$

COROLLARY 2.10. *Suppose (N, g) is a simply connected complete Riemannian manifold, and M is a connected taut submanifold of N. If γ_0 is an index 0 critical point of E_p with nullity k on $P(N, M \times p)$ then $E_p(\gamma_0)$ is the absolute minimum of E_p, and $C = E_p^{-1}(E_p(\gamma_0))$ is a connected k-dimensional critical submanifold of E_p.*

3. Variationally Complete and Polar Actions

DEFINITION 3.1. Let G act on a complete Riemannian manifold N isometrically. A Jacobi field J is called *G-transversal* if it is the variational field of a family of geodesics that are perpendicular to the orbits. The G-action is called *variationally complete* if any G-transversal Jacobi field that is tangent to orbits at two points is the restriction of some Killing field on N induced by the action.

One of the main results (Theorem I) of the paper [Bott and Samelson 1958], reformulated in our terminology, is the following theorem:

THEOREM 3.2. *Suppose G acts on N by isometries, and the G-action is variationally complete. Then the orbits of G are taut.*

Let (G, K) be a symmetric pair, and let $\mathcal{G} = \mathcal{K} \oplus \mathcal{P}$ be the corresponding Cartan decomposition. There are three natural actions associated to the symmetric pair:

(i) K acts on G/K by $g \cdot (hK) = (gh)K$.

(ii) The adjoint representation of G on $\mathcal{G}$ restricted to K leaves $\mathcal{P}$ invariant. So K acts on $\mathcal{P}$ by $\mathrm{Ad}(K)$, which is also the isotropy representation of G/K at eK.

(iii) The group $K \times K$ acts on G by $(k_1, k_2) \cdot g = k_1 g k_2^{-1}$.

Bott and Samelson apply the preceding theorem to symmetric spaces to prove the next result [Bott and Samelson 1958, Theorem II]:

THEOREM 3.3. *Let (G, K) be a symmetric pair, and let $\mathcal{G} = \mathcal{K} \oplus \mathcal{P}$ be the corresponding Cartan decomposition. Then the action of $K \times K$ on G, the action of K on G/K, and the action of K on $\mathcal{P}$ are variationally complete. Hence the orbits of these actions are taut.*

Bott had earlier [1956] proved important special cases of this theorem. The action of a compact Lie group G on itself by conjugation is variationally complete, and the same is true for the adjoint representation of G on its Lie algebra $\mathcal{G}$. In these two cases everything is much simpler and one does not really need Theorem 3.2 to prove the tautness of the orbits because all indices of critical points are even in these cases.

Let G be a compact Lie group equipped with a bi-invariant metric, and L a closed subgroup of $G \times G$. Then L acts on G isometrically by $(g_1, g_2) \cdot g = g_1 g g_2^{-1}$. Hermann [1960] generalized the first part of Theorem 3.3 as follows:

THEOREM 3.4. *Let (G, H) and (G, K) be two symmetric pairs of the compact Lie group G. Then the action of $H \times K$ on G and the action of H on G/K are variationally complete.*

Conlon [1971] found a geometric condition on isometric actions that implies variational completeness. We will use a terminology that differs somewhat from his.

DEFINITION 3.5 [Palais and Terng 1987]. Let G be a compact Lie group acting on the complete Riemannian manifold N by isometries. The G-action on N is said to be *polar* if there is a closed submanifold Σ of N that meets all orbits of G, and every intersection between Σ and an orbit is perpendicular. Such Σ is called a *section*. If the section is flat, the action is said to be *hyperpolar*.

It is easy to see that sections of a polar action are totally geodesic [Palais and Terng 1987]. So polar actions on flat Riemannian manifolds are hyperpolar.

THEOREM 3.6 [Conlon 1971]. *A hyperpolar action is variationally complete.*

COROLLARY 3.7. *The orbits of hyperpolar actions are taut.*

REMARK 3.8. All the variationally complete examples in Theorem 3.3 and 3.4 are hyperpolar.

EXAMPLE 3.9 [Heintze et al. 1995]. Recall that the *cohomogeneity* of a G-action on N is defined to be the codimension of the principal orbits. If an isometric G-action on N is of cohomogeneity one and the normal geodesics of principal orbits are closed, then the G-action is hyperpolar. Now suppose G/K is a rank-2 symmetric space, $\mathcal{G} = \mathcal{K} + \mathcal{P}$ is the corresponding Cartan decomposition, and the dimension of G/K is n. Then the K action on $\mathcal{P}$ leaves the unit sphere S^{n-1}

of $\mathcal{P}$ invariant, and the induced action of K on S^{n-1} is hyperpolar. Moreover, the action of $K \times \mathrm{SO}(n-1)$ on $\mathrm{SO}(n)$ is also hyperpolar. These examples of hyperpolar actions are different from those given in Theorem 3.3 and 3.4.

The following three problems arise naturally in the study of taut orbits in symmetric spaces:

PROBLEM 3.10. Classify all cohomogeneity one actions on symmetric spaces.

PROBLEM 3.11. Classify all hyperpolar actions on symmetric spaces.

PROBLEM 3.12. Classify all taut orbits in symmetric spaces.

When the symmetric space N is $\mathbb{R}^n$ or S^n, Problem 3.11 is solved by Dadok's Theorem [Dadok 1985]. In fact, he proved that if $\rho : H \to \mathrm{SO}(n)$ is polar then there exist a symmetric space G/K and a linear isometry $A : \mathbb{R}^n \to \mathcal{P}$ such that A maps H-orbits onto K-orbits, where $\mathcal{G} = \mathcal{K} \oplus \mathcal{P}$ is the corresponding Cartan decomposition. When the symmetric space N is the hyperbolic space H^n, Problem 3.11 is solved by Wu [1992].

Although there exist many examples, Problems 3.10 and 3.11 are far from being solved for general symmetric spaces. Next we explain the reduction of these problems to problems concerning Lie algebras. To explain this, we note that, since the group of isometries of a simply connected, compact symmetric space G/K is G, to classify hyperpolar actions on G/K it suffices to find all closed subgroups H of G such that the action of H on G/K is hyperpolar. It was proved in [Heintze et al. 1995] that for a closed subgroup H of G, the action of H on G/K is hyperpolar if and only if the action of $H \times K$ on G is hyperpolar. So to classify hyperpolar actions on compact symmetric spaces, it suffices to classify hyperpolar actions on simply connected, compact Lie groups. In fact, this was further reduced to a problem on Lie algebras:

THEOREM 3.13 [Heintze et al. 1995]. *Let G be a simply connected, compact Lie group equipped with the bi-invariant metric defined by the negative of the Killing form on $\mathcal{G}$, and let H be a closed subgroup of $G \times G$. Then the following statements are equivalent:*

(i) *The H-action on G is hyperpolar.*
(ii) *$g_0^{-1}\nu(H \cdot g_0)$ is abelian for some principal orbit $H \cdot g_0$.*
(iii) *There exists $g_0 \in G$ such that the orthogonal complement of*

$$\{g_0 x g_0^{-1} - y : (x, y) \in \mathcal{H}\}$$

is an abelian subalgebra of $\mathcal{G}$.

The problem of finding all $\mathcal{H}$ satisfying condition (iii) is still unsolved. For further results on hyperpolar actions, see [Alekseevskiĭ and Alekseevskĭ 1992; 1993; Heintze et al. 1994; 1995].

Problem 3.12 is not even solved for $\mathbb{R}^n$. In fact, it is not known what all the variationally complete actions on $\mathbb{R}^n$ are. Proposition 2.10 of [Terng 1991] asserted that an isometric action of G on $\mathbb{R}^n$ is variationally complete if and only if it is polar, but this must now be regarded as unproven: The proof depended on the main theorem of [Carter and West 1990], which stated that a totally focal submanifold of $\mathbb{R}^n$ is isoparametric, and which in turn is unsettled by the recent discovery of a gap in the demonstration of Theorem 5.1 of the same paper. (A submanifold M of $\mathbb{R}^n$ is *totally focal* if $\eta^{-1}(C)$ consists of all critical points of the endpoint map $\eta : \nu(M) \to \mathbb{R}^n$, where C is the set of all singular values of η. In other words, M is a totally focal submanifold of $\mathbb{R}^n$ if for any $a \in \mathbb{R}^n$ the critical points of the distance squared function f_a are either all nondegenerate or all degenerate.)

EXAMPLE 3.14. There are examples of orthogonal representations all of whose orbits are taut although the representations are neither polar nor variationally complete. For example, the orbits of the action of $\mathrm{SO}(n)$ on $\mathbb{R}^n \times \mathbb{R}^n$, for $n \geq 3$, defined by $g \cdot (v, w) = (gv, gw)$ are all taut by arguments as given in [Pinkall and Thorbergsson 1989]. First note that principal orbits of this action are diffeomorphic to the Stiefel manifold of orthonormal two-frames in $\mathbb{R}^n$, and the singular orbits are S^{n-1} and 0. To prove all orbits are taut, we note:

(i) If $\{v, w\}$ is orthonormal, then the $\mathrm{SO}(n)$-orbit through (v, w) is the standard embedding of the Stiefel manifold of orthonormal two-frames of $\mathbb{R}^n$ as a singular orbit of the isotropy representation of the symmetric space $\mathrm{Gr}(2, n)$. So it is taut.

(ii) If v and w are linearly independent and $v = ae_1 + be_2$ and $w = ce_1 + de_2$, where e_1, e_2 is an orthonormal two-frame, then the $\mathrm{SO}(n)$-orbit $M_{(v,w)}$ through (v, w) is the image of the orbit through (e_1, e_2) under the linear transformation $(x, y) \mapsto (ax + by, cx + dy)$. Since tightness is invariant under linear transformations and a taut submanifold in Euclidean space is tight, the orbit $M_{(v,w)}$ is tight. But $M_{(v,w)}$ lies in a sphere, so it is taut.

(iii) If (v, w) has rank one, then the orbit $M_{(v,w)}$ is a standard S^{n-1}, hence taut.

It is proved in [Heintze et al. 1994] that a polar representation cannot have repeated irreducible factors. So this action is not polar. To see this action is not variationally complete, we first note that a focal submanifold of an isoparametric submanifold is not totally focal. So $M_{(e_1,e_2)}$ is not totally focal. Now if the action of $\mathrm{SO}(n)$ on $\mathbb{R}^{2n}$ is variationally complete then its principal orbits must be totally focal [Terng 1991], which then implies that $M_{(e_1,e_2)}$ is totally focal, a contradiction.

Using a similar argument, we see that all orbits of the action of $\mathrm{SO}(n)$ on k copies of $\mathbb{R}^n$ by $g \cdot (v_1, \ldots, v_k) = (gv_1, \ldots, gv_k)$ with $k \leq n$ are taut. Similar constructions also work for other classical groups too. In fact, all orbits of k copies of the standard representation of $\mathrm{SU}(n)$ with $k < n$ on $\mathbb{C}^{kn}$ are taut, and all orbits of k (with $k \leq n$) copies of the standard representation of $\mathrm{Sp}(n)$ on $\mathbb{H}^{kn}$ are taut. (Here $\mathbb{H}$ denotes the quaternions.)

EXAMPLE 3.15. Examples of inhomogeneous isoparametric hypersurfaces in spheres were first given in [Ozeki and Takeuchi 1975], and then in a more systematic way in [Ferus et al. 1981]. It is shown in the latter paper that there is even an inhomogeneous isoparametric hypersurface in a sphere with a homogeneous focal manifold M_-. One sees easily that M_- cannot be an orbit of a polar representation. It is not difficult to show that

$$M_- = \{(u,v) \in \mathbb{H}^n \times \mathbb{H}^n : \|(u,v)\| = 1 \text{ and } u = \alpha v \text{ for some } \alpha \in \mathrm{Sp}(1)\}.$$

Clearly, M_- is an orbit of $\mathrm{Sp}(1) \times \mathrm{Sp}(n)$ acting on $\mathbb{H}^n \times \mathbb{H}^n$ by $(\alpha, A) \cdot (u_1, u_2) = (Au_1, \alpha A u_2)$.

The above examples explain the complexity of the following open problem:

PROBLEM 3.16. Classify all orthogonal representations all of whose orbits are taut.

It follows from the definition of a hyperpolar action that the cohomogeneity of such an action on a rank-k symmetric space has to be at most k. In particular, this implies that a hyperpolar action on S^n must be of cohomogeneity one. A polar action on S^n in general need not be variationally complete. To see this, first we recall that the set of focal points of a principal orbit of a variationally complete action is the set of all singular points of the action [Bott and Samelson 1958; Terng 1991]. Now suppose $\rho : G \to O(n)$ is irreducible, polar, and of cohomogeneity $k \geq 3$, and that M is a principal orbit in S^{n-1}. Then the action of G on S^{n-1} is polar and of cohomogeneity $k - 1 \geq 2$. Moreover, $\tau(M)$ is again a principal orbit in S^{n-1}, where $\tau : S^{n-1} \to S^{n-1}$ is the antipodal map. But $\tau(x)$ is a focal point of multiplicity $k - 2$ of M with respect to x for all $x \in M$. So the set of all focal points of M as a submanifold of S^{n-1} is the union of the set of singular points and $\tau(M)$. This implies that the action of G on S^{n-1} is not variationally complete. But G-orbits are taut in S^{n-1}. These examples lead us naturally to the following question:

QUESTION 3.17. Are orbits of polar actions on a symmetric space taut?

4. Equifocal and Weakly Equifocal Submanifolds

The notions of equifocal and weakly equifocal submanifolds in symmetric spaces were introduced in [Terng and Thorbergsson 1995]. These submanifolds give new examples of nonhomogeneous taut submanifolds, and are geometric analogues of the principal orbits of hyperpolar actions on symmetric spaces. In this section, we will review some results on these submanifolds proved in [Terng and Thorbergsson 1995].

First, we summarize some geometric and topological properties of principal orbits of hyperpolar actions. Suppose the action of G on a compact Riemannian manifold N is hyperpolar, and M is a principal G-orbit in N. Then [Bott and Samelson 1958; Palais and Terng 1987] we can say that:

(a) $\nu(M)$ is flat and has trivial holonomy. In fact, given $v \in \nu(M)_p$, let $\tilde{v}(g \cdot p) = g_*(v)$. Then $\tilde{v}$ gives a well-defined equivariant normal vector field on M, and $\tilde{v}$ is parallel with respect to the induced normal connection.

(b) $\exp(\nu(M)_x)$ is a closed flat submanifold of N for all $x \in M$.

(c) $v \in \nu(M)_p$ is a focal normal of multiplicity k of M with respect to p if and only if $\tilde{v}(x)$ is a focal normal of multiplicity k of M with respect to x for all $x \in M$.

(d) $y \in N$ is a focal point of multiplicity k of M if and only if $G \cdot y$ is a singular orbit and $k = \dim(G \cdot p) - \dim(G \cdot y)$.

(e) Let $\eta : \nu(M) \to N$ denote the endpoint map, take $v \in \nu(M)_p$, and let

$$M_v = \{\eta(\tilde{v}(x)) : x \in M\}.$$

Then $M_v = G \cdot \exp(v)$. Moreover, the map $\eta_v : M \to M_v$ defined by

$$\eta_v(x) = \exp(\tilde{v}(x)) = \exp(g_*(v)) = g \cdot \exp(v)$$

is a fibration, and the fiber $\eta_v^{-1}(y)$ is diffeomorphic to a principal orbit of the slice representation of G at y. (Recall that the slice representation at y is the representation of G_y on $\nu(G \cdot y)_y$ defined by $g * v = g_*(v)$).

(f) A point $q \in N$ is called *subregular* if there is no singular point $x \in N$ such that $G_x \subset G_q$ and $G_x \neq G_q$. Suppose $q = \exp(v)$ for some $v \in \nu(M)_x$ and q is subregular. Then there exists an integer m_q such that q is a focal point of multiplicity m_q of M with respect to all $y \in \eta_v^{-1}(q)$, and $\eta_v^{-1}(q)$ is diffeomorphic to the sphere S^{m_q}. This follows since the slice representation at q is polar with only one nontrivial orbit type.

(g) Bott and Samelson proved that E_a on $P(N, M \times a)$ is perfect for generic $a \in N$ (see Section 3). By definition, therefore, M is taut in N. They proved this by constructing a linking cycle at every critical point of E_a. We now give a geometric sketch of their construction. Assume that $a \in N$ is not a focal point of M and all focal points on each critical point γ of E_a on $P(N, M \times a)$ are subregular. Let γ be a critical point of E_a, and let $\gamma(0) = p \in M$. Then there exists $v \in \nu(M)_p$ such that $\gamma(t) = \exp(tv)$ and $\gamma(1) = a$. Let $p_1 = \gamma(t_1)$, ..., $p_r = \gamma(t_r)$ be focal points on γ with $0 < t_r < \cdots < t_2 < t_1 < 1$, and let $m_i = \dim(G_{p_i}) - \dim(G_p)$ be the dimension of $\eta_{t_i v}^{-1}(p_i)$. Now construct an iterated sphere bundle ξ_r as follows:

$$\xi_r = \{(g_1 p, g_2 g_1 p, \ldots, g_r g_{r-1} \cdots g_1 p) : g_j \in G_{p_j}, \ 1 \leq j \leq r\}.$$

Define a smooth map $\phi : \xi_r \to P(N, M \times a)$ by setting $\phi(y_1, \ldots, y_r)$ to the curve that restricted to the interval $[i/n, (i+1)/n]$ is the image of $\gamma | [i/n, (i+1)/n]$ under $g_{i+1} g_i \cdots g_1$. The image of ϕ lies on a constant energy level E_a. By cutting off corners of the broken geodesics in $\phi(\xi_r)$ we can deform $\phi(\xi_r)$ into a linking cycle of E_a at γ, i.e., $\phi(\xi_r)$ a completion of a local unstable manifold at γ below the energy level γ; see Section 2.

We realized that at least in symmetric spaces the properties (d)–(g) only depend on properties (a)–(c). This led us to the definition of equifocal submanifolds as being those submanifolds of symmetric spaces that have these three properties.

To make the definition more precise, we recall that an r-flat in a rank-k symmetric space $N = G/K$ is an r-dimensional, totally geodesic, flat submanifold. Let $\mathcal{G} = \mathcal{K} \oplus \mathcal{P}$ be the corresponding Cartan decomposition. Then every flat is contained in some k-flat, and every k-flat is of the form $\pi(g \exp(\mathcal{A}))$, where $g \in G$ and $\mathcal{A}$ is a maximal abelian subalgebra in $\mathcal{P}$. If N is a compact Lie group of rank k, then a k-flat in N is just a maximal torus. But an r-flat need not be closed in general.

Let M be an immersed submanifold of a symmetric space N. The normal bundle $\nu(M)$ is called *abelian* if $\exp(\nu(M)_x)$ is contained in some flat of N for each $x \in M$. It is called *globally flat* if the induced normal connection is flat and has trivial holonomy.

Let v be a globally defined normal field on M, and $\eta_v : M \to N$ denote the endpoint map associated to v defined by setting $\eta_v(x) = \exp(v(x))$.

DEFINITION 4.1. A connected, immersed submanifold M in a symmetric space N is called *equifocal* if

(1) $\nu(M)$ is globally flat and abelian, and
(2) if v is a parallel normal field on M such that $\eta_v(x_0)$ is a focal point of multiplicity k of M with respect to x_0, then $\eta_v(x)$ is a focal point of multiplicity k of M with respect to x for all $x \in M$. (Equivalently, the focal data $\Gamma(M)$ is "invariant under normal parallel translation").

EXAMPLES 4.2. (i) Principal orbits of a hyperpolar action on a compact symmetric space are equifocal since they satisfy properties (a), (b), and (c) of page 199.

(ii) A distance sphere in an irreducible compact symmetric space N is equifocal if and only if N has rank one. This follows from the fact that a geodesic normal to an equifocal hypersurface in an irreducible compact symmetric space is closed (see Theorem 4.8(b) below). As a consequence, if a distance sphere is equifocal, then all geodesics in N are closed and the rank of N is one. However, we will see in Section 6 that distance spheres in compact symmetric spaces are always taut.

REMARKS 4.3. (i) It is proved in [Terng 1985] that if M is isoparametric in $\mathbb{R}^n$ then $\nu(M)$ is globally flat. Given a unit vector $v \in \nu(M)_x$, then $t_0 v$ is a focal normal of multiplicity k if and only if $1/t_0$ is a principal curvature of multiplicity k of M in the direction v. So it follows that a submanifold M in $\mathbb{R}^n$ is isoparametric if and only if it is equifocal.

(ii) A hypersurface M in a sphere S^n is called *isoparametric* if it has constant principal curvatures. It turns out that M is isoparametric in S^n if and only if M is equifocal in S^n. The study of isoparametric hypersurfaces in S^n has a long history, and these hypersurfaces have many remarkable properties [Münzner 1980; 1981]. We will make some remarks about equifocal hypersurfaces and isoparametric hypersurfaces towards the end of this section.

(iii) Notice that an isoparametric hypersurface of the real hyperbolic space H^n is equifocal, but the converse is not true. In fact, an equifocal hypersurface in H^n can be characterized by the property that the principal curvatures whose absolute values are greater than one are constant.

Ewert [1997] has proved the following two results:

THEOREM 4.4. *A complete hypersurface in a symmetric space of noncompact type is equifocal if and only if it is a tube around a submanifold without focal points. Furthermore, such an equifocal hypersurface is taut.*

THEOREM 4.5. *Suppose M is a submanifold of a simply connected complete Riemannian manifold N such that M has no focal points. Then M is taut in N.*

Wu [1994] defined a submanifold M in N to be *hyper-isoparametric* if it satisfies the following conditions:

(1) M is curvature adapted, i.e., the operator $B_v(u) = R(v, u)(v)$ leaves TM
    invariant and commutes with the shape operator A_v for all $v \in \nu(M)$,
(2) $\nu(M)$ is globally flat and abelian,
(3) the principal curvatures along any parallel normal field are constant.

Note that M is hyper-isoparametric if and only if M is curvature adapted and equifocal. Wu independently obtained some of our results by using the method of moving frames. But an equifocal submanifold is in general neither curvature adapted nor has constant principal curvatures. For example, there are many such equifocal hypersurfaces in $\mathbb{CP}^n$ [Wang 1982].

Henceforth, we will assume that $N = G/K$ is a compact, rank-k symmetric space of semisimple type, that $\mathcal{G} = \mathcal{K} + \mathcal{P}$ is a Cartan decomposition, and that N is equipped with the G-invariant metric given by the restriction of the negative of the Killing form of $\mathcal{G}$ to $\mathcal{P}$.

To simplify the terminology we make the following definition:

DEFINITION 4.6. Let M be a submanifold in N, and $v \in \nu(M)_x$. Then t_0 is called a *focal radius* of M with multiplicity m along v if $\exp_x(t_0 v)$ is a focal point of multiplicity m of M with respect to v.

Then a submanifold M with globally flat abelian normal bundle of a compact symmetric space N is equifocal if the focal radii of M along any parallel normal field are constant.

A smooth normal field v on a submanifold M is called a *focal normal field* if $v/\|v\|$ is parallel and there exists an integer k such that $\exp(v(x))$ is a focal point of multiplicity k of M with respect to x for all $x \in M$. If v is a smooth focal normal field of an equifocal submanifold M in N, then $\|v\|$ is constant on

M, and the endpoint map $\eta_v : M \to N$ has constant rank. So the kernel of $d\eta_v$ defines an integrable distribution $\mathcal{F}_v$ with $\eta_v^{-1}(y)$ as leaves, and $M_v = \eta_v(M)$ is an immersed submanifold of N. We will call $\mathcal{F}_v$, $\eta_v^{-1}(y)$ and M_v respectively the *focal distribution, focal leaf* and the *focal manifold* defined by the focal normal field v.

THEOREM 4.7 [Terng and Thorbergsson 1995]. *Let v be a parallel normal field on an equifocal submanifold M of a compact, simply connected, symmetric space N. Then $M_v = \eta_v(M)$ is an embedded submanifold. Moreover:*

(a) *M_v is taut;*
(b) *if $\exp(v(x))$ is not a focal point then M_v is again equifocal and the endpoint map $\eta_v : M \to M_v$ is a diffeomorphism; and*
(c) *if $\exp(v(x))$ is a focal point then $\eta_v : M \to M_v$ is a fibration and the fiber $\eta_v^{-1}(y)$ is diffeomorphic to a finite dimensional isoparametric submanifold in the Euclidean space $\nu(M_v)_y$.*

If M is a principal orbit of a hyperpolar G-action on N, then (b) and (c) are consequences of the following facts:

(i) *If v is a parallel normal field on M, then $\eta_v(g \cdot x_0) = g \cdot \exp(v(x_0))$.*
(ii) *The focal points of M are the set of singular points of the G-action.*
(iii) *The slice representation of a polar action is hyperpolar. So in particular, the orbits of the slice representation are isoparametric.*

The basic ideas in the proof of Theorem 4.7(a) are:

(1) There is a geometric analogue of subregular points for equifocal submanifolds.
(2) The focal leaves corresponding to "subregular" points are diffeomorphic to standard spheres.
(3) There is a construction of linking cycles for critical points of E_a that is similar to the one sketched in (g) on page 199.

We also associated to each equifocal submanifold M of N an affine Weyl group W and a marked affine Dynkin diagram. We proved that the critical points of $E_p : P(N, M \times p) \to \mathbb{R}$ and their indices can be described in terms of W and the marked Dynkin diagram. We describe this situation more precisely in the following theorem. Notice that the results are to a large extent analogous to the rich structure theory of isoparametric submanifolds in Euclidean spaces [Terng 1985].

THEOREM 4.8. *Suppose M is a codimension-r equifocal submanifold of a simply connected, compact symmetric space N. Then:*

(a) *For a focal normal field v, the leaf of the focal distribution $\mathcal{F}_v$ through $x \in M$ is diffeomorphic to an isoparametric submanifold in $\nu(M_v)_{\eta_v(x)}$.*
(b) *$\exp(\nu(M)_x) = T_x$ is an r-dimensional flat torus in N for all $x \in M$.*

(c) *There exists an affine Weyl group W with $r + 1$ nodes in its affine Dynkin diagram such that, for $x \in M$,*

   (i) *W acts isometrically on $\nu(M)_x$, and the set of singular points of the W-action on $\nu(M)_x$ is the set of all $v \in \nu(M)_x$ such that $\exp(v)$ is a focal point of M with respect to x, and*

   (ii) *$M \cap T_x = \exp_x(W \cdot 0)$.*

(d) *Let D_x denote the Weyl chamber of the W-action on $\nu(M)_x$ containing 0, and let $\triangle_x = \exp(D_x)$. Then*

   (i) *$\exp_x$ maps the closure of D_x isometrically onto the closure of $\triangle_x$, and*

   (ii) *there is a labeling of the open faces of $\triangle_x$ by $\sigma_1(x)$, ..., $\sigma_{r+1}(x)$ and integers m_1, ..., m_{r+1} independent of x such that if $y \in \partial\triangle_x$, then y is a focal point with respect to x of multiplicity m_y, where m_y is the sum of m_i such that y is in the closure of $\sigma_i(x)$.*

(e) *Let $p \in N$, let v be a parallel normal field on M, and let E be the energy functional on the path space $P(M, p \times M_v)$. Then the $\mathbb{Z}_2$-homology of $P(M, p \times M_v)$ can be computed explicitly in terms of W and m_1, ..., m_{r+1}; moreover,*

   (i) *if p is not a focal point of M then E is a perfect Morse function, and*

   (ii) *if p is a focal point of M then E is nondegenerate in the sense of Bott and perfect.*

REMARK 4.9. Theorems 4.7 and 4.8 are invalid when N is not simply connected. To see this, let N be the real projective space $\mathbb{RP}^n$ and M a distance sphere in N centered at x_0. Then M is certainly equifocal. Let v be a unit normal field on M. Then there exists $t_0 \in \mathbb{R}$ such that $\exp(t_0 v(x)) = x_0$ for all $x \in M$. Let T_x be the normal circle at a point x in M. Then D_x is an interval, and $\triangle_x = T_x \setminus \{x_0\}$. Moreover, there exists t_1 such that the parallel set M_{t_1} is the cut locus of the center x_0, which is a $\mathbb{Z}_2$-quotient of M, i.e., a projective hyperplane. Notice that the focal variety of M consists of only one point $(x_0, n-1)$ and M_{t_1} is not diffeomorphic to M. In fact M_{t_1} has the same dimension as M and satisfies all the conditions in the definition of an equifocal submanifold except that the normal bundle does not have trivial holonomy. Although a parallel manifold M_v of M in a simply connected compact symmetric space N is either equifocal or a focal submanifold, this need not be the case if N is not simply connected.

DEFINITION 4.10. A connected, compact, immersed submanifold M with a globally flat and abelian normal bundle in a symmetric space N is called *weakly equifocal* if, given a parallel normal field v on M, the following conditions are satisfied:

(1) The multiplicities of the focal radius functions along v are constant, i.e., the focal radius functions t_j are smooth functions on M that can be ordered,

$$\cdots < t_{-2}(x) < t_{-1}(x) < 0 < t_1(x) < t_2(x) < \cdots,$$

and the multiplicities m_j of the focal radii $t_j(x)$ are constant on M.

(2) the focal radius function t_j is constant on $\eta_{t_j v}^{-1}(x)$ for all $x \in M_{t_j v}$.

REMARKS 4.11. (i) It is proved in [Terng and Thorbergsson 1995] that condition (2) on the focal radii in the definition of weakly equifocal submanifolds is always satisfied if the dimension of $\eta_{t_j v}^{-1}(x)$ is at least two.

(ii) It follows from the definitions that a (weakly) equifocal submanifold in a rank k symmetric space has codimension less than or equal to k, and that equifocal implies weakly equifocal.

In the following theorem we bring our main results on weakly equifocal submanifolds.

THEOREM 4.12 [Terng and Thorbergsson 1995]. *Suppose M is an immersed, weakly equifocal compact submanifold of a simply connected symmetric space N of compact type. Then*

(a) *M is embedded,*

(b) *M is taut,*

(c) *for a focal normal field v, the set $\eta_v^{-1}(x)$ is diffeomorphic to a taut submanifold of a finite-dimensional Euclidean space for all $x \in M$.*

We now discuss how equifocal and weakly equifocal submanifolds relate to submanifolds of Euclidean space.

DEFINITION 4.13 [Terng 1987]. A submanifold M in $\mathbb{R}^n$ is called *weakly isoparametric* if

(1) $\nu(M)$ is globally flat,

(2) the multiplicities of the principal curvatures λ along a parallel normal field v are constant, and

(3) if $d\lambda(X) = 0$ for X in the eigenspace $E_\lambda(v)$ corresponding to the principal curvature λ.

A submanifold of $\mathbb{R}^n$ is therefore clearly weakly isoparametric if and only if it is weakly equifocal. Pinkall [1985] called a hypersurface in $\mathbb{R}^n$ or S^n *proper Dupin* if the multiplicities of the principal curvatures are constant and $d\lambda(X) = 0$ for X in the eigenspace $E_\lambda(v)$ corresponding to the principal curvature λ. A hypersurface M in $\mathbb{R}^n$ or S^n is therefore weakly equifocal if and only if it is proper Dupin. It was proved in [Thorbergsson 1983] that proper Dupin hypersurfaces are taut. In [Terng 1987] this was generalized to weakly isoparametric submanifolds. These results are of course special cases of Theorem 4.12.

Assume that M is an isoparametric hypersurface of S^n with g distinct constant principal curvatures $\lambda_1 > \cdots > \lambda_g$ along the unit normal field v with multiplicities $m_1, \ldots, m_g$. Let E_j denote the curvature distribution defined by λ_j, i.e., $E_j(x)$ is equal to the eigenspace of $A_{v(x)}$ with respect to the eigenvalue $\lambda_j(x)$. Then the focal distributions of M are the curvature distributions E_j. It follows from the structure equations of S^n that there exists $0 < \theta < \pi/g$ such that the principal curvatures are $\lambda_j = \cot(\theta + (j-1)\pi/g)$ with $j = 1, \ldots, g$, and the parallel set $M_t = M_{tv}$ for $-\pi/g + \theta < t < \theta$ is again an isoparametric hypersurface. The focal sets $M^+ = M_{\theta v}$ and $M^- = M_{\theta - \pi/g}$ are embedded submanifolds of S^n with codimension $m_1 + 1$ and $m_g + 1$, respectively, and the focal variety of M in S^n is equal to

$$\{(x, m_1) : x \in M^+\} \cup \{(x, m_g) : x \in M^-\}.$$

Another consequence of the structure equations is that the leaves of each E_j are standard spheres. Using topological methods, Münzner proved that

(1) g has to be $1, 2, 3, 4$ or 6,
(2) $m_i = m_1$ if i is odd, and $m_i = m_2$ if i is even,
(3) S^n can be written as the union $D_1 \cup D_2$, where D_1 is the normal disk bundle of M^+, D_2 is the normal disk bundle of M^- and $D_1 \cap D_2 = M$, and
(4) the $\mathbb{Z}_2$-homology of M can be given explicitly in terms of g and m_1, m_2; in particular, the sum of the $\mathbb{Z}_2$-Betti numbers of M is $2g$.

It is proved in [Thorbergsson 1983] that proper Dupin hypersurfaces have the above properties (1)–(4).

We end this section by restricting ourselves to equifocal hypersurfaces in symmetric spaces to see to which extent our results on equifocal submanifolds generalize the theory of isoparametric hypersurfaces in spheres that we have been sketching.

Assume that M is an immersed compact equifocal hypersurface in a simply connected, compact, semisimple symmetric space N. Then the following statements follow from Theorems 4.7 and 4.8:

(a) The normal geodesics to M are circles of constant length, which will be denoted by l.
(b) There exist integers m_1, m_2, an even number $2g$ and $0 < \theta < l/(2g)$ such that

    (1) the focal points on the normal circle $T_x = \exp(\nu(M)_x)$ are

$$x(j) = \exp((\theta + (j - 1)l/(2g))v(x)) \quad \text{for } 1 \le j \le 2g,$$

    with multiplicity m_1 if j is odd and m_2 if j is even, and

    (2) the group generated by reflections in pairs of focal points $x(j)$, $x(j + g)$ on the normal circle T_x is isomorphic to the dihedral group W with $2g$ elements, and hence W acts on T_x.

(c) $M \cap T_x = W \cdot x$.
(d) Let $\eta_{tv} : M \to N$ denote the endpoint map defined by tv, where v is a unit normal field, and let $M_t = \eta_{tv}(M) = \{\exp(tv(x)) : x \in M\}$ denote the set parallel to M at distance t; then M_t is an equifocal hypersurface and η_{tv} maps M diffeomorphically onto M_t if $t \in (-l/(2g) + \theta, \theta)$.

(e) $M^+ = M_\theta$ and $M^- = M_{-l/(2g)+\theta}$ are embedded submanifolds of codimension $m_1 + 1$ and $m_2 + 1$ in N, and the maps $\eta_{\theta v} : M \to M^+$ and $\eta_{(-l/(2g)+\theta)v} : M \to M^-$ are S^{m_1}- and S^{m_2}-bundles, respectively.

(f) The focal variety $\mathcal{V}(M)$ equals $(M^+, m_1) \cup (M^-, m_2)$.

(g) $\{M_t : t \in [-l/(2g)+\theta, \theta]\}$ gives a singular foliation of N, which is analogous to the orbit foliation of a cohomogeneity one isometric group action on N.

(h) $N = D_1 \cup D_2$ and $D_1 \cap D_2 = M$, where D_1 and D_2 are diffeomorphic to the normal disk bundles of M^+ and M^-, respectively.

(i) M_t is taut in N for all $t \in \mathbb{R}$.

(j) The $\mathbb{Z}_2$-homology of $P(N, p \times M_t)$ can be computed explicitly in terms m_1 and m_2.

This generalizes most of the theory of isoparametric hypersurfaces in spheres to equifocal hypersurfaces in simply connected compact symmetric spaces. There is only one important result that we have not been able to generalize: Münzner's celebrated restriction on the possible values of g. Bing-le Wu [1995] solves this problem for the rank-one symmetric spaces except the Cayley plane:

THEOREM 4.14 [Wu 1995]. *Suppose M is an equifocal hypersurface of a projective space $\mathbb{CP}^n$ or $\mathbb{HP}^n$, and g is the number of focal points along a normal geodesic of M. Then g is either 2, 4 or 6.*

5. Dupin Submanifolds in a Complete Riemannian Manifold

In this section, we will introduce the notion of Dupin submanifolds in a general Riemannian manifold and study its relation to tautness. First we review some results concerning Dupin submanifolds in $\mathbb{R}^n$ and in Hilbert spaces. Then we explain a linearization technique developed in [Terng and Thorbergsson 1995], which lifts submanifolds in symmetric spaces to submanifolds in Hilbert space. This lifting technique allows us to apply the extensive theory developed for taut submanifolds in Hilbert spaces to taut submanifolds in symmetric spaces. These results then motivate our definition of Dupin submanifolds in an arbitrary Riemannian manifold.

The spectral theory of the shape operators and the Morse theory of the Euclidean distance squared functions of submanifolds in $\mathbb{R}^n$ are closely related, and they play essential roles in the study of the geometry and topology of submanifolds in $\mathbb{R}^n$. Given a submanifold M of $\mathbb{R}^n$, the shape operator A is a smooth bundle morphism from $\nu(M)$ to the bundle of self-adjoint operators $L_s(TM, TM)$. So there is an open dense subset $\mathcal{V}_0$ of $\nu^1(M)$ such that the principal curvatures are differentiable functions on $\mathcal{V}_0$ with locally constant multiplicities. Reckziegel [1979] proved that given any $v \in \mathcal{V}_0 \cap \nu(M)_{x_0}$ and an eigenspace E_0 of A_v with respect to a nonzero eigenvalue, there exist a connected submanifold S of M through x_0 and a parallel normal field ξ of M along S such that $\xi(x_0) = v$, $TS_{x_0} = E_0$, and TS_x is equal to an eigenspace of $A_{\xi(x)}$ for all $x \in S$. Moreover, if $d = \dim S > 1$, then S is a standard sphere. However, for $v \notin \mathcal{V}_0$ and E_0 an eigenspace of A_v of dimension > 1, such S may not exist and even when it exists it need not be a standard sphere. This leads to the definition in [Reckziegel 1979] that a connected submanifold S of M is called a *curvature surface* if there is a parallel normal field ξ of M along S such that at each point x of S the tangent plane TS_x is a full eigenspace of $A_{\xi(x)}$. A one dimensional curvature surface is called a *line of curvature*.

DEFINITION 5.1 [Pinkall 1985]. A submanifold M of $\mathbb{R}^n$ is called *Dupin* if every line of curvature is a standard circle.

Recall that a Dupin hypersurface is proper Dupin if the multiplicities of the principal curvatures are constant; see Section 4.

THEOREM 5.2 [Pinkall 1986]. *A taut submanifold M in $\mathbb{R}^n$ is Dupin, i.e., the lines of curvature of M are standard circles.*

QUESTION 5.3. Let M be a complete Dupin submanifold of $\mathbb{R}^n$. Is M taut?

THEOREM 5.4 [Pinkall 1986]. *Let M be a compact submanifold of $\mathbb{R}^n$, and let*

$$\nu_\varepsilon(M) = \{x + v : v \in \nu(M)_x, \|v\| = \varepsilon\}$$

denote the ε-tube of M. For small ε, the hypersurface $\nu_\varepsilon(M)$ is taut if and only if M is taut. Moreover, the set of focal points of $\nu_\varepsilon(M)$ is $\Gamma(M) \cup M$, where $\Gamma(M)$ is the set of focal points of M in $\mathbb{R}^n$.

If M is Dupin with constant multiplicities, then $\nu_\varepsilon(M)$ is a proper Dupin hypersurface. So, by [Thorbergsson 1983], $\nu_\varepsilon(M)$ and hence also M are taut. Notice that a weakly isoparametric submanifold of $\mathbb{R}^n$ is taut, but the multiplicities of the principal curvatures are in general not constant.

The three basic local invariants (the induced metric, the second fundamental form, and the induced normal connection), the endpoint map, and the focal points and focal radii of an immersed Hilbert manifold M in a Hilbert space V are defined exactly the same way as for immersed submanifolds in $\mathbb{R}^n$. But the spectral theory for the shape operators of M is complicated and the infinite-dimensional differential topology and Morse theory cannot be applied easily without further restrictions. In order to develop a good theory of submanifold geometry in Hilbert space, proper Fredholm submanifolds were introduced in [Terng 1989]. Recall that an immersed finite codimension submanifold M of a Hilbert space V is called *proper Fredholm* (PF) if the endpoint map η restricted to any finite radius normal disk bundle is proper and Fredholm [Terng 1989]. Properness implies that the map $f_a : M \to \mathbb{R}$ defined by $f_a(x) = \|x - a\|^2$ satisfies the Palais–Smale condition C, so we can apply Morse theory to these functions. The Fredholm condition allows us to use infinite-dimensional differential topology of Fredholm maps. Note that, if $V = \mathbb{R}^n$, then M is PF if and only if the immersion is a proper map.

A PF submanifold M of V has the following general properties [Terng 1989]:

(a) The shape operators are compact.
(b) All nonzero principal curvatures have finite multiplicities.
(c) The set of nonfocal points of M is open and dense in V (by the transversality theorem for Fredholm maps).
(d) The set of focal points along the normal ray $\{x + tv : t \in R\}$ is locally finite for any $v \in \nu(M)_x$.

A submanifold M of V is *taut* if it is PF and f_a is perfect for generic $a \in V$. Curvature surfaces of a PF submanifold M in V can be defined exactly the same way as in the finite-dimensional case. Using the same proof as in that case we obtain the following:

THEOREM 5.5. *Suppose M is a PF submanifold of V. Then there exists an open dense subset $\mathcal{V}_0$ of $\nu^1(M)$ such that given any $v_0 \in \mathcal{V}_0 \cap \nu(M)_{x_0}$ and multiplicity k principal curvature λ_0 of M in the direction of v_0 there exists a curvature surface S through x_0 satisfying the following conditions:*

(i) *TS_{x_0} is the eigenspace of A_{v_0} with eigenvalue λ_0.*
(ii) *If $k > 1$ then S is a standard sphere.*
(iii) *If M is taut and $k = 1$, then S is a standard circle.*

It is easy to see that linear subspaces of V with finite codimension are PF and taut. But if the dimension of V is infinite, the unit sphere S centered at 0 is not PF. This is because $\eta^{-1}(0) = S$ and S is not compact, contradicting the condition that η is proper. Since S is contractible, the Euclidean distance squared functions are not perfect.

Using exactly the same proof as in [Pinkall 1986], we can generalize Theorem 5.4 to Hilbert spaces:

THEOREM 5.6. *Suppose M is a PF submanifold of a Hilbert space V. Then the ε-tube $\nu_\varepsilon(M)$ is taut if and only if M is taut.*

A PF submanifold M of a Hilbert space V is called *isoparametric* if $\nu(M)$ is globally flat and the principal curvatures along any parallel normal field are constant. If the multiplicities of principal curvatures (but not necessarily the curvatures themselves) are constant and lines of curvatures are circles, we call M *weakly isoparametric*. A theory of isoparametric and weakly isoparametric submanifolds of Hilbert spaces is developed in [Terng 1989]. In particular, the next result is proved there:

THEOREM 5.7. *If M is a (weakly) isoparametric submanifold of a Hilbert space V, then M is taut.*

To get geometrically and topologically interesting taut submanifolds in V, we need to use infinite-dimensional transformation groups. First we review some definitions. Let G be a Hilbert Lie group and M a Riemannian Hilbert manifold. A smooth isometric G-action on M is called *proper* if $g_n \cdot x_n \to x_0$ and $x_n \to x_0$

in M implies that $\{g_n\}$ has a convergent subsequence in G; and the G-action is called *Fredholm* if every orbit map $G \to G \cdot x$ is a Fredholm map. An isometric, proper Fredholm (PF) action is called *polar* if there exists a smooth, closed submanifold Σ of M such that every orbit meets Σ, and all intersections between Σ and orbits are perpendicular. Such a Σ is called a *section* for the action.

THEOREM 5.8 [Terng 1989]. *The orbits of a polar action on a Hilbert space are isoparametric submanifolds or focal manifolds of isoparametric submanifolds, and hence taut.*

THEOREM 5.9 [Terng 1995]. *Let G be a compact Lie group, H a closed subgroup of $G \times G$, $P(G, H)$ the group of H^1-paths $g : [0, 1] \to G$ such that $(g(0), g(1)) \in H$, and $V = H^0([0, 1], \mathcal{G})$. Suppose the action of H on G is hyperpolar. Then the action of $P(G, H)$ on V by gauge transformations*

$$g \cdot u = gug^{-1} - g_x g^{-1}$$

is polar.

Therefore the hyperpolar actions in Section 3 provide many examples of taut orbits in Hilbert space. The simplest example of this kind comes from the action of the diagonal group $H = \Delta(\mathrm{SU}(2))$ on $G = \mathrm{SU}(2)$, i.e., the Adjoint action. Let M be a principal Adjoint orbit of $\mathrm{SU}(2)$. Then a principal orbit of $P(G, H)$ is a hypersurface of V with Poincaré polynomial

$$1 + 2 \sum_{i > 0} t^{2i},$$

and is diffeomorphic to $P(\mathrm{SU}(2), e \times M)$. There are exactly two singular orbits in V, which are codimension 3 submanifolds of V with Poincaré polynomial

$$\sum_{i \geq 0} t^{2i}.$$

The singular orbits are diffeomorphic to the group of based loops in $\mathrm{SU}(2)$.

Since f_a satisfies condition C, we can prove an analogue of Ozawa's theorem [Ozawa 1986] (see also Section 2):

THEOREM 5.10 [Terng and Thorbergsson 1995]. *Suppose M is a taut submanifold in a Hilbert space V, and take $a \in V$. Then the distance squared function $f_a : M \to \mathbb{R}$ is a perfect Morse–Bott function. Moreover, if x_0 is a critical point of f_a with nullity k, the k-dimensional critical submanifold of f_a at x_0 is a taut submanifold in some finite-dimensional affine subspace of V.*

Next we will explain the relation between tautness in symmetric spaces and in Hilbert spaces. Let $\pi : G \to G/K$ denote the natural fibration defined by $\pi(g) = gK$, and $\phi : L^2([0, 1], \mathcal{G}) \to G$ the parallel translation from 0 to 1, i.e., $\phi(u) = g(1)$, where g satisfies $g^{-1}g_x = u$ and $g(0) = e$. Now for a submanifold M of G/K, we set $M^* = \pi^{-1}(M)$ and $\tilde{M} = \phi^{-1}(\pi^{-1}(M))$. One of our main steps in developing the theory of equifocal submanifolds in symmetric spaces is to relate the focal structures of M, M^* and $\tilde{M}$. In fact, we proved in [Terng and Thorbergsson 1995] that the following three statements are equivalent:

(i) M is (weakly) equifocal in G/K.

(ii) $M^* = \pi^{-1}(M)$ is (weakly) equifocal in G.

(iii) $\tilde{M} = \phi^{-1}(M^*)$ is (weakly) isoparametric in $L^2([0,1], \mathcal{G})$.

Hence we can apply the theory of (weakly) isoparametric submanifolds in Hilbert space to obtain many of the properties described in Theorems 4.8 and 4.12. This trick of lifting submanifolds of a symmetric space G/K to submanifolds in Hilbert space can be viewed as a useful linearization process. In fact, we can do the same for tautness:

THEOREM 5.11. *Let $V = L^2([0,1], \mathcal{G})$, and let $\phi : V \to G$ the parallel translation from 1 to 0, i.e., $\phi(u) = g(0)$, where g is the solution of $g^{-1}g_x = u$ and $g(1) = e$. Let M be a submanifold of G, and set $\tilde{M} = \phi^{-1}(M)$. Then M is taut in G if and only if $\tilde{M}$ is taut in V.*

PROOF. Suppose $p = e^{-a}$. Given $g \in P(G, M \times p)$, let $v = g^{-1}g_x$ and define $\tilde{g}(x) = g(x)e^{ax}$. Then $\tilde{g} \in P(G, M \times e)$ and

$$\tilde{g}^{-1}\tilde{g}_x = e^{-ax}g^{-1}g_x e^{ax} + a = e^{-ax}v(x)e^{ax} + a \in \tilde{M}.$$

Let $F_a : V \to V$ be the isometry defined by $F_a(v) = e^{-ax}ve^{ax} + a$, and $\psi : P(G, G \times p) \to V$ the diffeomorphism defined by $\psi(g) = g^{-1}g_x$. Then $(F_a \circ \psi)(P(G, M \times p)) = \tilde{M}$ and $E_p = f_a \circ F_a \circ \psi$. Since both F_a and ψ are diffeomorphisms, E_p is perfect if and only if f_a is perfect. □

THEOREM 5.12. *Let G/K be a symmetric space, $\pi : G \to G/K$ the natural projection, M a submanifold of G/K, and $M^* = \pi^{-1}(M)$. Then M is taut in G/K if and only if M^* is taut in G.*

PROOF. Fix $g_0 \in G$, and set $p_0 = \pi(g_0)$. Let F denote the diffeomorphism from $P(G/K, p_0 \times M) \times P(K, e \times K)$ to $P(G, g_0 \times M^*)$ defined by $F(x, k)(t) = \tilde{x}(t)k(t)$, where $\tilde{x}(t)$ is the horizontal lift of $x(t)$ to G with $\tilde{x}(0) = g_0$. Note that if α is a horizontal curve then $\alpha^{-1}\alpha_x \in \mathcal{P}$. Since $\mathcal{K} \perp \mathcal{P}$,

$$\|(\tilde{x}k)'\|^2 = \|\tilde{x}'k + \tilde{x}k'\|^2 = \|\tilde{x}^{-1}\tilde{x}' + k'k^{-1}\|^2 = \|\tilde{x}'\|^2 + \|k'\|^2.$$

So we have

$$E_{g_0}(F(x,k)) = E_{p_0}(x) + E_e(k).$$

Notice that $P(K, e \times K)$ is contractible and that the only critical point of $E_e : P(K, e \times K) \to \mathbb{R}$ is the constant path. This shows that $E_e : P(K, e \times K) \to \mathbb{R}$ is perfect. So E_{p_0} on $P(G/K, p_0 \times M)$ is perfect if and only if E_{g_0} is perfect on $P(G, g_0 \times M^*)$. □

If M is a taut submanifold of $N = \mathbb{R}^n$ and if C is a critical submanifold of the distance squared function f_a, then C is also taut in $\mathbb{R}^n$ [Ozawa 1986]. But it is not known whether this is true for general N.

QUESTION 5.13. Suppose M is a taut submanifold of a complete Riemannian manifold N and $\tilde{C}$ is a critical submanifold of the energy functional E_p on $P(N, p \times M)$, and let $C = \{\gamma(1) : \gamma \in \tilde{C}\}$. Is C taut in N?

Now we are ready to define the notion of Dupin submanifolds in any complete Riemannian manifold.

DEFINITION 5.14. Let M be a submanifold of a complete Riemannian manifold N, let $v_0 \in \nu(M)_{x_0}$ be a focal normal of multiplicity k of M, and set $p_0 = \exp(v_0)$. A k-dimensional submanifold S of N is called a *focal leaf* of M at x_0 if there exists a parallel normal field v of M along S such that

(i) $\exp(v(x)) = p_0$ for all $x \in S$,

(ii) $v(x_0) = v_0$,

(iii) $v(x)$ is a focal normal of multiplicity k of M for all $x \in S$.

DEFINITION 5.15. A submanifold M of a nonnegatively curved complete Riemannian manifold N is called *Dupin* if there exists an open dense subset $\mathcal{V}_0$ of the set $\mathcal{V}$ of all focal normals of M such that given any multiplicity-k focal normal $v_0 \in \mathcal{V}_0 \cap \nu(M)_{x_0}$ there exists a focal leaf S of M through x_0 such that S is a k-dimensional submanifold of the distance sphere centered at $\exp(v_0)$ with radius $\|v_0\|$, and S is diffeomorphic to S^k.

DEFINITION 5.16. A submanifold M of a nonnegatively curved complete Riemannian manifold N is called *proper Dupin* if it is Dupin and the multiplicities are locally constant on the set of focal normals of M.

REMARKS 5.17. (i) When $N = \mathbb{R}^n$, Definitions 5.14, 5.15 and 5.16 agree with the original definitions.

(ii) We restrict ourselves to nonnegatively curved manifolds in Definitions 5.15 and 5.16 for the following reason: If the ambient space N is the real hyperbolic space, then the conditions on M in the definitions would not imply anything for principal curvatures whose absolute value is less than or equal to one. Consequently, the definitions would not agree with the existing definition of Dupin and proper Dupin submanifolds in hyperbolic space. However, Definition 5.16 does make sense for arbitrary complete Riemannian manifold regardless of the sign of the curvature, and "Dupin" in our sense is closely related with the notion of tautness.

(iii) The local theory of Dupin submanifolds in hyperbolic space is equivalent to the one in Euclidean space and the sphere. The reason for this is that "stereographic projections" from the sphere and the hyperbolic space into the Euclidean space respects the Dupin property. The theory of compact Dupin submanifolds is therefore also the same in the real space forms. In hyperbolic space the local theory coincides with the theory of complete Dupin submanifolds for the following reason: The image of the hyperbolic space under the "stereographic projection" is a ball in the Euclidean space. Hence a local Dupin submanifold in hyperbolic space can be mapped to Euclidean space where by a homothety one can arrange that its boundary lies outside of the ball model. Now we can map it back to a complete Dupin submanifold in hyperbolic space.

From Theorems 5.10, 5.11, and 5.12 follows:

PROPOSITION 5.18. *Suppose M is a taut submanifold of the compact symmetric space G/K, and $v_0 \in \nu(M)_{x_0}$ is a focal normal of multiplicity k of M in G/K. Then there exists a k-dimensional focal leaf S through x_0. Moreover, S lies in the distance sphere of radius $\|v_0\|$ centered at $\exp(v_0)$.*

The following proposition follows from Theorems 5.5, 5.11 and 5.12.

PROPOSITION 5.19. *If M is a taut submanifold of a nonnegatively curved symmetric space, then M is Dupin.*

CONJECTURE 5.20. If M is taut in a nonnegatively curved complete Riemannian manifold N, then M is Dupin.

CONJECTURE 5.21. If M is a compact submanifold of a complete Riemannian manifold N, then there exists a subset $\mathcal{F}_0$ of $\nu(M)$ such that

(i) $\mathcal{F}_0$ is an open and dense subset of the set of all focal normals of M, and
(ii) if $v_0 \in \nu(M)_{x_0} \cap \mathcal{F}_0$ is a focal normal of multiplicity k and $k > 1$, there exists a focal leaf S through x_0 such that S is a k-dimensional submanifold of the distance sphere centered at $\exp(v_0)$ with radius $\|v_0\|$ and S is diffeomorphic to S^k.

If Conjecture 5.21 is true and M is a proper Dupin submanifold of N, then using a construction similar to the one explained in (g) on page 199, we can construct a linking cycle for each critical point of the energy functional E_p on $P(N, M \times p)$, i.e., M is taut in N.

The study of equifocal submanifolds (Section 4) indicates that the following more general class of submanifolds in complete Riemannian manifolds might have very rich geometric and topological properties.

DEFINITION 5.22. A submanifold M of a complete Riemannian manifold N is said to have *parallel focal structure* if

(i) $\nu(M)$ is globally flat, and
(ii) given a parallel normal field v on M such that $v(x_0)$ is a focal normal of multiplicity k of M, the vector $v(x)$ is a focal normal of multiplicity k of M for all $x \in M$.

CONJECTURE 5.23. Suppose M is a submanifold of a complete Riemannian manifold N and M has parallel focal structure. Then M is taut.

We end this section with a question of a somewhat different nature.

QUESTION 5.24. Is a totally geodesic submanifold in a compact symmetric space taut? Assume M is taut in N_1 which is totally geodesic in the compact symmetric space N_0. Is then M taut in N_0? If the answer to the second question is yes, it is sufficient to study whether maximal totally geodesic submanifolds are taut in the first question. Notice that it is very easy to find counterexamples to both questions if the ambient space is not symmetric.

6. Taut Points and Spheres

One of the main topics of this section is to study the geometric and topological properties of a Riemannian manifold containing taut points. We discuss the relation between Blaschke manifolds and manifolds all of whose points are taut. We also show that a point is taut if and only if a distance sphere centered at that point is taut. At the end of the section, we study the question whether a taut S^{n-1} in an n-dimensional complete Riemannian manifold is a distance sphere.

Let (N, g) be a complete Riemannian manifold, $p \in N$, and $\gamma_v(t) = \exp(tv)$ for $v \in TN_p$. Let

$$\mu(v) = \sup\{r > 0 : \gamma_v \mid [0, r] \text{ is a minimizing geodesic}\}.$$

If $\mu(v)$ is finite, $\exp(\mu(v)v)$ is called the *cut point* of p along γ_v.

PROPOSITION 6.1. *Suppose (N, g) is a complete simply connected Riemannian manifold of dimension n, with a taut point p. Then the first conjugate point of p along a geodesic coincides with the cut point of p along that geodesic and vise versa.*

PROOF. Since N is simply connected, $P(N, p \times q)$ is connected. By Corollary 2.10, any critical point of index 0 of E_q is an absolute minimum.

Suppose $\gamma_v(t) = \exp(tv)$, $\mu(v) = t_0$ and $\gamma(t_1)$ is the first conjugate point along γ. We will prove that $t_0 = t_1$. First notice that $t_0 \le t_1$ since a geodesic is not minimizing after its first conjugate point. Assume that $t_1 > t_0$. By definition of $\mu(v)$, for $t_2 \in (t_0, t_1)$, $\gamma = \gamma_v \mid [0, t_2]$ does not minimize the distance between p and $\gamma_v(t_2)$. Since there are no conjugate points on γ, γ is a critical point of $E_{\gamma(t_2)}$ with index zero. By the tautness of p, γ is therefore an absolute minimum, a contradiction. This proves that $t_0 = t_1$.

Assume that $\gamma(t_0)$ is a cut point of p along the geodesic γ and that there is no conjugate point of p along γ. Then, for $s > t_0$, $\gamma \mid [0, s]$ is not a minimum point of $E_{\gamma(s)}$ on $P(n, p \times \gamma(s))$. But $\gamma \mid [0, s]$ has index 0. Since p is taut, $E_{g(s)}(\gamma \mid [0, s])$ is an absolute minimum, a contradiction. So there must be conjugate point along γ, say at $\gamma(t_1)$. So $t_0 = t_1$.   □

COROLLARY 6.2. *Suppose (N, g) is a compact simply connected Riemannian manifold of dimension n with a taut point p. Then there is a conjugate point along every geodesic starting in p.*

PROOF. Since N is compact, every geodesic γ starting at p has a cut point of p. By Proposition 6.1, this cut point is a conjugate point along γ.   □

THEOREM 6.3. *If (N, g) is a symmetric space, then every point of N is taut.*

PROOF. Suppose $N = G/K$. By Theorem 3.3, orbits of the K-action on N are taut. But the K-orbit at eK is $\{eK\}$. So eK is a taut point. Using Proposition 2.3 and the fact that N is homogeneous, all points in N are taut.   □

This gives rise to the following question.

CONJECTURE 6.4. Suppose all points of (N, g) are taut, and N is homotopy equivalent to a compact symmetric space. Then (N, g) is a symmetric space.

We will prove this conjecture when N is a torus or a sphere by using deep results in Riemannian geometry. Then we will show that it is equivalent to the Blaschke conjecture when N is homotopy equivalent to a compact rank-one symmetric space. So this conjecture is more general than the Blaschke conjecture which is still not settled.

THEOREM 6.5. *Let (N, g) be a complete Riemannian manifold without conjugate points. Then all points in (N, g) are taut. Now let $\tilde{g}$ be another Riemannian metric on N with respect to which all points are taut. Then also $\tilde{g}$ has no conjugate points.*

PROOF. Let p and q be points in N. Then all critical points of the energy functional E_p on $P(N, p \times q)$ have index zero and are nondegenerate since there are no conjugate points. It follows that all points are taut. It also follows that all connected components of $P(N, p \times q)$ are contractible. Now let $\tilde{g}$ be a metric on N such that all points are taut. Let γ be a geodesic starting in p. If there is a conjugate point on γ, then we can find a q on γ such that E_q has a nondegenerate critical point with positive index i. By tautness, $P(N, p \times q)$ would have a nonvanishing Betti number in a positive dimension contradicting the contractibility of the connected components of $P(N, p \times q)$. Hence $\tilde{g}$ has no conjugate points.   □

E. Hopf [1948] proved that a Riemannian metric on a two-torus without conjugate points is flat. The generalization of this result to higher dimensions was one of the well-known open problems in Riemannian geometry. It was finally solved by Burago and Ivanov [1994].

THEOREM 6.6 [Burago and Ivanov 1994]. *Suppose (T^n, g) has no conjugate points. Then g is flat.*

It follows from Theorem 6.5 that all points on a flat torus are taut. As a consequence of Theorems 6.6 and 6.5, we have:

THEOREM 6.7. *Let g be a metric on T^n so that all points are taut. Then g is flat.*

We now come to Blaschke manifolds [Besse 1978]. We first review some definitions. Let p and q be distinct points in a complete Riemannian manifold M. Set $d = d(p, q)$. Then the *link from p to q* is defined to be the set $\Lambda(p, q)$ of unit vectors $X = \gamma'(d)$ in $T_q M$ where $\gamma : [0, d] \to M$ is a length-minimizing geodesic between p and q. A compact Riemannian manifold is said to be a *Blaschke manifold at p* if for every point q in the cut locus $C(p)$ of p the link $\Lambda(p, q)$ is a totally geodesic sphere in the unit sphere of $T_q M$. A Riemannian manifold is said to be *Blaschke* if it is Blaschke at all of its points. One says that a compact n-dimensional Riemannian manifold M is *Allamigeon–Warner* at a point $p \in M$ if there is a number $l > 0$ and an integer k between 1 and $n - 1$ such that the first conjugate point on any geodesic starting in p comes after distance l and has multiplicity k. It is proved in [Besse 1978, Theorem 5.43, p. 137] that if (M, g) is Blaschke at p, then M is Allamigeon–Warner at p. Moreover, if M is simply connected, then M is Blaschke at p if and only if it is Allamigeon–Warner at p.

It is well-known that if (N^n, g) is a simply connected compact, rank-one symmetric space, it is Blaschke and hence also Allamigeon–Warner at every point. The integer k in the definition of N being Allamigeon–Warner at a point is 1, 3, 7 and $n - 1$ if N is a complex projective space, a quaternionic projective space, a Cayley plane and a standard sphere respectively. Moreover, the first three nonvanishing Betti numbers of the space of based loops in N are b_0, b_k, b_{k+n-1}.

BLASCHKE CONJECTURE 6.8. *Every Blaschke manifold is isometric to a compact symmetric space of rank one.*

In spite of the name, the Blaschke conjecture was never made by Blaschke in this generality. It was solved for the two-sphere by L. Green [1963] and for the n-sphere by Berger and Kazdan:

THEOREM 6.9 [Besse 1978]. *If (S^n, g) is a Blaschke manifold, then (S^n, g) is the standard sphere with constant sectional curvature.*

The Blaschke Conjecture is still not settled for the other simply connected, compact rank-one symmetric spaces. But a lot of progress has been made; see the references in [Reznikov 1994]. For example, if (N, g) is a Blaschke manifold, then:

(i) There exists $l > 0$ such that all geodesics of N are closed with length l [Besse 1978].

(ii) N is homeomorphic to a compact rank-one symmetric space [Sato 1984; Yang 1990].

(iii) We may assume that all geodesics of N are closed with length 2π by rescaling the metric. Then the number $i(N) = \text{vol}(N)/\text{vol}(S^n)$ is an integer [Weinstein 1974].

(iv) If N is homeomorphic to a rank-one symmetric space N_0 then $i(N) = i(N_0)$ [Reznikov 1994].

We now want to apply our generalization of Ozawa's theorem 2.8 to study compact manifolds with taut points. In fact, we will show that under certain assumptions on the Betti numbers of the space of based loops, these manifolds are Blaschke manifolds. First we consider the sphere case.

THEOREM 6.10. *Let (N, g) be a simply connected compact Riemannian manifold of dimension n such that the first two nonvanishing Betti numbers of the based loop space are b_0 and b_{n-1}. If $p \in N$ is a taut point, then (N, g) is Blaschke at p and N is homeomorphic to S^n.*

PROOF. Let γ be a geodesic starting in p. By Proposition 6.1 and Corollary 6.2 there is a point q on γ that is the first conjugate point on γ and is also a cut point. Since the path space $P(N, p \times q)$ has trivial homology in dimensions $0 < i < n - 1$, it follows from the tautness of p that the multiplicity of q as a conjugate point along γ is $n - 1$. We assume that $\gamma(1) = q$. Then $\gamma_0 = \gamma \mid [0, 1]$ is a critical point of E_q on $P(N, p \times q)$ with index 0 and nullity $n - 1$. It now follows from Theorem 2.8 that there is an $(n - 1)$-dimensional nondegenerate critical manifold in $P(N, p \times q)$ through γ_0. As a consequence, all geodesics starting in p first meet in q after constant distance l, have their first conjugate point at constant distance l and this conjugate point is q. It follows easily that N is homeomorphic to a sphere.   □

As consequence of Theorems 6.9 and 6.10, we have:

COROLLARY 6.11. *Suppose all points of (N^n, g) are taut, and N is homotopy equivalent to S^n. Then (N, g) is isometric to the standard sphere.*

COROLLARY 6.12. *If all points of (S^n, g) are taut, then g is the standard metric.*

REMARK 6.13. There are Riemannian metrics on spheres with some, but not all, points taut. For example, let (S^2, g) be a surface of revolution whose curvature is not constant. Then the $SO(2)$-action is hyperpolar, and the north and the south poles are fixed points of the action. So both the north and the south poles are taut. Not all points of (S^2, g) are taut, because otherwise g would be the standard metric on S^2 by Corollary 6.12.

THEOREM 6.14. *Let (N, g) be a simply connected compact Riemannian manifold of dimension n. Suppose the first three nonvanishing Betti numbers b_i of the based loop space of N are b_0, b_a and b_{a+n-1} for some $1 \le a < n - 1$ and $b_a = 1$. If p is taut in N, then (N, g) is Blaschke at p.*

PROOF. Let $\gamma : [0, \infty) \to N$ be a unit-speed geodesic starting at p. By Corollary 6.2, γ has a conjugate point. Let q be the first conjugate point along γ. Using the tautness of p and the assumption on the Betti numbers of the based loop space, the multiplicity of q must be a. It now follows again from the assumption on the Betti numbers of the loop space and the tautness of p that the multiplicity of the second conjugate point q_2 on γ must be $n-1$. It follows from Theorem 2.8 that all geodesics starting in p will meet at q_2 after distance l, where l is the arc length of γ from p to q_2. Furthermore, all the geodesics of length l from p to q_2 have index a with q_2 as a second conjugate point of multiplicity $n-1$. We claim that $q_2 = p$. To see this, notice that the energy functional $E_p : P(N, p{\times}p) \to \mathbb{R}$ has a critical point γ of index a. It follows that γ is degenerate, since otherwise both γ and γ^{-1} would be nondegenerate critical points of index a contradicting the tautness of p and $b_a = 1$. This implies that $q_2 = p$. Using Theorem 2.8, we obtain that every geodesic γ starting in p will be back to p after length l, the first conjugate point has multiplicity a, and the second conjugate point along γ is $\gamma(l) = p$ with multiplicity $n-1$.

We would now like to show that the first conjugate point on γ comes at distance $l/2$. Assume it comes earlier at distance t_0. We look at γ^{-1} between p and $q = \gamma(t_0 + \varepsilon)$, where ε is a small number such that neither γ nor γ^{-1} have a conjugate point in q and $t_0 + \varepsilon \le l/2$. Then γ^{-1} from p to q is not a minimizing geodesic between p and q. By tautness it must therefore have a conjugate point. This conjugate point must have multiplicity a by what we have already proved. It follows that $E_q : P(N, p{\times}q) \to \mathbb{R}$ has at least two nondegenerate critical points of index a contradicting tautness and $b_a = 1$.

Now assume that the first conjugate point on γ comes later than $l/2$, say at $\gamma(t_0) = q$. Then $\gamma\,|\,[0, t_0]$ is an index 0 critical point of E_q. So γ is a minimizing geodesic from p to q by tautness of p. But γ^{-1} from p to q is shorter, a contradiction. It follows that the first conjugate point of a geodesic starting in p comes at distance $l/2$ and has multiplicity a.

So we have proved that every geodesic starting in p has its first conjugate point after distance $l/2$ and its multiplicity is a. This is exactly the definition of N being Allamigeon–Warner at p. Since N is simply connected it follows from [Besse 1978, Theorem 5.43] that N is Blaschke at p. This finishes the proof. $\square$

COROLLARY 6.15. *Suppose all points of (N, g) are taut, and the first three nonzero Betti numbers of the based loop space of N are b_0, b_a, b_{a+n-1} for some $1 \le a \le n-1$ and $b_a = 1$. Then (N, g) is Blaschke. In particular, N is homeomorphic to a compact rank-one symmetric space.*

COROLLARY 6.16. *Suppose N is homotopy equivalent to a simply connected rank-one symmetric space, and g is a Riemannian metric on N such that all points of (N, g) are taut. Then (N, g) is Blaschke.*

Next we recall a definition in [Besse 1978]:

DEFINITION 6.17. *A Riemannian manifold (N, g) is called a L_l^p-manifold if all geodesics starting at p return to p after length l.*

It follows from the proof of Theorem 6.14 that:

COROLLARY 6.18. *Let (N, g) be as in Theorem 6.14. If (N, g) has a taut point p, then (N, g) is a L_l^p-manifold.*

One can say a lot about the topology of L_l^p-manifolds. Using Corollary 6.18 and results from [Bott 1954; Samelson 1963] (compare [Besse 1978, Theorem 7.23]) on L_l^p-manifolds, we have:

THEOREM 6.19. *Let (N, g) be as in Theorem 6.14. Assume that N has a taut point. Then a must be 1, 3, 7, or $n - 1$. Moreover:*

(1) *If $a = 1$, then $n = 2m$ and N has the homotopy type of $\mathbb{CP}^m$.*
(2) *If $a = 3$, then $n = 4m$ and N has the integral cohomology ring of $\mathbb{HP}^n$.*
(3) *If $a = 7$, then $n = 16$ and N has the integral cohomology ring of CaP^2.*
(4) *If $a = n - 1$, then N is homeomorphic to S^n.*

The next theorem gives a converse of Theorem 6.14, which shows that Conjecture 6.4 is more general than the Blaschke Conjecture 6.8.

THEOREM 6.20. *Let N be a simply connected Blaschke manifold. Then all points in N are taut.*

PROOF. There is a number l such that every unit-speed geodesic in N is a simple closed geodesic with least period l [Besse 1978, Corollary 5.42]. Here simple means that there is no self-intersection in one period. Moreover, for every p and a point q in the cut locus of p we have $d(p, q) = l/2$ [Besse 1978, Proposition 5.39]. The conjugate points on a closed geodesics come after distance $l/2$, l, $3l/2$ and so on. Now fix p and let $q \neq p$ be a point in N that is not at distance $l/2$ from p. Then every geodesic between p and q has image on the same closed geodesic. It follows that $E_q : P(N, p \times q) \to \mathbb{R}$ is a Morse function whose critical points have indices a, $a + n - 1$, $2a + n - 1$, $2(a + n - 1)$ and so on. It is clear that E_q is perfect if $a > 1$ (and hence $n > 2$), since there are no critical points with indices differing by 1. If $a = 1$, then by [Besse 1978, Theorem 7.23], N is homotopy equivalent to a $\mathbb{CP}^m$, where $2m = n$. The loop spaces of N and $\mathbb{CP}^m$ are therefore also homotopy equivalent. It follows that the loop space of N has nontrivial homology in dimensions a, $a + n - 1$, $2a + n - 1$, $2(a + n - 1)$ and so on, and hence that E_q is perfect.   □

Combining Corollary 6.16 and Theorem 6.20 we get:

THEOREM 6.21. *Let (N, g) be a simply connected compact Riemannian manifold that is homotopy equivalent to a compact rank one symmetric space. Then (N, g) is Blaschke if and only if all points of (N, g) are taut.*

We assume that N is topologically or homotopically a rank-one symmetric space in the discussion above for simplicity. Of course, the following problem is one of our main interests:

PROBLEM 6.22. Suppose (N, g) is a simply connected complete Riemannian manifold such that all points of N are taut. What can we say about the geometry and topology of (N, g)?

The last topic we will deal with in this section is tautness of spheres. First we prove a theorem that relates tautness of a point p and tautness of the distance spheres centered at p.

THEOREM 6.23. *Let N be a complete Riemannian manifold and p a point in N. Set $\phi = \exp_p : S_\varepsilon(0) \to N$, where $S_\varepsilon(0)$ is the sphere of radius ε around the origin in TM_p and ε is smaller than the conjugate radius of N at p. Then ϕ is taut if and only if p is taut.*

PROOF. Let B denote the geodesic disk $B_\varepsilon(p)$. Since B is contractible, the fibration $P(N, B \times p) \to B$ defined by $\gamma \mapsto \gamma(0)$ is trivial. Hence the fibration restricts to $P(N, \phi \times p)$ is trivial, and we have

$$P(N, \phi \times p) \sim S \times P(N, p \times p).$$

So the Poincaré polynomial of $P(N, \phi \times p)$ is

$$(1 + t^{n-1}) \sum b_k t^k, \tag{6.1}$$

where $n = \dim N$ and b_k is the k-th Betti number of $P(N, p \times p)$. Now let q be some point in N. Then the critical points of $E_q : P(N, \phi \times q) \to \mathbb{R}$ are pairs (r, γ) where γ is a geodesic starting perpendicularly to $\phi(S_\varepsilon(p))$ in $\phi(r)$ and ending in q. There are two possibilities: either after adding to γ or deleting from γ a geodesic segment of length ε we end up in p. A critical point of $E_q : P(N, p \times q) \to \mathbb{R}$ of index k therefore gives rise two critical points of $E_p : P(N, \phi \times q) \to R$ of index k and $k + (n - 1)$ respectively, and vise versa. The equivalence of the tautness of S and p now follows from Equation (6.1). □

As a consequence of Theorems 6.3 and 6.23, we have:

COROLLARY 6.24. *The distance spheres in a symmetric space N are taut if their radius is smaller than the conjugate radius of N.*

Next we study whether a null-homotopic taut sphere S^{n-1} in an n-dimensional manifold (N, g) must be a distance sphere. First remember that if S is an embedded hypersurface of N that is null-homotopic, then $N \setminus S$ has two components. If one of the components is diffeomorphic to an n-dimensional ball, then we say that S bounds a ball on one side.

THEOREM 6.25. *Let $S = S^{n-1}$ be a null-homotopic embedded taut hypersurface in a complete Riemannian manifold N of dimension n. Assume that S^{n-1} bounds a ball on one side. Then S is a distance sphere.*

PROOF. We denote by B the component of $M \setminus S$ that is diffeomorphic to an n-ball. We look at the parallel hypersurfaces of S in B. There must be a singularity in one of them. There is therefore a geodesic γ starting perpendicularly in S, going into B and having a first focal point p in B at distance l. We assume l is minimal with this property. We denote the multiplicity of this focal point by k and want to prove that it is equal to $n - 1$. By Theorem 2.8, there is a k-dimensional critical submanifold C of minima of E_p through γ in $P(N, S \times p)$, which, by tautness represents a nontrivial homology class of $P(N, S \times p)$. Now C is contained in $P(B, S \times p)$. This implies that C represents a nontrivial homology class of $P(B, S \times p)$. Since $P(B, S \times p)$ does not have any nontrivial homology class of positive dimension less than $n - 1$, it follows that $k = n - 1$. The focal point p is therefore of index $n - 1$ and it follows that C is an $(n - 1)$-dimensional family of geodesics starting perpendicularly to S, going into B and meeting in p at distance l. Hence all geodesics starting perpendicularly to S and going into B meet in p at distance l. It follows that S is a geodesic sphere. $\square$

The Differentiable Schoenfliess Theorem says that if $n \geq 5$ an embedded S^{n-1} in S^n always bounds a ball [Milnor 1965, Proposition D, p. 112]. The question whether an embedded and null-homotopic S^{n-1} in an arbitrary smooth n-manifold N bounds a ball is more complicated. It was answered by Ruberman [Ruberman 1997]:

THEOREM 6.26. *Suppose that $i : S^{n-1} \to N$ is a null-homotopic smooth embedding, where N has dimension n. Let $S = i(S^{n-1})$. Then one of the following statements must hold:*

(1) *S bounds a ball on one side.*
(2) *N is a rational homology sphere, the fundamental groups of both components of $N \setminus S$ are finite, and at least one of them is trivial.*

It follows from [Chern and Lashof 1957] that any taut hypersurface in S^n that is homeomorphic to a sphere, is a distance sphere. Together with Theorems 6.25 and 6.26, this implies the following results:

COROLLARY 6.27. *A null-homotopic, embedded taut hypersphere in a symmetric space must be a distance sphere.*

COROLLARY 6.28. *Suppose N is homotopy equivalent to a compact symmetric space which is not a sphere. Then a null-homotopic, embedded, taut hypersphere of (N, g) must be a distance sphere.*

THEOREM 6.29. *Suppose the n-dimensional manifold N is not a rational homology sphere. Then a null-homotopic, embedded taut hypersphere S^{n-1} of (N, g) must be a distance sphere.*

Appendix: Applications of Infinite-Dimensional Morse Theory to Section 2

Let $\phi : M \to N$ be an immersion, and take $p \in N$. It is known that $E_p : P(N, \phi \times p) \to \mathbb{R}$ satisfies the Palais–Smale condition. So we can apply infinite-dimensional Morse theory to E_p. As in Section 2 we set

$$\mathcal{M}_p = P(N, \phi \times p) \quad \text{and} \quad \mathcal{M}_p^r = \{\gamma \in \mathcal{M}_p : E_p(\gamma) \leq r\},$$

and let $R(E_p)$ and $C(E_p)$ denote the set of regular values and the set of singular values of E_p. In this appendix, we will explain how one can use infinite-dimensional Morse theory to prove the following theorem.

THEOREM A.1. *If M is an immersed taut submanifold of N and $p \in N$ and r is a regular value of E_p, then the map*

$$i_* : H_*(\mathcal{M}_p^r) \to H_*(\mathcal{M}_p)$$

induced by the inclusion of $\mathcal{M}_p^r$ in $\mathcal{M}_p$ is injective.

The proof follows the same method as the proof of [Terng 1989, Proposition 5.8], except that here the function E_p changes its domain as p changes in N. But it is easy to see that it suffices to prove analogues of [Terng 1989, Propositions 5.6 and 5.7], here called Lemmas A.3 and A.4, and the rest of the proof is exactly the same.

We will make some remarks at the end of this appendix on how we can prove injectivity of i_* for all r with these methods if we use Čech homology instead of singular homology. Notice that we proved in Section 2 that i_* is injective for all r in singular homology, but that proof relies on Theorem 2.8 and is therefore more difficult. It would be nice to have a proof of Theorem 2.8 that does not use finite-dimensional approximations. The computations below should also be useful for writing such a proof.

Recall that the tangent space of $\mathcal{M}_p$ at γ is the set of all absolutely continuous vector fields v along γ such that $v(0) \in TM_{\gamma(0)}$ and $v(1) = 0$. A Riemannian metric is defined on $\mathcal{M}_p$ by

$$\langle u, v \rangle_1 = \int_0^1 (\nabla_{\gamma'} u(t), \nabla_{\gamma'} v(t))\, dt.$$

The gradient $\nabla E_p(\gamma) \in T(\mathcal{M}_p)_\gamma$ of E_p is implicitly defined by

$$d(E_p)_\gamma(u) = \langle \nabla E_p(\gamma), u \rangle_1$$

for all $u \in T(\mathcal{M}_p)_\gamma$. We next prove a formula for $\nabla E_p(\gamma)$.

PROPOSITION A.2. *Let $\gamma \in \mathcal{M}_p$, and let $\{e_i\}$ be a parallel orthonormal frame field along γ such that $e_1(0), \dots, e_m(0)$ span $TM_{\gamma(0)}$. Write $\gamma'(t) = \sum_i a_i(t) e_i(t)$.*

Then

$$\|\nabla E_p(\gamma)\|_1^2 = \sum_{i \leq m} \int_0^1 a_i^2 \, dt + \sum_{i > m} \int_0^1 (a_i - \alpha_i)^2 = E(\gamma) - \sum_{i > m} \alpha_i^2,$$

where $\alpha_i = \int_0^1 a_i(t) \, dt$ denotes the mean value of a_i.

PROOF. Let F denote $\nabla E_p(\gamma)$, and write $F = \sum_i f_i e_i$. Note that

$$d(E_p)_\gamma(u) = \int_0^1 (\nabla_{\gamma'} u, \gamma') \, dt = \int_0^1 (\gamma', u)' \, dt - \int_0^1 (\nabla_{\gamma'} \gamma', u) \, dt$$

$$= \langle F, u \rangle_1 = \int_0^1 (\nabla_{\gamma'} F, \nabla_{\gamma'} u) \, dt$$

$$= \int_0^1 (\nabla_{\gamma'} F, u)' \, dt - \int_0^1 (\nabla_{\gamma'} \nabla_{\gamma'} F, u) \, dt,$$

for all $u \in T(\mathcal{M}_p)_\gamma$. So we have $f_i'' = a_i'$, $f_i(1) = 0$ for all i, $f_i(0) = 0$ for all $i > m$, and

$$\int_0^1 \sum_i (f_i' - a_i)'(t) \, dt = 0.$$

The proposition now follows by a direct computation. □

LEMMA A.3. *If $[r, s] \subset R(E_p)$, there exists $\delta > 0$ such that $[r, s] \subset R(E_q)$ if the distance $d(p, q) < \delta$.*

PROOF. If the claim is not true, there exist a sequence $q_n \in N$ converging to p and critical points γ_n of E_{q_n} on $P(N, M \times q_n)$ such that $E_{q_n}(\gamma_n) \in [r, s]$. Since γ_n is a critical point of E_{p_n}, there is a $v_n \in TN_{q_n}$ such that

$$\gamma_n(t) = \exp((1-t)v_n) \quad \text{and} \quad \|v_n\|^2 = E(\gamma_n).$$

But $q_n \to p$ and $\|v_n\| \leq \sqrt{s}$. So there exists a subsequence, still denoted by v_n, converging to some $v_0 \in TN_p$. Set $\gamma(t) = \exp((1-t)v_0)$. Then γ is a critical point of E_p and $E(\gamma) \in [r, s]$, a contradiction. □

LEMMA A.4. *If $[r, s] \subset R(E_p)$, then there exist $\delta_1 > 0, \delta_2 > 0$ such that $d(p, q) < \delta_1$ and $E_q(\gamma) \in [r, s]$ imply that $\|\nabla E_q(\gamma)\|_1 \geq \delta_2$.*

PROOF. Suppose the claim is false. First choose a $\delta > 0$ as in Lemma A.3 that is also less than the injectivity radius at p. Then there exist a sequence $q_n \in N$ and $\gamma_n \in \mathcal{M}_{q_n}$ such that $d(q_n, p) < \delta$, $q_n \to p$, $E_{q_n}(\gamma_n) \in [r, s]$ and $\|\nabla E_{q_n}(\gamma_n)\|_1 \to 0$.

Set $\delta_n = d(p, q_n)$ and let $\beta_n : [1-\delta_n, 1] \to N$ denote the geodesic joining q_n to p parametrized by arc length, and

$$\tilde{\gamma}_n(t) = \begin{cases} \gamma_n\left(\dfrac{t}{1-\delta_n}\right), & \text{if } t \in [0, 1-\delta_n], \\ \beta_n(t), & \text{if } t \in [1-\delta_n, 1]. \end{cases}$$

Let $e_i(t)$ be a parallel orthonormal frame along $\tilde{\gamma}_n$ as in Proposition A.2. Write $\gamma_n' = \sum_i a_i^n e_i$, and $\tilde{\gamma}_n' = \sum_i \tilde{a}_i^n e_i$ and $\beta_n' = \sum_i d_i^n e_i$. Then the d_i^n are constant and $\sum_i (d_i^n)^2 = 1$,

$$\tilde{a}_i^n(t) = \begin{cases} \dfrac{1}{1 - \delta_n} a_i^n \left(\dfrac{t}{1 - \delta_n} \right), & \text{if } t \in [0, 1 - \delta_n], \\ d_i^n, & \text{if } t \in [1 - \delta_n, 1]. \end{cases}$$

A direct computation shows that the mean $\tilde{\alpha}_i^n$ of $\tilde{a}_i^n$ is related to the mean α_i^n of a_i^n by

$$\tilde{\alpha}_i^n = \alpha_i^n + \delta_n d_i^n.$$

Using Proposition A.2, we obtain

$$\|\nabla E_p(\tilde{\gamma}_n)\|_1^2 = \frac{\|\nabla E_{q_n}(\gamma_n)\|_1^2}{1 - \delta_n} + \delta_n + \sum_{i > m} \left(\frac{\delta_n}{1 - \delta_n} (d_i^n)^2 - 2\alpha_i^n d_i^n \delta_n - (d_i^n)^2 \delta_n^2 \right).$$

But $\|\nabla E_{q_n}(\gamma_n)\|_1 \to 0$, $\|d^n\| = 1$ and $\delta_n \to 0$. So we have $\|\nabla E_p(\tilde{\gamma}_n)\|_1 \to 0$. Since E_p satisfies condition C, there is a convergent subsequence $\tilde{\gamma}_{n_k} \to \gamma_0$ and γ_0 is a critical point of E_p. It is clear that $E_p(\gamma_0) \in [r, s]$, contradicting the assumption that $[r, s] \subset R(E_p)$. □

REMARK A.5. We now explain how Theorem A.1 can be proved for all values r, not only regular ones, if we replace singular homology by Čech homology. We will denote Čech homology by $\check{H}_*$ and singular homology by H_*.

Let r be any real number and let (r_k) be a decreasing sequence of regular values of E_p that converge to r. Notice that $\check{H}_*(\mathcal{M}_p) = H_*(\mathcal{M}_p)$ and $\check{H}_*(\mathcal{M}_p^{r_k}) = H_*(\mathcal{M}_p^{r_k})$ for all k since $\mathcal{M}_p$ and $\mathcal{M}_p^{r_k}$ are manifolds (with or without boundary). By Theorem A.1 we have

$$\check{H}_*(\mathcal{M}_p^{r_k}) \to \check{H}_*(\mathcal{M}_p)$$

is injective for all k. We have

$$\check{H}_*(\mathcal{M}_p^r) = \lim_{k \to \infty} \check{H}_*(\mathcal{M}_p^{r_k})$$

by continuity of Čech homology. We also know that $\check{H}_*(\mathcal{M}_p^r) = \check{H}_*(\mathcal{M}_p^{r_k})$ for $k > k_0$ for some k_0 since $\check{H}_*(\mathcal{M}_p^{r_l}) \to \check{H}_*(\mathcal{M}_p^{r_m})$ is injective for $l > m$ and $H_*(\mathcal{M}_p^{r_k})$ is finite-dimensional for all k so that the sequence $(H_*(\mathcal{M}_p^{r_k}))$ must stabilize for big k. It now follows that

$$\check{H}_*(\mathcal{M}_p^r) \to \check{H}_*(\mathcal{M}_p)$$

is injective as we wanted to show.

The above argument is quite typical in the theory of tight and taut immersions; see [Kuiper 1980], for example.

It follows from Theorem 2.8 that $\mathcal{M}_p^r$ is homotopy equivalent to a CW-complex for all r. Consequently, Čech and singular homology coincide, and we have injectivity in singular homology in Theorem A.1 for all r (see Section 2).

Acknowledgemnt

We thank Professor D. Ruberman for proving a topological result we need in this paper: A null-homotopic embedded S^{n-1} in N^n bounds a ball on one side if N is not a rational homology sphere (Theorem 6.26). The proof is given in [Ruberman 1997] in this volume.

References

[Alekseevskiĭ and Alekseevskĭ 1992] A. V. Alekseevskiĭ and D. V. Alekseevskĭ, "G-manifolds with one-dimensional orbit space", pp. 1–31 in *Lie groups, their discrete subgroups, and invariant theory*, edited by E. B. Vinberg., Adv. Soviet Math. **8**, Amer. Math. Soc., Providence, RI, 1992.

[Alekseevskiĭ and Alekseevskĭ 1993] A. V. Alekseevskiĭ and D. V. Alekseevskĭ, "Riemannian G-manifold with one-dimensional orbit space", *Ann. Global Anal. Geom.* **11**:3 (1993), 197–211.

[Banchoff 1970] T. F. Banchoff, "The spherical two-piece property and tight surfaces in spheres", *J. Differential Geometry* **4** (1970), 193–205.

[Besse 1978] A. L. Besse, *Manifolds all of whose geodesics are closed*, Ergebnisse der Mathematik und ihrer Grenzgebiete **93**, Springer, Berlin, 1978.

[Bott 1954] R. Bott, "On manifolds all of whose geodesics are closed", *Ann. of Math.* (2) **60** (1954), 375–382.

[Bott 1956] R. Bott, "An application of the Morse theory to the topology of Lie-groups", *Bull. Soc. Math. France* **84** (1956), 251–281.

[Bott and Samelson 1958] R. Bott and H. Samelson, "Applications of the theory of Morse to symmetric spaces", *Amer. J. Math.* **80** (1958), 964–1029. Corrections in **83** (1961), pp. 207–208.

[Burago and Ivanov 1994] D. Burago and S. Ivanov, "Riemannian tori without conjugate points are flat", *Geom. Funct. Anal.* **4**:3 (1994), 259–269.

[Carter and West 1972] S. Carter and A. West, "Tight and taut immersions", *Proc. London Math. Soc.* (3) **25** (1972), 701–720.

[Carter and West 1985] S. Carter and A. West, "Isoparametric systems and transnormality", *Proc. London Math. Soc.* (3) **51**:3 (1985), 520–542.

[Carter and West 1990] S. Carter and A. West, "Isoparametric and totally focal submanifolds", *Proc. London Math. Soc.* (3) **60**:3 (1990), 609–624.

[Cecil and Ryan 1985] T. E. Cecil and P. J. Ryan, *Tight and taut immersions of manifolds*, Research Notes in Mathematics **107**, Pitman, Boston, 1985.

[Chern and Lashof 1957] S.-s. Chern and R. K. Lashof, "On the total curvature of immersed manifolds", *Amer. J. Math.* **79** (1957), 306–318.

[Chern and Lashof 1958] S.-s. Chern and R. K. Lashof, "On the total curvature of immersed manifolds II", *Michigan Math. J.* **5** (1958), 5–12.

[Conlon 1971] L. Conlon, "Variational completeness and K-transversal domains", *J. Differential Geometry* **5** (1971), 135–147.

[Dadok 1985] J. Dadok, "Polar coordinates induced by actions of compact Lie groups", *Trans. Amer. Math. Soc.* **288**:1 (1985), 125–137.

[Ewert 1997] H. Ewert, *Equifocal submanifolds in Riemannian symmetric spaces*, Ph.D. thesis, Universität zu Köln, 1997.

[Ferus et al. 1981] D. Ferus, H. Karcher, and H. F. Münzner, "Cliffordalgebren und neue isoparametrische Hyperflächen", *Math. Z.* **177** (1981), 479–502.

[Gluck et al. 1983] H. Gluck, F. Warner, and C. T. Yang, "Division algebras, fibrations of spheres by great spheres and the topological determination of space by the gross behavior of its geodesics", *Duke Math. J.* **50**:4 (1983), 1041–1076.

[Green 1963] L. W. Green, "Auf Wiedersehensflächen", *Ann. of Math.* (2) **78** (1963), 289–299.

[Harle 1982] C. E. Harle, "Isoparametric families of submanifolds", *Bol. Soc. Brasil. Mat.* **13**:2 (1982), 35–48.

[Heintze et al. 1994] E. Heintze, R. S. Palais, C.-L. Terng, and G. Thorbergsson, "Hyperpolar actions and k-flat homogeneous spaces", *J. Reine Angew. Math.* **454** (1994), 163–179.

[Heintze et al. 1995] E. Heintze, R. S. Palais, C.-L. Terng, and G. Thorbergsson, "Hyperpolar actions on symmetric spaces", pp. 214–245 in *Geometry, topology, and physics for Raoul Bott*, edited by S.-T. Yau, Conf. Proc. Lecture Notes Geom. Topology 4, Internat. Press, Cambridge, MA, 1995.

[Hermann 1960] R. Hermann, "Variational completeness for compact symmetric spaces", *Proc. Amer. Math. Soc.* **11** (1960), 544–546.

[Hopf 1948] E. Hopf, "Closed surfaces without conjugate points", *Proc. Nat. Acad. Sci. USA* **34** (1948), 47–51.

[Hsiang et al. 1988] W.-Y. Hsiang, R. S. Palais, and C.-L. Terng, "The topology of isoparametric submanifolds", *J. Differential Geom.* **27**:3 (1988), 423–460.

[Kobayashi 1967] S. Kobayashi, "Imbeddings of homogeneous spaces with minimum total curvature", *Tôhoku Math. J.* (2) **19** (1967), 63–70.

[Kuiper 1959] N. H. Kuiper, "Immersions with minimal total absolute curvature", pp. 75–88 in *Colloque Géom. Diff. Globale* (Bruxelles, 1958), Centre Belge Rech. Math., Louvain, 1959.

[Kuiper 1980] N. H. Kuiper, "Tight embeddings and maps: Submanifolds of geometrical class three in $E^{n''}$", pp. 97–145 in *The Chern Symposium* (Berkeley, 1979), edited by W.-Y. Hsiang et al., Springer, New York, 1980.

[Milnor 1963] J. Milnor, *Morse theory*, Annals of Mathematical Studies **51**, Princeton U. Press, Princeton, 1963.

[Milnor 1965] J. Milnor, *Lectures on the h-cobordism theorem*, Princeton University Press, Princeton, N.J., 1965. Notes by L. Siebenmann and J. Sondow.

[Morse and Cairns 1969] M. Morse and S. S. Cairns, *Critical point theory in global analysis and differential topology: An introduction*, Pure and Applied Mathematics **33**, Academic Press, New York, 1969.

[Münzner 1980] H. F. Münzner, "Isoparametrische Hyperflächen in Sphären", *Math. Ann.* **251** (1980), 57–71.

[Münzner 1981] H. F. Münzner, "Isoparametrische Hyperflächen in Sphären II: Über die Zerlegung der Sphäre in Ballbündel", *Math. Ann.* **256** (1981), 215–232.

[Ozawa 1986] T. Ozawa, "On critical sets of distance functions to a taut submanifold", *Math. Ann.* **276**:1 (1986), 91–96.

[Ozeki and Takeuchi 1975] H. Ozeki and M. Takeuchi, "On some types of isoparametric hypersurfaces in spheres I", *Tôhoku Math. J.* (2) **27**:4 (1975), 515–559.

[Palais 1963] R. S. Palais, "Morse theory on Hilbert manifolds", *Topology* **2** (1963), 299–340.

[Palais and Smale 1964] R. S. Palais and S. Smale, "A generalized Morse theory", *Bull. Amer. Math. Soc.* **70** (1964), 165–172.

[Palais and Terng 1987] R. S. Palais and C.-L. Terng, "A general theory of canonical forms", *Trans. Amer. Math. Soc.* **300**:2 (1987), 771–789.

[Palais and Terng 1988] R. S. Palais and C.-L. Terng, *Critical point theory and submanifold geometry*, Lecture Notes in Math. **1353**, Springer, Berlin, 1988.

[Pinkall 1985] U. Pinkall, "Dupin hypersurfaces", *Math. Ann.* **270**:3 (1985), 427–440.

[Pinkall 1986] U. Pinkall, "Curvature properties of taut submanifolds", *Geom. Dedicata* **20**:1 (1986), 79–83.

[Pinkall and Thorbergsson 1989] U. Pinkall and G. Thorbergsson, "Deformations of Dupin hypersurfaces", *Proc. Amer. Math. Soc.* **107**:4 (1989), 1037–1043.

[Reckziegel 1979] H. Reckziegel, "On the eigenvalues of the shape operator of an isometric immersion into a space of constant curvature", *Math. Ann.* **243** (1979), 71–82.

[Reznikov 1985a] A. G. Reznikov, "The volume of certain manifolds with closed geodesics", *Ukrain. Geom. Sb.* **28** (1985), 102–106. In Russian.

[Reznikov 1985b] A. G. Reznikov, "The weak Blaschke conjecture for $\mathbb{HP}^n$", *Dokl. Akad. Nauk SSSR* **283**:2 (1985), 308–312.

[Reznikov 1994] A. G. Reznikov, "The weak Blaschke conjecture for $\mathbb{CP}^n$", *Invent. Math.* **117**:3 (1994), 447–454.

[Ruberman 1997] D. Ruberman, "Null-homotopic, codimension-one embedded spheres", pp. 229–232 in *Tight and Taut Submanifolds*, edited by T. E. Cecil and S.-s. Chern, Cambridge U. Press, 1997.

[Samelson 1963] H. Samelson, "On manifolds with many closed geodesics", *Portugal. Math.* **22** (1963), 193–196.

[Sato 1984] H. Sato, "On topological Blaschke conjecture, I: Cohomological complex projective spaces", pp. 231–238 in *Geometry of geodesics and related topics* (Tokyo, 1982), edited by K. Shiohama, Adv. Stud. Pure Math. **3**, North-Holland, Amsterdam, 1984.

[Sharpe 1988] R. W. Sharpe, "Total absolute curvature and embedded Morse numbers", *J. Differential Geom.* **28**:1 (1988), 59–92.

[Smale 1964] S. Smale, "Morse theory and a non-linear generalization of the Dirichlet problem", *Ann. of Math.* (2) **80** (1964), 382–396.

[Smale 1965] S. Smale, "An infinite dimensional version of Sard's theorem", *Amer. J. Math.* **87** (1965), 861–866.

[Takeuchi and Kobayashi 1968] M. Takeuchi and S. Kobayashi, "Minimal imbeddings of R-spaces", *J. Differential Geometry* **2** (1968), 203–215.

[Terng 1985] C.-L. Terng, "Isoparametric submanifolds and their Coxeter groups", *J. Differential Geom.* **21**:1 (1985), 79–107.

[Terng 1987] C.-L. Terng, "Submanifolds with flat normal bundle", *Math. Ann.* **277**:1 (1987), 95–111.

[Terng 1989] C.-L. Terng, "Proper Fredholm submanifolds of Hilbert space", *J. Differential Geom.* **29**:1 (1989), 9–47.

[Terng 1991] C.-L. Terng, "Variational completeness and infinite-dimensional geometry", pp. 279–293 in *Geometry and topology of submanifolds, III* (Leeds, 1990), edited by L. Verstraelen and A. West, World Sci. Publishing, River Edge, NJ, 1991.

[Terng 1995] C.-L. Terng, "Polar actions on Hilbert spaces", *J. Geom. Anal.* **5**:1 (1995), 129–150.

[Terng and Thorbergsson 1995] C.-L. Terng and G. Thorbergsson, "Submanifold geometry in symmetric spacs", *J. Differential Geom.* **42**:3 (1995), 665–718.

[Thorbergsson 1983] G. Thorbergsson, "Dupin hypersurfaces", *Bull. London Math. Soc.* (2) **15**:5 (1983), 493–498.

[Thorbergsson 1991] G. Thorbergsson, "Isoparametric foliations and their buildings", *Ann. of Math.* (2) **133**:2 (1991), 429–446.

[Wang 1982] Q. M. Wang, "Isoparametric hypersurfaces in complex projective spaces", pp. 1509–1523 in *Symposium on Differential Geometry and Differential Equations* (Beijing, 1980), vol. 3, Science Press, Beijing, 1982.

[Warner 1965] F. W. Warner, "The conjugate locus of a Riemannian manifold", *Amer. J. Math.* **87** (1965), 575–604.

[Warner 1967] F. W. Warner, "Conjugate loci of constant order", *Ann. of Math.* (2) **86** (1967), 192–212.

[Weinstein 1974] A. Weinstein, "On the volume of manifolds all of whose geodesics are closed", *J. Differential Geometry* **9** (1974), 513–517.

[Wolter 1979] F.-E. Wolter, "Distance function and cut loci on a complete Riemannian manifold", *Arch. Math. (Basel)* **32** (1979), 92–96.

[Wu 1992] B.-L. Wu, "Isoparametric submanifolds of hyperbolic spaces", *Trans. Amer. Math. Soc.* **331**:2 (1992), 609–626.

[Wu 1994] B.-L. Wu, "Hyper-isoparametric submanifolds in Riemannian symmetric spaces", preprint, Univ. of Pennsylvania, 1994.

[Wu 1995] B.-L. Wu, "Equifocal focal hypersurfaces of rank one symmetric spaces", preprint, Northeastern Univ., 1995.

[Yang 1990] C. T. Yang, "Smooth great circle fibrations and an application to the topological Blaschke conjecture", *Trans. Amer. Math. Soc.* **320**:2 (1990), 507–524.

CHUU-LIAN TERNG
DEPARTMENT OF MATHEMATICS
NORTHEASTERN UNIVERSITY
360 HUNTINGTON AVENUE
BOSTON, MA 02115
  terng@neu.edu

GUDLAUGUR THORBERGSSON
MATHEMATISCHES INSTITUT
UNIVERSITÄT ZU KÖLN
WEYERTAL 86–90
D-50931 KÖLN
GERMANY
  gthorbergsson@mi.uni-koeln.de

Tight and Taut Submanifolds
MSRI Publications
Volume 32, 1997

Null-Homotopic Embedded Spheres
of Codimension One

DANIEL RUBERMAN

ABSTRACT. Let S be an $(n-1)$-sphere smoothly embedded in a closed, orientable, smooth n-manifold M, and let the embedding be null-homotopic. We show that, if S does not bound a ball, then M is a rational homology sphere, the fundamental groups of both components of $M \setminus S$ are finite, and at least one of them is trivial.

Let M be a closed, oriented n-manifold, and suppose that $\iota : S^{n-1} \to M$ is a smooth embedding that is null-homotopic. It follows easily that the image $\iota(S^{n-1}) = S$ separates M into two pieces: $M = X_0 \cup_S Y_0$, or $M = X \# Y$ with $X = X_0 \cup B^n$ and $Y = Y_0 \cup B^n$. An obvious instance is when X_0 or Y_0 is diffeomorphic to B^n; we then say that S bounds a ball on one side. The question as to whether this is the only possibility arises in [Terng and Thorbergsson 1997]. The following theorem describes what can happen; there are examples in every dimension to show that this is (more or less) the best possible. The only qualification is that it is perhaps possible to show that both X and Y must be simply connected; all of the examples constructed at the end of this article have this property.

THEOREM 1. *Suppose that $\iota : S^{n-1} \to M^n$ is a null-homotopic smooth embedding. Then either S bounds a homotopy ball on one side, or the following statements hold:*

(i) *M is a rational homology sphere, and therefore X and Y are as well.*

(ii) *The fundamental groups of both X and Y are finite, and at least one of them is trivial.*

For $n > 4$, if S bounds a homotopy ball then it bounds a (smooth) ball, while if $n = 4$ it bounds a topological ball.

The basic ingredient in the proof is the well-known principle that a manifold admitting a map from a sphere of nonzero degree must be a rational homology sphere:

LEMMA 2. *Suppose that M is an n-dimensional oriented manifold, and that $f : S^n \to M$ has degree $k > 0$. Then M is a rational homology sphere, and $H_*(M; \mathbb{Z})$ has no m-torsion if $\gcd(m, k) = 1$. Moreover, $\pi_1(M)$ is finite, and its order divides k.*

The first part follows by Poincaré duality with rational or $\mathbb{Z}/m$ coefficients. The second part follows by considering the lift of f to the universal cover of M.

To apply the lemma, note that there are maps $\pi_X : M \to X$ and $\pi_Y : M \to Y$ collapsing Y_0 and X_0, respectively, to a point. These maps induce an isomorphism from $H_n(M, S)$ to the direct sum $H_n(X) \oplus H_n(Y)$. Here X and Y are oriented by the image of $H_n(M)$ in $H_n(M, S)$, and also X_0 and Y_0 acquire orientations as manifolds with boundary. The inverse of the isomorphism $(\pi_X)_* \oplus (\pi_Y)_*$ is then induced by the inclusions ι_X, ι_Y of X_0, Y_0 into M.

Suppose now that $F : B^n \to M$ is an extension of ι coming from the null-homotopy of ι. Composing with the projections π_X and π_Y gives maps $F_X : S^n \to X$ and $F_Y : S^n \to Y$.

LEMMA 3. *The degrees of F_X and F_Y satisfy $\deg F_X - \deg F_Y = \pm 1$.*

PROOF. This is a small diagram chase. The point is that (with suitable orientation conventions) the boundary map $\partial : H_n(M, S) \to H_{n-1}(S)$, takes the class $\iota_*([X_0])$ to $+1$ and $\iota_*([Y_0])$ to -1. $\qquad \Box$

PROOF OF THEOREM 1. Suppose that S is null-homotopic, and that neither X nor Y is a homotopy ball. The fact that one of X and Y must be simply connected follows from the van Kampen theorem, which implies that $\pi_1(M)$ is the free product $\pi_1(X) * \pi_1(Y)$. It is easily seen that a lift of S to the universal cover $\tilde{M}$ intersects a properly embedded line, and is thus essential (in homology). But the covering homotopy theorem implies that any lift of S is null-homotopic. In dimension $n = 3$, a standard argument implies that a simply connected manifold with boundary S^2 is a homotopy ball; hence one of X_0 or Y_0 is a homotopy ball.

Suppose now that $n > 3$, and that one of the degrees, say $\deg(F_Y)$, is zero. By the preceding lemma, the other one must be ± 1. By the first lemma, X must be a homotopy sphere, i.e., X_0 is a homotopy ball. In all dimensions except 4, X_0 is then known to be diffeomorphic to a ball [Milnor 1965]; in dimension 4, all one can say at present is that X_0 is homeomorphic to a ball [Freedman and Quinn 1990].

If neither degree is zero, both X and Y are rational homology spheres, by the first lemma. $\qquad \Box$

We remark that a simply connected four-manifold has no torsion in its homology, so a simply connected rational homology four-sphere must be homotopy equivalent to, and thus homeomorphic to, a sphere. In dimension four, therefore, a null-homotopic sphere must bound a ball, and the new phenomena must be in higher dimensions.

We now construct examples that show that in some sense the theorem gives as much information as possible. Clearly, by the theorem, one needs a source of simply connected manifolds that arise as the target of a map of nonzero degree from a sphere. We use the following two lemmas to put such manifolds together to give examples of manifolds M containing a null-homotopic sphere.

LEMMA 4. *Suppose that X is a simply connected n-manifold whose homology in dimensions $0 < m < n$ is all k-torsion, for some integer k. Then the image of the Hurewicz map $\pi_n(M) \to H_n(M)$ is given by $k^r \mathbb{Z}$ for some r. In particular, there is a map $S^n \to M$ of degree k^r.*

PROOF. This follows from the mod $\mathcal{C}$ Hurewicz theorem [Serre 1953], where $\mathcal{C}$ is the class of finite abelian groups.   □

LEMMA 5. *Suppose that X and Y are oriented simply connected manifolds, admitting maps from S^n of degrees k and l, respectively. Then the connected sum $M = X \# Y$ admits a map from B^n such that the restriction takes S^{n-1} to the sphere separating X from Y, and the induced map has degree $k + l$.*

PROOF. Choose regular values $x \in X$ and $y \in Y$. By a homotopy of the maps, if necessary, we can assume that the local degree at some point p in the preimage of x is positive, and that the local degree at some point q in the preimage of y is negative. Remove small ball neighborhoods of x and y, and form the connected sum $X \# Y$ using an orientation reversing diffeomorphism of S^{n-1}. There is an obvious map of a punctured sphere to X_0, and another one to Y_0, that fit together (near p and q) to give a map of a punctured sphere to M. All of the boundary S^{n-1}'s map to S, and the total degree of all the maps is clearly $k + l$.

Choose one of the boundary components S_0 of the punctured sphere, and for each of the other boundary components, choose an arc joining it to S_0. The arcs become loops in M, which can be contracted to lie in a neighborhood of S. (This is where the simple connectivity gets used.) Remove a neighborhood of each of the arcs, to get a map of B^n, with boundary lying in $S \times I$. The map on the boundary can be homotoped to lie in S; the homotopy extension theorem says that this homotopy extends to a homotopy of the map of the ball as well.   □

REMARK 6. The simple connectivity of at least one of X and Y is essential, as the proof of the theorem shows. It is not known if X and Y both have to be one-connected. There is also some possible confusion about orientations: the sphere S gets its orientation as the boundary of the submanifold X_0 of M.

To apply these lemmas, suppose X and Y admit degree-k and degree-l maps from the sphere. By precomposing with maps of the sphere to itself, of degrees a and b, we can get a map from the ball to $X \# Y$ sending S^{n-1} to S with degree $ak + bl$. If $\gcd(k, l) = 1$, we can choose $ak + bl = 1$, so the map is homotopic to the embedding of S in M. So all we need is a collection of rational homology spheres, in each dimension $n \geq 5$, with only k-torsion in their homology.

EXAMPLE 7. For $n \geq 5$, start with the manifold $S^2 \times B^{n-2}$. Add a three-handle to $S^2 \times B^{n-2}$ where the attaching two-sphere in the boundary $S^2 \times S^{n-3}$ represents k times the generator of $H_2(S^2 \times S^{n-3})$. (When $n = 5$, some care must be taken, as not every homology class is represented by an embedded sphere. But in the case at hand, this is not a problem; tube together k parallel copies of the obvious sphere $S^2 \times$ pt.) Double the resulting manifold with boundary, to obtain a simply connected manifold X_k. If $n > 5$, the only homology in X_k (apart from dimensions 0 and n) is $\mathbb{Z}/k$ in dimensions 2 and $n - 2$. For dimension 5, the homology is $\mathbb{Z}/k \oplus \mathbb{Z}/k$ in dimension 2.

The 5-manifolds X_k were constructed by D. Barden [1965] by a somewhat different method. As an alternative to the previous paragraph, one could obtain higher-dimensional examples inductively, starting from Barden's manifolds, as follows: From an n-dimensional X_k, form the product $X_k \times S^1$, and then surger the circle (this is called spinning X) to get an $(n+1)$-manifold X_k with nontrivial homology ($\mathbb{Z}/k \oplus \mathbb{Z}/k$) only in dimensions 2 and $n - 2$.

EXAMPLE 8. Start with the Hopf map $p : S^7 \to S^4$. As a (linear) S^3 bundle over S^4, it has an Euler class that is easily seen to be a generator of $H^4(S^4)$. Now let $g : S^4 \to S^4$ have nonzero degree, say k, and let $p_k : X_k \to S^4$ be the pull back bundle $g^*(p)$. By naturality, p_k has Euler class k; it is easy to compute (with a Gysin sequence) that the homology of X_k is $\mathbb{Z}/k$ in dimension 3, $\mathbb{Z}$ in dimensions 0 and 7, and trivial otherwise. Using the naturality of the Gysin sequence, or a geometric argument, it is easy to see that the degree of the map $X_k \to S^7$ covering g is exactly k. From properties of the Hopf invariant, it is not hard to check that there is a map $S^7 \to X_k$ of degree exactly k. By spinning as in the previous example, one gets examples in every dimension ≥ 7.

References

[Barden 1965] D. Barden, "Simply connected five-manifolds", *Ann. of Math.* (2) **82** (1965), 365–385.

[Freedman and Quinn 1990] M. H. Freedman and F. Quinn, *Topology of 4-manifolds*, Princeton Mathematical Series **39**, Princeton University Press, Princeton, NJ, 1990.

[Milnor 1965] J. Milnor, *Lectures on the h-cobordism theorem*, Princeton University Press, Princeton, N.J., 1965. Notes by L. Siebenmann and J. Sondow.

[Serre 1953] J.-P. Serre, "Groupes d'homotopie et classes de groupes abéliens", *Ann. of Math.* (2) **58** (1953), 258–294.

[Terng and Thorbergsson 1997] C.-L. Terng and G. Thorbergsson, "Taut immersions into complete Riemannian manifolds", pp. 181–228 in *Tight and Taut Submanifolds*, edited by T. E. Cecil and S.-s. Chern, Cambridge U. Press, 1997.

DANIEL RUBERMAN
DEPARTMENT OF MATHEMATICS
BRANDEIS UNIVERSITY
WALTHAM, MA 02254

Tight and Taut Submanifolds
MSRI Publications
Volume **32**, 1997

Real Hypersurfaces
in Complex Space Forms

ROSS NIEBERGALL AND PATRICK J. RYAN

ABSTRACT. The study of real hypersurfaces in complex space forms has
been an active field of study over the past decade. This article attempts to
give the necessary background material to access this field, as well as a detailed construction of the important examples of hypersurfaces in complex
projective and complex hyperbolic space. Following this we give a survey
of the major classification results, including such topics as restrictions on
the shape operator, the η-parallel condition, and restrictions on the Ricci
tensor. We conclude with a brief discussion of some additional areas of
study and some open problems. A comprehensive bibliography is included.

Introduction

The study of real hypersurfaces in complex projective space $\mathbb{C}P^n$ and complex
hyperbolic space $\mathbb{C}H^n$ has been an active field over the past decade. Although
these ambient spaces might be regarded as the simplest after the spaces of constant curvature, they impose significant restrictions on the geometry of their
hypersurfaces. For instance, they do not admit umbilic hypersurfaces and their
geodesic spheres do not have constant curvature. They also do not admit Einstein hypersurfaces. M. Okumura [1978] remarked that there was a poverty of
vocabulary for describing the differential geometric properties of the hypersurfaces that can arise. That situation has since been improved.

One can regard $\mathbb{C}P^n$ as a projection from S^{2n+1} with fibre S^1. H. B. Lawson
[1970] was the first to exploit this idea to study a hypersurface in $\mathbb{C}P^n$ by
lifting it to an S^1-invariant hypersurface of the sphere. He identified certain
hypersurfaces called *equators* of $\mathbb{C}P^n$ which are minimal and lift to Clifford
minimal hypersurfaces of the sphere. Subsequently, other investigators explored
properties that lifted to familiar properties of hypersurfaces in S^{2n+1}.

1991 *Mathematics Subject Classification*. 53C40.

Partially supported by the Natural Sciences and Engineering Research Council of Canada.

R. Takagi's classification [1973] of the homogeneous real hypersurfaces of $\mathbb{C}P^n$ was important in its own right, but it also identified a whole list of hypersurfaces, gave them names (type A1, type B, etc.), and focused attention on them. Other geometers began to study them and to derive new characterizations of various subsets of the list.

Another important notion that developed was the relevance of the *structure vector* of a hypersurface. It is defined by $W = -J\xi$ where J is the complex structure and ξ is the unit normal field. In early investigations, it was found that computations were more tractable when W was a principal vector. Further, it was observed that W is principal for all homogeneous hypersurfaces in $\mathbb{C}P^n$. Later, geometric characterizations of this property were found and hypersurfaces that satisfy it are now called *Hopf hypersurfaces*.

It has also developed that certain interesting classes of hypersurfaces can be characterized by simply stated conditions on the so-called holomorphic distribution $W^\perp$. For example, the notions of η-umbilical and pseudo-Einstein have arisen. These are the appropriate analogues of umbilical and Einstein, respectively, and essentially say that the indicated property holds on $W^\perp$.

The homogeneous hypersurfaces of $\mathbb{C}P^n$ all have constant principal curvatures, and Hopf hypersurfaces with constant principal curvatures have been determined, both for $\mathbb{C}P^n$ and for $\mathbb{C}H^n$. In real space forms, constant principal curvature hypersurfaces are isoparametric and have many nice properties related to parallel families and focal sets. The various equivalent definitions of *isoparametric* that can be used in real space forms lead to distinct classes of hypersurfaces in $\mathbb{C}P^n$. Several of these alternatives have been investigated but the constant principal curvature hypersurfaces have been studied most intensively.

The study of hypersurfaces in $\mathbb{C}H^n$ has followed developments in $\mathbb{C}P^n$, often with similar results, but sometimes with differences. For example, a Hopf hypersurface with constant principal curvatures in $\mathbb{C}P^n$ must have 2, 3, or 5 distinct principal curvatures. For $\mathbb{C}H^n$, 2 and 3 are the only possibilities. On the other hand, $\mathbb{C}H^n$ admits a wider variety of hypersurfaces with a specific number (say 2) of principal curvatures.

In this article, we will present the fundamental definitions and results necessary for reaching the frontiers of research in the field. We will state the known classification results and provide proofs of many of them. For those proofs that we cannot include because of time and space limitations, we provide appropriate pointers to the literature.

In Section 1 we construct the standard models of spaces of constant holomorphic sectional curvature, and give the essential background for studying real hypersurfaces. In Section 2 we discuss the notion of Hopf hypersurfaces, and show that the shape operator satisfies rather stringent conditions for these hypersurfaces. In Section 3 we list the standard examples of real hypersurfaces that occur in spaces of constant holomorphic sectional curvature. The classification results are discussed in Sections 4–7. Finally, in Sections 8 and 9, we discuss areas for further study.

We conclude this section by mentioning a few notational conventions. In addition to the usual end-of-proof symbol □, we use ◁ to conclude the statement of a theorem whose proof is to be omitted.

For Hopf hypersurfaces, the shape operator A preserves the holomorphic distribution $W^\perp$. Rather than say that λ is a principal curvature whose corresponding principal vectors lie in $W^\perp$, we often say that λ is a principal curvature "on $W^\perp$". This will allow us to avoid repeating a wordy and awkward phrase.

When X and Y are vectors, $X \wedge Y$ will denote the linear transformation satisfying

$$(X \wedge Y)Z = \langle Y, Z \rangle X - \langle X, Z \rangle Y$$

where $\langle \, , \, \rangle$ is the inner product. Finally, since covariant differentiation acts as a derivation on the algebra of tensor fields, and commutes with contractions, the curvature operator $R(X,Y)$ can operate in the same way. For any tensor field T, the tensor field defined by

$$R(X,Y) \cdot T = \nabla_X \nabla_Y T - \nabla_Y \nabla_X T - \nabla_{[X,Y]} T$$

is abbreviated $R \cdot T$. For instance, if T is a tensor field of type $(1,1)$,

$$(R \cdot T)(X,Y,Z) = (R(X,Y) \cdot T)Z = R(X,Y)(TZ) - T(R(X,Y)Z).$$

1. Preliminaries

In this section we construct the standard models of spaces of constant holomorphic sectional curvature. We first construct $\mathbb{C}P^n$ and then $\mathbb{C}H^n$, and then we take a unified approach to the discussion of spaces of constant holomorphic sectional curvature. We then discuss the geometry of hypersurfaces and their lifts.

Complex Projective Space. We first introduce the complex projective space $\mathbb{C}P^n$ and the basic equations for studying its hypersurfaces. For $z = (z_0, \ldots, z_n)$, $w = (w_0, \ldots, w_n)$ in $\mathbb{C}^{n+1}$, write

$$F(z,w) = \sum_{k=0}^{n} z_k \bar{w}_k$$

and let $\langle z, w \rangle = \mathrm{Re}\, F(z,w)$, the real part of $F(z,w)$. The $(2n+1)$-sphere $S^{2n+1}(r)$ of radius r is defined by

$$S^{2n+1}(r) = \{z \in \mathbb{C}^{n+1} : \langle z, z \rangle = r^2\}.$$

We may consider $\mathbb{C}^{n+1}$ as $\mathbb{R}^{2n+2}$ and define $u, v \in \mathbb{R}^{2n+2}$ by

$$z_k = u_{2k} + u_{2k+1} i, \qquad w_k = v_{2k} + v_{2k+1} i.$$

Then

$$\langle z, w \rangle = \langle u, v \rangle = \sum_{k=0}^{2n+1} u_k v_k.$$

We will use $\langle z, w \rangle$ and $\langle u, v \rangle$ interchangeably. When desired, we can work exclusively in real terms by introducing the operator J for multiplication by the complex number i. Note that for $z \in S^{2n+1}(r)$,

$$T_z S^{2n+1}(r) = \{ w \in \mathbb{C}^{n+1} : \langle z, w \rangle = 0 \}.$$

Restricting $\langle \, , \, \rangle$ to $S^{2n+1}(r)$ gives a Riemannian metric whose Levi-Civita connection $\widetilde{\nabla}$ satisfies

$$D_X Y = \widetilde{\nabla}_X Y - \langle X, Y \rangle \frac{z}{r^2}$$

for X, Y tangent to $S^{2n+1}(r)$ at z, where D is the Levi-Civita connection of $\mathbb{R}^{2n+2}$. The usual calculations of the Gauss equation yield that the curvature tensor $\tilde{R}$ of $\widetilde{\nabla}$ satisfies

$$\tilde{R}(X, Y) = \frac{1}{r^2} \, X \wedge Y. \tag{1.1}$$

Let $V = Jz$ and write down the orthogonal decomposition into so-called vertical and horizontal components,

$$T_z S^{2n+1}(r) = \mathrm{span}\{V\} \oplus V^\perp.$$

Let π be the canonical projection of $S^{2n+1}(r)$ to complex projective space $\mathbb{C}P^n$,

$$\pi : S^{2n+1}(r) \to \mathbb{C}P^n.$$

Complex Hyperbolic Space. Next, we introduce the complex hyperbolic space $\mathbb{C}H^n$. The construction is parallel to that of $\mathbb{C}P^n$ with some important differences. For z, w in $\mathbb{C}^{n+1}$, write

$$F(z, w) = -z_0 \bar{w}_0 + \sum_{k=1}^{n} z_k \bar{w}_k$$

and let $\langle z, w \rangle = \mathrm{Re}\, F(z, w)$. The anti–de Sitter space of radius r in $\mathbb{C}^{n+1}$ is defined by

$$H_1^{2n+1}(r) = \{ z \in \mathbb{C}^{n+1} : \langle z, z \rangle = -r^2 \}.$$

We denote $H_1^{2n+1}(r)$ by $\mathbb{H}$ for short. We use the same identification of $\mathbb{C}^{n+1}$ with $\mathbb{R}^{2n+2}$ so that

$$\langle z, w \rangle = \langle u, v \rangle = -u_0 v_0 - u_1 v_1 + \sum_{k=2}^{2n+1} u_k v_k.$$

For $z \in \mathbb{H}$,

$$T_z \mathbb{H} = \{ w \in \mathbb{C}^{n+1} : \langle z, w \rangle = 0 \}.$$

Restricting $\langle \, , \, \rangle$ to $\mathbb{H}$ gives a Lorentz metric whose Levi-Civita connection $\widetilde{\nabla}$ satisfies

$$D_X Y = \widetilde{\nabla}_X Y + \langle X, Y \rangle \frac{z}{r^2}$$

for X, Y tangent to $\mathbb{H}$ at z. The Gauss equation takes the form

$$\tilde{R}(X,Y) = -\frac{1}{r^2} X \wedge Y. \qquad (1.2)$$

Again take $V = Jz$ and we get the analogous orthogonal decomposition

$$T_z\mathbb{H} = \text{span}\{V\} \oplus V^\perp.$$

Denote by $\mathbb{C}H^n$ the image of $\mathbb{H}$ by the canonical projection π to complex projective space,

$$\pi : \mathbb{H} \to \mathbb{C}H^n \subset \mathbb{C}P^n.$$

Thus, topologically, $\mathbb{C}H^n$ is an open subset of $\mathbb{C}P^n$. However, as Riemannian manifolds, they have quite different structures.

Complex Space Forms From here we make a uniform exposition covering both $\mathbb{C}P^n$ and $\mathbb{C}H^n$. When convenient, we make use of the letter ε to distinguish the two cases. It denotes the sign of the constant holomorphic sectional curvature $4c = 4\varepsilon/r^2$. For example, (1.1) and (1.2) could be written as

$$\tilde{R}(X,Y) = \frac{\varepsilon}{r^2} X \wedge Y.$$

We also use $\tilde{M}$ to stand for either $\mathbb{C}P^n$ or $\mathbb{C}H^n$ and $\tilde{M}'$ for $S^{2n+1}(r)$ or $\mathbb{H}$.

Note that $\pi_* V = 0$ but that π_* is an isomorphism on $V^\perp$. Let z be any point of $\tilde{M}'$. For $X \in T_{\pi z}\tilde{M}$, let X^L be the vector in $V_z^\perp$ that projects to X. X^L is called the *horizontal lift* of X to z. Define a Riemannian metric on $\tilde{M}$ by $\langle X, Y \rangle = \langle X^L, Y^L \rangle$. It is well-defined since the metric on $\tilde{M}'$ is invariant by the fibre S^1. Since $V^\perp$ is J-invariant, $\tilde{M}$ can be assigned a complex structure (also denoted by J) by $JX = \pi_*(JX^L)$. It is easy to check that $\langle \ , \ \rangle$ is Hermitian on $\tilde{M}$ and that its Levi-Civita connection $\tilde{\nabla}$ satisfies

$$\tilde{\nabla}_X Y = \pi_*(\tilde{\nabla}_{X^L} Y^L). \qquad (1.3)$$

We also note that on $\tilde{M}'$

$$\tilde{\nabla}_{X^L} V = \tilde{\nabla}_V X^L = JX^L = (JX)^L \qquad (1.4)$$

while

$$\tilde{\nabla}_V V = 0.$$

See [O'Neill 1966; Gray 1967] for background on Riemannian submersions.

THEOREM 1.1. *The curvature tensor $\tilde{R}$ of $\tilde{M}$ satisfies*

$$\tilde{R}(X,Y)Z = \frac{\varepsilon}{r^2}(X \wedge Y + JX \wedge JY + 2\langle X, JY \rangle J)Z.$$

In particular, the sectional curvature of a holomorphic plane spanned by X and JX is $4\varepsilon/r^2$ so that $\tilde{M}$ is a space of constant holomorphic sectional curvature. ◁

The Riemannian metrics we have just constructed are known as the *Fubini–Study metric* on $\mathbb{C}P^n$ and the *Bergman metric* on $\mathbb{C}H^n$ respectively. See [Kobayashi and Nomizu 1969, Chapter IX] for additional information on these metrics.

Hypersurfaces in Complex Space Forms. Let $\tilde{M}(c)$ be a space of constant holomorphic sectional curvature $4c$ with real dimension $2n$ and Levi-Civita connection $\tilde{\nabla}$. For an immersed manifold $f : M^{2n-1} \to \tilde{M}$, the Levi-Civita connection ∇ of the induced metric and the shape operator A of the immersion are characterized respectively by

$$\tilde{\nabla}_X Y = \nabla_X Y + \langle AX, Y \rangle \xi$$

and

$$\tilde{\nabla}_X \xi = -AX$$

where ξ is a local choice of unit normal. We omit mention of the immersion f for brevity of notation. Let

$$J : T\tilde{M} \to T\tilde{M}$$

be the complex structure with properties $J^2 = -I$, $\tilde{\nabla} J = 0$, and $\langle JX, JY \rangle = \langle X, Y \rangle$. Define the *structure vector*

$$W = -J\xi.$$

Clearly $W \in TM$, and $|W| = 1$. Write $a = \langle AW, W \rangle$. We reserve the symbols W and a for these purposes throughout.

Define a skew-symmetric $(1,1)$-tensor φ from the tangential projection of J by

$$JX = \varphi X + \langle X, W \rangle \xi. \tag{1.5}$$

Then, since $-X = J^2 X = J(\varphi X + \langle X, W \rangle \xi) = \varphi^2 X + \langle \varphi X, W \rangle \xi + \langle X, W \rangle J\xi$, we see that

$$\varphi^2 X = -X + \langle X, W \rangle W. \tag{1.6}$$

It is easy to check that

$$\langle \varphi X, \varphi Y \rangle = \langle X, Y \rangle - \langle X, W \rangle \langle Y, W \rangle. \tag{1.7}$$

Putting $X = W$ in (1.5) gives $\varphi W = 0$. Noting that $\varphi^2 = -I$ on $W^\perp = \{X \in TM : \langle X, W \rangle = 0\}$ we see that φ has rank $2n - 2$ and that

$$\ker \varphi = \operatorname{span}\{W\}.$$

Such a φ determines an almost contact metric structure [Blair 1976, pp. 19–21] and $W^\perp$ is called the *holomorphic distribution*.

In the usual way, we derive the Gauss and Codazzi equations:

$$R(X, Y) = AX \wedge AY + c(X \wedge Y + \varphi X \wedge \varphi Y + 2\langle X, \varphi Y \rangle \varphi), \tag{1.8}$$

$$(\nabla_X A)Y - (\nabla_Y A)X = c(\langle X, W \rangle \varphi Y - \langle Y, W \rangle \varphi X + 2\langle X, \varphi Y \rangle W). \tag{1.9}$$

COROLLARY 1.2. *We have* $\langle (\nabla_X A)Y - (\nabla_Y A)X, W \rangle = 2c\langle X, \varphi Y \rangle$ *and*

$$\langle (\nabla_X A)W, W \rangle = \langle (\nabla_W A)X, W \rangle = \langle (\nabla_W A)W, X \rangle.$$

PROOF. The first equation follows by taking the inner product of the Codazzi equation with W, and the second follows by letting $Y = W$. □

From equation (1.8) we get the Ricci tensor S of type $(1,1)$ defined by

$$\langle SX, Y \rangle = \text{trace}\{Z \to R(Z, X)Y\} \tag{1.10}$$

as

$$SX = (2n+1)cX - 3c\langle X, W \rangle W + (\text{trace } A)AX - A^2 X. \tag{1.11}$$

The scalar curvature is

$$s = \text{trace } S = 4(n^2 - 1)c + (\text{trace } A)^2 - \text{trace } A^2.$$

The mean curvature is $\mathfrak{m} = \text{trace } A$ and we reserve the symbol $\mathfrak{m}$ for this purpose throughout.

PROPOSITION 1.3. *We have* $\nabla_X W = \varphi AX$ *and*

$$(\nabla_X \varphi)Y = \langle Y, W \rangle AX - \langle AX, Y \rangle W.$$

PROOF. For the first equality,

$$\begin{aligned}
\nabla_X W &= -\nabla_X (J\xi) = -\tilde{\nabla}_X (J\xi) + \langle AX, J\xi \rangle \xi \\
&= -J\tilde{\nabla}_X \xi + \langle AX, J\xi \rangle \xi = JAX - \langle JAX, \xi \rangle \xi = \varphi AX.
\end{aligned}$$

For the second,

$$\begin{aligned}
(\nabla_X \varphi)Y &= \nabla_X(\varphi Y) - \varphi \nabla_X Y = \nabla_X(JY - \langle Y, W \rangle \xi) - \varphi \nabla_X Y \\
&= \tilde{\nabla}_X(JY - \langle Y, W \rangle \xi) - \langle A\varphi Y, X \rangle \xi - \varphi \nabla_X Y \\
&= J(\nabla_X Y + \langle AX, Y \rangle \xi) - X\langle Y, W \rangle \xi + \langle Y, W \rangle AX - \langle A\varphi Y, X \rangle \xi - \varphi \nabla_X Y \\
&= \varphi \nabla_X Y + \langle \nabla_X Y, W \rangle \xi - \langle AX, Y \rangle W - \langle \nabla_X Y, W \rangle \xi \\
&\qquad\qquad - \langle Y, \varphi AX \rangle \xi + \langle Y, W \rangle AX - \langle A\varphi Y, X \rangle \xi - \varphi \nabla_X Y \\
&= \langle Y, W \rangle AX - \langle AX, Y \rangle W.
\end{aligned}$$
□

PROPOSITION 1.4. *If* $c \neq 0$ *then* ∇W *cannot be identically zero. Equivalently,* φA *cannot be identically zero.*

PROOF. By Proposition 1.3, $\nabla_X W = 0$ if and only if $\varphi AX = 0$. Suppose that this condition holds for all X. Then $AX = \langle AX, W \rangle W$. Thus, for all X and Y,

$$(\nabla_X A)Y = \nabla_X(AY) - A\nabla_X Y \in \text{span}\{W\}.$$

Applying the Codazzi equation, we have

$$c(\langle X, W \rangle \varphi Y - \langle Y, W \rangle \varphi X) \in \text{span}\{W\}.$$

In particular, put $Y = W$ to get that $-c\varphi X$ lies in the span of W for all X. This is clearly impossible since $c \neq 0$. □

THEOREM 1.5. *Let M^{2n-1}, where $n \geq 2$, be a hypersurface in a complex space form of constant holomorphic sectional curvature $4c \neq 0$. Then the shape operator A cannot be parallel. Also, no identity of the form $A = \lambda I$ can hold, even with λ nonconstant. In particular, umbilic hypersurfaces cannot occur.*

PROOF. Let us first assume that $A = \lambda I$. The Codazzi equation (1.9) becomes

$$(X\lambda)Y - (Y\lambda)X = c(\langle X, W \rangle \varphi Y - \langle Y, W \rangle \varphi X + 2\langle X, \varphi Y \rangle W).$$

If we put $Y = W$ in this equation, it simplifies to

$$(X\lambda)W - (W\lambda)X = -c\varphi X.$$

For $X \neq 0$ orthogonal to W, the set $\{X, \varphi X, W\}$ is linearly independent, and so $c = 0$ which contradicts the hypothesis. Now suppose that $\nabla A = 0$. Take $X \neq 0$ orthogonal to W and $Y = W$ in the Codazzi equation, to get $-c\varphi X = 0$, another contradiction. □

The nonexistence of umbilic hypersurfaces was proved by Tashiro and Tachibana [1963].

Lifts of Hypersurfaces in $\tilde{M}$ to $\tilde{M}'$. Once again, we let $\tilde{M}$ represent $\mathbb{C}P^n$ or $\mathbb{C}H^n$ and $\tilde{M}'$ represent $S^{2n+1}(r)$ or $\mathbb{H}$ respectively, with the canonical projection

$$\pi : \tilde{M}' \to \tilde{M}.$$

Now consider a hypersurface M in $\tilde{M}$. Then $M' = \pi^{-1}M$ is an S^1-invariant hypersurface in $\tilde{M}'$ (Lorentzian in the case $\tilde{M}' = \mathbb{H}$). If ξ is a unit normal for M, then $\xi' = \xi^L$ is a unit normal for M'. The induced connection ∇' and the shape operator A' for M' satisfy

$$\tilde{\nabla}_X Y = \nabla'_X Y + \langle A'X, Y \rangle \xi', \qquad \tilde{\nabla}_X \xi' = -A'X,$$

and the more familiar form of the Codazzi equation

$$(\nabla'_X A')Y = (\nabla'_Y A')X.$$

There is also a Gauss equation, but we will not have occasion to use it. We also have $W^L = U = -J\xi'$ where $W = -J\xi$ is the structure vector introduced earlier.

LEMMA 1.6. *For X and Y tangent to $\tilde{M}$,*

$$(\tilde{\nabla}_X Y)^L = \tilde{\nabla}_{X^L} Y^L + \varepsilon \langle JX^L, Y^L \rangle \frac{1}{r^2} V. \tag{1.12}$$

PROOF. By (1.3), π_* applied to each side of (1.12) yields the same result. Thus, it only remains to check that the right side is horizontal. However,

$$\langle \tilde{\nabla}_{X^L} Y^L, V \rangle + \varepsilon \langle JX^L, Y^L \rangle \frac{1}{r^2} \langle V, V \rangle = -\langle Y^L, \tilde{\nabla}_{X^L} V \rangle + \langle JX^L, Y^L \rangle \varepsilon^2$$

$$= -\langle Y^L, JX^L \rangle + \langle JX^L, Y^L \rangle = 0. \quad \square$$

LEMMA 1.7. (i) $\tilde{\nabla}_V \xi' = J\xi' = -U$, *so $A'V = U$.*
(ii) *For X tangent to M, $(AX)^L = A'X^L - \langle X^L, U \rangle \varepsilon r^{-2} V$.*
(iii) *In particular, $(AW)^L = A'U - \varepsilon r^{-2} V$.*
(iv) *If $AW = aW$, then $A'U = aU + \varepsilon r^{-2} V$.*

PROOF. To verify the first assertion, note that $\tilde{\nabla}_V \xi' = \tilde{\nabla}_V \xi^L = J\xi^L = -U$. For the second, we compute

$$-A'X^L = \tilde{\nabla}_{X^L} \xi^L = (\tilde{\nabla}_X \xi)^L - \varepsilon \langle JX^L, \xi^L \rangle r^{-2} V,$$

and hence

$$A'X^L = (AX)^L + \varepsilon \langle X^L, U \rangle r^{-2} V.$$

Assertions (iii) and (iv) are special cases of (ii). $\quad \square$

We now look at the relationship between the covariant derivatives of the respective shape operators of M and M'.

THEOREM 1.8. *Let M^{2n-1}, where $n \geq 2$, be a real hypersurface in a complex space form of constant holomorphic sectional curvature $4c \neq 0$. Then the shape operator A' of $M' = \pi^{-1} M$ satisfies*

$$\pi_*((\nabla'_{X^L} A')Y^L) = (\nabla_X A)Y + c(\langle \varphi X, Y \rangle W + \langle Y, W \rangle \varphi X)$$

for all X, Y tangent to M.

PROOF. We begin with the fundamental identity,

$$(\nabla_X A)Y = \nabla_X(AY) - A(\nabla_X Y),$$

and take the horizontal lift of each side. In the following equation, all equalities are to be understood mod V. We freely use (1.3), (1.4) and the results of

Lemmas 1.6 and 1.7.

$$
\begin{aligned}
\left((\nabla_X A)Y\right)^L &= \left(\tilde{\nabla}_X(AY) - \langle AX, AY\rangle \xi\right)^L - A'(\nabla_X Y)^L\\
&= \tilde{\nabla}_{X^L}(AY)^L - \langle (AX)^L, (AY)^L\rangle \xi' - A'\left((\tilde{\nabla}_X Y)^L - \langle (AX)^L, Y^L\rangle \xi'\right)\\
&= \tilde{\nabla}_{X^L}(A'Y^L - \varepsilon r^{-2}\langle Y^L, U\rangle V)\\
&\quad - \left(\langle A'X^L, A'Y^L\rangle + r^{-4}\langle X^L, U\rangle \langle Y^L, U\rangle \langle V, V\rangle\right)\xi'\\
&\quad + 2\varepsilon r^{-2}\langle X^L, U\rangle \langle Y^L, U\rangle \xi'\\
&\quad - A'\left(\tilde{\nabla}_{X^L}Y^L + \varepsilon r^{-2}\langle JX^L, Y^L\rangle V - \langle (AX)^L, Y^L\rangle \xi'\right)\\
&= \nabla'_{X^L}(A'Y^L) + \langle A'X^L, A'Y^L\rangle \xi' - \varepsilon r^{-2}\langle Y^L, U\rangle JX^L\\
&\quad - \langle A'X^L, A'Y^L\rangle \xi' + \varepsilon r^{-2}\langle X^L, U\rangle \langle Y^L, U\rangle \xi'\\
&\quad - A'\left(\nabla'_{X^L}Y^L + \langle A'X^L, Y^L\rangle \xi' + \varepsilon r^{-2}\langle JX^L, Y^L\rangle V - \langle (AX)^L, Y^L\rangle \xi'\right)\\
&= (\nabla'_{X^L}A')Y^L - \varepsilon r^{-2}\langle Y^L, U\rangle (JX)^L\\
&\quad + \varepsilon r^{-2}\langle X^L, U\rangle \langle Y^L, U\rangle \xi' - \varepsilon r^{-2}\langle JX, Y\rangle U\\
&= (\nabla'_{X^L}A')Y^L - \varepsilon r^{-2}\langle Y, W\rangle (\varphi X)^L - \varepsilon r^{-2}\langle Y, W\rangle \langle X, W\rangle \xi^L\\
&\quad + \varepsilon r^{-2}\langle X, W\rangle \langle Y, W\rangle \xi^L - \varepsilon r^{-2}\langle \varphi X, Y\rangle W^L,
\end{aligned}
$$

from which the result follows.   □

Note that the Codazzi equation (1.9) for M in $\tilde{M}$ is a consequence of Theorem 1.8 together with the Codazzi equation for M' in $\tilde{M}'$. We also look at the vertical component of the covariant derivative of A' and observe the following nice relationship. The proof is a straightforward application of the same methods used in Theorem 1.8.

PROPOSITION 1.9. *Under the hypothesis of Theorem 1.8,*

$$
\langle (\nabla'_{X^L}A')Y^L, V\rangle = \langle (\varphi A - A\varphi)X, Y\rangle.
$$

Therefore $(\nabla'_{X^L}A')Y^L$ *is horizontal for all* X *and* Y *if and only if* φ *and* A *commute.*   ◁

A $(1,1)$ tensor A is said to be a *Codazzi tensor* with respect to a semi-Riemannian metric $\langle\ ,\ \rangle$ if, for all tangent vectors X and Y,

$$
\langle AX, Y\rangle = \langle X, AY\rangle \qquad \text{and} \qquad (\nabla_X A)Y = (\nabla_Y A)X,
$$

where ∇ is the Levi-Civita connection.

LEMMA 1.10. *Let* A *be a Codazzi tensor. Assume that there are constants* α *and* β *such that* $A^2 = \alpha A + \beta I$. *If* $\alpha^2 + 4\beta \neq 0$, *then* $\nabla A = 0$. *Furthermore, if* $\alpha^2 + 4\beta < 0$, *the tangent space splits into spacelike and timelike subspaces of equal dimensions.*

PROOF. Differentiating the quadratic condition yields

$$(\nabla_X A)A + A(\nabla_X A) = \alpha(\nabla_X A) \tag{1.13}$$

so that

$$A(\nabla_X A)A = (\alpha A - A^2)\nabla_X A = -\beta\nabla_X A. \tag{1.14}$$

On the other hand, we can write (1.13) in the form

$$\langle(\nabla_Z A)Y, AX\rangle + \langle(\nabla_Z A)X, AY\rangle = \alpha\langle(\nabla_Z A)X, Y\rangle$$

where we have used the symmetry of A and $\nabla_Z A$. Applying the Codazzi equation and then replacing Z by AZ yields

$$\langle(\nabla_Y A)AZ, AX\rangle + \langle(\nabla_X A)AZ, AY\rangle = \alpha\langle(\nabla_X A)AZ, Y\rangle.$$

Again using symmetry and the Codazzi equation, along with (1.14), we have

$$-2\beta\langle(\nabla_X A)Y, Z\rangle = \alpha\langle A(\nabla_X A)Y, Z\rangle$$
$$= \alpha\langle(\nabla_X A)AZ, Y\rangle.$$

Since the left side is symmetric in Y and Z, we have

$$\alpha(\nabla_X A)A = \alpha A(\nabla_X A) = -2\beta\nabla_X A$$

so that, in view of (1.13),

$$(\alpha^2 + 4\beta)\nabla_X A = 0.$$

If $\alpha^2 + 4\beta < 0$, there is a constant γ such that $P = \gamma(A - \frac{1}{2}\alpha I)$ is a symmetric transformation satisfying $P^2 = -I$. In fact, $\gamma = \left(-(\beta + \frac{1}{4}\alpha^2)\right)^{-1/2}$. The result follows immediately. $\square$

Using Theorem 1.8, we can strengthen the result that the shape operator cannot be parallel. In fact, its covariant derivative cannot vanish even at one point. Specifically:

THEOREM 1.11. *Let M^{2n-1}, where $n \geq 2$, be a real hypersurface in a complex space form of constant holomorphic sectional curvature $4c \neq 0$. Then the shape operator A satisfies $|\nabla A|^2 \geq 4c^2(n-1)$.*

PROOF. Let

$$T(X, Y) = (\nabla_X A)Y + c(\langle\varphi X, Y\rangle W + \langle Y, W\rangle\varphi X),$$

the right side of the equation in Theorem 1.8. Then

$$0 \leq |T|^2 = |\nabla A|^2 + 2ck_1 + c^2 k_2,$$

where k_1 is the sum of

$$\langle(\nabla_X A)Y, \langle\varphi X, Y\rangle W + \langle Y, W\rangle\varphi X\rangle$$

as X and Y range over an orthonormal basis, while k_2 is the sum of

$$\langle\varphi X, Y\rangle^2 + \langle Y, W\rangle^2|\varphi X|^2$$

over the same range of X and Y. Summing over Y and using the fact that $\nabla_X A$ is symmetric, we see that k_1 is equal to the sum over X of

$$\langle \varphi X, (\nabla_X A) W \rangle + \langle W, (\nabla_X A) \varphi X \rangle = 2\langle \varphi X, (\nabla_X A) W \rangle.$$

By the Codazzi equation,

$$\langle \varphi X, (\nabla_X A) W \rangle = \langle \varphi X, (\nabla_W A) X \rangle - c\langle \varphi X, \varphi X \rangle.$$

Now note that $(\nabla_W A)\varphi$ has zero trace while the trace of φ^2 is $-2(n-1)$. Thus we can calculate that $k_1 = -4c(n-1)$, while $k_2 = 4(n-1)$. Therefore

$$|\nabla A|^2 - 8c^2(n-1) + 4c^2(n-1) \geq 0,$$

which gives the desired result. $\square$

As a byproduct of this proof, we also see that equality holds if and only if $T = 0$. This means that

$$(\nabla_X A) Y = -c(\langle \varphi X, Y \rangle W + \langle Y, W \rangle \varphi X).$$

We will discuss this further at the beginning of Section 4. See Corollary 4.4.

2. Hopf Hypersurfaces: When W Is Principal

If W is a principal vector, M is called a *Hopf hypersurface*. Hopf hypersurfaces have several nice characterizations. The notion makes sense in any Kähler ambient space, and corresponds to the property that the integral curves of W are geodesics. Tubes over complex submanifolds are known to be Hopf.

A fundamental fact about Hopf hypersurfaces is that the principal curvature a corresponding to W is constant for complex space forms of nonzero curvature. For $c > 0$, the proof is fairly direct, but for $c < 0$, it is rather lengthy, and involves formidable computation. Nevertheless, it is of significance for the geometry of real hypersurfaces in complex space forms, and so we will include it here. When $c = 0$, a need not be constant, but nonconstancy puts rather strong restrictions on A.

THEOREM 2.1. *Let M^{2n-1}, where $n \geq 2$, be a Hopf hypersurface in a complex space form of constant holomorphic sectional curvature $4c$, and let a be the principal curvature corresponding to W.*

(i) *If $c \neq 0$, then a must be constant.*

(ii) *If $c = 0$ and $\operatorname{grad} a \neq 0$ at some point, then $A|_{W^\perp} = 0$ in a neighborhood of this point. Consequently, the number of distinct principal curvatures is 1 or 2 in this neighborhood.*   ◁

We first remark that $a = \langle AW, W \rangle$ is a smooth function whether or not W is principal. The first few lemmas establish a relationship between the shape operator A and the structure tensor φ. A consequence of these preliminary results is the proof of the theorem in the case when $c \geq 0$. The remainder of the section is necessary to establish the result when $c < 0$.

Throughout this section, M^{2n-1}, where $n \geq 2$, will be a real hypersurface in a space of constant holomorphic sectional curvature $4c$, and X, Y, and Z will be vectors tangent M.

LEMMA 2.2. *Let M^{2n-1}, where $n \geq 2$, be a Hopf hypersurface in a complex space form of constant holomorphic sectional curvature $4c$. Then*

$$A\varphi A - \frac{a}{2}(A\varphi + \varphi A) - c\varphi = 0 \tag{2.1}$$

and

$$\operatorname{grad} a = (Wa)W. \tag{2.2}$$

PROOF. Expand $(\nabla_X A)W$, by making use of Proposition 1.3 to calculate that

$$(\nabla_X A)W = \nabla_X(AW) - A\nabla_X W = (Xa)W + (aI - A)\varphi AX. \tag{2.3}$$

Then, by Corollary 1.2,

$$Xa = \langle (\nabla_X A)W, W \rangle = \langle (\nabla_W A)X, W \rangle, \tag{2.4}$$

and $\langle (\nabla_W A)W, X \rangle = (Wa)\langle W, X \rangle$. Since $\langle \operatorname{grad} a, X \rangle = Xa$, we have $\operatorname{grad} a = (Wa)W$. Now, using (2.3) and (2.4),

$$\langle (\nabla_X A)Y, W \rangle = \langle (\nabla_X A)W, Y \rangle$$
$$= (Wa)\langle W, X \rangle\langle W, Y \rangle + \langle (aI - A)\varphi AX, Y \rangle. \tag{2.5}$$

Interchanging X and Y in (2.5) and subtracting, we calculate

$$\langle (\nabla_X A)Y, W \rangle - \langle (\nabla_Y A)X, W \rangle = \langle (aI - A)\varphi AX, Y \rangle - \langle (aI - A)\varphi AY, X \rangle.$$

Comparing this with Corollary 1.2 we see that

$$2c\langle X, \varphi Y \rangle = \langle (aI - A)\varphi AX, Y \rangle - \langle (aI - A)\varphi AY, X \rangle$$
$$= -\langle X, A\varphi(aI - A)Y \rangle - \langle X, (aI - A)\varphi AY \rangle.$$

Since this is true for all tangent X and Y, we get

$$2c\varphi Y = -a(A\varphi Y + \varphi AY) + 2A\varphi AY,$$

and so

$$A\varphi A - \frac{a}{2}(A\varphi + \varphi A) - c\varphi = 0. \qquad \square$$

Here is an immediate consequence of this lemma.

COROLLARY 2.3. (i) *If $X \in W^{\perp}$ and $AX = \lambda X$, then*

$$(\lambda - \frac{a}{2})A\varphi X = (\frac{\lambda a}{2} + c)\varphi X.$$

(ii) *If a nonzero $X \in W^{\perp}$ satisfies $AX = \lambda X$ and $A\varphi X = \mu \varphi X$, then*

$$\lambda \mu = \frac{\lambda + \mu}{2} a + c.$$

(iii) *If T_{λ} is φ-invariant, then $\lambda^2 = a\lambda + c$. (The notation T_{λ} is used for the set of principal vectors for a principal curvature λ).*   ◁

LEMMA 2.4. *Let M^{2n-1}, where $n \geq 2$, be a Hopf hypersurface in a complex space form of constant holomorphic sectional curvature $4c$. Then $(Wa)(\varphi A + A\varphi) = 0$.*

PROOF. Let $\beta = Wa$, so that $\operatorname{grad} a = \beta W$. Then

$$\langle \nabla_X(\operatorname{grad} a), Y \rangle - \langle \nabla_Y(\operatorname{grad} a), X \rangle$$
$$= X\langle \operatorname{grad} a, Y \rangle - \langle \operatorname{grad} a, \nabla_X Y \rangle - Y\langle \operatorname{grad} a, X \rangle + \langle \operatorname{grad} a, \nabla_Y X \rangle$$
$$= XYa - YXa - \operatorname{grad} a\langle \nabla_X Y - \nabla_Y X \rangle = ([X, Y] - (\nabla_X Y - \nabla_Y X))a = 0.$$

(This is, of course, true for any function a and expresses the symmetry of the Hessian.) So, since $\beta W = \operatorname{grad} a$,

$$0 = \langle \nabla_X(\beta W), Y \rangle - \langle \nabla_Y(\beta W), X \rangle$$
$$= X\beta\langle W, Y \rangle + \beta\langle \varphi AX, Y \rangle - Y\beta\langle W, X \rangle - \beta\langle \varphi AY, X \rangle$$
$$= (X\beta)\langle W, Y \rangle - (Y\beta)\langle W, X \rangle + \beta\langle(\varphi A + A\varphi)X, Y \rangle. \qquad (2.6)$$

If we set $Y = W$ in this equation, we get

$$0 = X\beta - (W\beta)\langle W, X \rangle + \beta\langle A\varphi X, W \rangle,$$

where the last term is zero since $AW = aW$ and $\varphi W = 0$. Thus

$$X\beta = (W\beta)\langle W, X \rangle.$$

Using this to simplify equation (2.6) and noticing that this is true for all X proves the lemma.   □

COROLLARY 2.5. *Let M^{2n-1}, where $n \geq 2$, be a real hypersurface in a complex space form of constant holomorphic sectional curvature $4c$. If $\varphi A = A\varphi$, then $AW = aW$. If, in addition, $c \neq 0$, then a must be constant.*

PROOF. Start with $a = \langle AW, W \rangle$. Because $\varphi AW = A\varphi W = 0$, AW lies in the span of W; that is, $AW = \langle AW, W \rangle W = aW$. Now suppose that $\beta = Wa \neq 0$ at some point. From Lemma 2.4, we have $A\varphi = -\varphi A = -A\varphi$ so that $A\varphi = 0$, and hence $c\varphi = 0$ by (2.1). If $c \neq 0$, we have a contradiction. If $c = 0$, there is no contradiction, but A must vanish on $W^{\perp}$ in a neighborhood of the point in question.   □

A cylinder of the form $\Gamma \times \mathbb{C}^n$, where Γ is any plane curve other than a circle or line, shows that a need not be constant when $c = 0$.

COROLLARY 2.6. *Let M^{2n-1}, where $n \geq 2$, be a real hypersurface in a complex space form of constant holomorphic sectional curvature $4c$. Suppose $AW = aW$. If $c > 0$, then a is constant. If $c = 0$, then either a is constant or $g \leq 2$, where g is the number of distinct eigenvalues of A.*

PROOF. By Lemma 2.4, if $\beta \neq 0$ at some point, then $\varphi A + A\varphi = 0$ in some neighborhood of this point, and (2.1) reduces to

$$\varphi A^2 + c\varphi = 0.$$

Now, for any eigenvector of A, say X, orthogonal to W, we have

$$0 = \varphi(A^2 + cI)X = \varphi(\lambda^2 + c)X,$$

where λ is the eigenvalue of A for X. Hence $\lambda^2 + c = 0$. For $c > 0$, this is a contradiction. If $c = 0$, then $\lambda = 0$, and there can be no eigenvalues other than 0 and a. □

To complete the proof of Theorem 2.1, the main result of this section, we must verify the theorem for the case when $c < 0$ The purpose of this next series of lemmas is to compute the explicit form of $\nabla_X A$. This is accomplished in Lemma 2.9. We can then compute $(R(X,Y) \cdot A)Z$ and use this to prove that $A\varphi + \varphi A$ cannot be zero and hence that $Wa = 0$. Our proof follows [Ki and Suh 1990].

LEMMA 2.7. *Let M^{2n-1}, where $n \geq 2$, be a real hypersurface in a complex space form of constant holomorphic sectional curvature $4c$. If $A\varphi + \varphi A = 0$, then*

$$(\nabla_X A)AY + A(\nabla_X A)Y = 2a\beta\langle X, W\rangle\langle Y, W\rangle W$$
$$+ (a^2 + c)(\langle \varphi AX, Y\rangle W + \langle Y, W\rangle \varphi AX), \quad (2.7)$$
$$(\nabla_X A)AY - (\nabla_Y A)AX = 2ca\langle \varphi X, Y\rangle W + a^2(\langle Y, W\rangle \varphi AX - \langle X, W\rangle \varphi AY). \quad (2.8)$$

PROOF. First note that $A\varphi = -\varphi A$ implies that $\varphi AW = 0$ and hence that W is principal, so we can write $AW = aW$. Also, (2.1) simplifies to $\varphi(A^2 + cI) = 0$, and from this we compute

$$0 = (\nabla_X(\varphi(A^2 + cI)))Y$$
$$= (\nabla_X \varphi)(A^2 + cI)Y + \varphi(\nabla_X A)AY + \varphi A(\nabla_X A)Y. \quad (2.9)$$

Proposition 1.3 allows us to rewrite the first term, and equation (2.9) becomes

$$0 = (a^2 + c)\langle Y, W\rangle AX - \langle(A^3 + cA)X, Y\rangle W + \varphi(\nabla_X A)AY + \varphi A(\nabla_X A)Y.$$

Applying φ to this equality gives

$$\varphi((a^2 + c)\langle Y, W\rangle AX) + \varphi^2((\nabla_X A)AY) + \varphi^2(A(\nabla_X A)Y) = 0. \quad (2.10)$$

Using (1.6), the second term of (2.10) is

$$\varphi^2(\nabla_X A)AY = -(\nabla_X A)AY + \langle(\nabla_X A)AY, W\rangle W.$$

From Proposition 1.3 and the fact that $\varphi A^2 = -c\varphi$, we calculate directly that

$$
\begin{aligned}
\langle (\nabla_X A)AY, W \rangle &= \langle AY, (Xa)W + (aI - A)\varphi AX \rangle \\
&= a\beta \langle X, W \rangle \langle Y, W \rangle + a\langle A\varphi AX, Y \rangle - \langle A^2 \varphi AX, Y \rangle \\
&= a\beta \langle X, W \rangle \langle Y, W \rangle + ca\langle \varphi X, Y \rangle + c\langle \varphi AX, Y \rangle.
\end{aligned}
$$

Again using (1.6), we can rewrite the third term of (2.10) as

$$
\varphi^2 A(\nabla_X A)Y = -A(\nabla_X A)Y + \langle A(\nabla_X A)Y, W \rangle W,
$$

and we can calculate directly that

$$
\langle A(\nabla_X A)Y, W \rangle = a\langle (\nabla_X A)Y, W \rangle a\beta \langle X, W \rangle \langle Y, W \rangle + a^2 \langle \varphi AX, Y \rangle - ac\langle \varphi X, Y \rangle.
$$

All of the information allows us to rewrite (2.10) as

$$
\begin{aligned}
(\nabla_X A)AY &+ A(\nabla_X A)Y \\
&= (a^2 + c)\langle Y, W \rangle \varphi AX + 2a\beta \langle X, W \rangle \langle Y, W \rangle W + (a^2 + c)\langle \varphi AX, Y \rangle W,
\end{aligned}
$$

which is equation (2.7). Then, interchanging X and Y in (2.7) and subtracting gives

$$
\begin{aligned}
(\nabla_X A)AY &- (\nabla_Y A)AX + c(\langle X, W \rangle A\varphi Y - \langle Y, W \rangle A\varphi X + 2a\langle X, \varphi Y \rangle W) \\
&= (a^2 + c)(\langle Y, W \rangle \varphi AX - \langle X, W \rangle \varphi AY + \langle \varphi AX, Y \rangle W - \langle \varphi AY, X \rangle W).
\end{aligned}
$$

Equation (2.8) results from this. $\qquad\square$

LEMMA 2.8. *Let M be a real hypersurface in a complex space form of constant holomorphic sectional curvature $4c$. If $\varphi A + A\varphi = 0$, then*

$$
\begin{aligned}
(\nabla_X A)AY - A(\nabla_X A)Y &= (a^2 - c)(\langle Y, W \rangle \varphi AX - \langle \varphi AX, Y \rangle W) \\
&\quad - 2ac(\langle W, Y \rangle \varphi X + \langle W, X \rangle \varphi Y + \langle \varphi Y, X \rangle W).
\end{aligned}
$$

PROOF. Taking the inner product of Z with $(\nabla_X A)AY$ and using the Codazzi equation results in

$$
\begin{aligned}
\langle (\nabla_X A)AY, Z \rangle &= \langle AY, (\nabla_X A)Z \rangle \\
&= \langle AY, (\nabla_Z A)X \rangle + c\langle AY, \langle X, W \rangle \varphi Z - \langle Z, W \rangle \varphi X + 2\langle X, \varphi Z \rangle W \rangle \\
&= \langle AY, (\nabla_Z A)X \rangle \\
&\quad + c(\langle X, W \rangle \langle A\varphi Y, Z \rangle - \langle Z, W \rangle \langle A\varphi X, Y \rangle + 2a\langle W, Y \rangle \langle X, \varphi Z \rangle).
\end{aligned}
$$

Reversing X and Y and subtracting the two equations, we get

$$
\begin{aligned}
\langle (\nabla_X A)AY, Z \rangle &- \langle (\nabla_Y A)AX, Z \rangle = \langle AY, (\nabla_Z A)X \rangle - \langle AX, (\nabla_Z A)Y \rangle \\
&+ c(\langle X, W \rangle \langle A\varphi Y, Z \rangle - \langle Y, W \rangle \langle A\varphi X, Z \rangle) + 2ac(\langle W, Y \rangle \langle X, \varphi Z \rangle - \langle W, X \rangle \langle Y, \varphi Z \rangle).
\end{aligned}
$$

Then the coefficient of X on the right hand side is

$$
\begin{aligned}
(\nabla_Z A)AY - A(\nabla_Z A)Y &+ c(\langle A\varphi Y, Z \rangle W - \langle Y, W \rangle A\varphi Z) \\
&+ 2ac(\langle W, Y \rangle \varphi Z - \langle Y, \varphi Z \rangle W). \quad (2.11)
\end{aligned}
$$

Now consider equation (2.8). If we take the inner product of this equation with Z, the coefficient of X on the right hand side is

$$-2ca\langle W, Z\rangle\varphi Y + a^2(\langle Y, W\rangle\varphi AZ - \langle\varphi AY, Z\rangle W).$$

Since this represent the same quantity as (2.11), the two expressions can be equated, and we get

$$(\nabla_Z A)AY - A(\nabla_Z A)Y = (a^2 - c)\langle Y, W\rangle\varphi AZ$$
$$- 2ac(\langle W, Y\rangle\varphi Z + \langle W, Z\rangle\varphi Y + \langle\varphi Y, Z\rangle W) - (a^2 - c)\langle\varphi AY, Z\rangle W.$$

The statement of the lemma can be obtained by replacing Z with X.   □

LEMMA 2.9. *Let M^{2n-1}, where $n \geq 2$, be a real hypersurface in a complex space form of constant holomorphic sectional curvature $4c \neq 0$. If $\varphi A + A\varphi = 0$, then*

$$(\nabla_X A)Y = \beta\langle X, W\rangle\langle Y, W\rangle W + a(\langle X, W\rangle\varphi AY + \langle Y, W\rangle\varphi AX + \langle\varphi AX, Y\rangle W)$$
$$+ c(\langle\varphi Y, X\rangle W - \langle W, Y\rangle\varphi X).$$

PROOF. Adding the result of Lemma 2.8 to the first equation from Lemma 2.9 we get

$$(\nabla_X A)AY = a\beta\langle X, W\rangle\langle Y, W\rangle W + c\langle\varphi AX, Y\rangle W + a^2\langle Y, W\rangle\varphi AX$$
$$- ac(\langle Y, W\rangle\varphi X + \langle X, W\rangle\varphi Y + \langle\varphi Y, X\rangle W). \quad (2.12)$$

Notice that $(A^2 + cI)Y = (a^2 + c)\langle Y, W\rangle W$, since $\varphi(A^2 + cI) = 0$. Also recall that $A\varphi A = c\varphi$. Replacing Y by AY in (2.12) we get

$$(\nabla_X A)A^2Y = a^2\beta\langle X, W\rangle\langle Y, W\rangle W + c^2\langle\varphi X, Y\rangle W$$
$$+ a^3\langle Y, W\rangle\varphi AX - a^2c\langle Y, W\rangle\varphi X - ac\langle X, W\rangle\varphi AY - ac\langle\varphi AY, X\rangle W. \quad (2.13)$$

On the other hand, by computing it directly we see that

$$(\nabla_X A)A^2Y = -c(\nabla_X A)Y + (a^2 + c)\langle Y, W\rangle(\nabla_X A)W$$
$$= -c(\nabla_X A)Y + (a^2 + c)\beta\langle Y, W\rangle\langle X, W\rangle W$$
$$+ (a^2 + c)\langle Y, W\rangle a\varphi AX - (a^2 + c)c\langle Y, W\rangle\varphi X$$
$$= -c(\nabla_X A)Y + a^2\beta\langle X, W\rangle\langle Y, W\rangle W + \beta c\langle X, W\rangle\langle Y, W\rangle W$$
$$+ a^3\langle Y, W\rangle\varphi AX + ac\langle Y, W\rangle\varphi AX - a^2c\langle Y, W\rangle\varphi X - c^2\langle Y, W\rangle\varphi X.$$

Now after equating this with (2.13), we can make several cancellations to get

$$c(\nabla_X A)Y = c\beta\langle X, W\rangle\langle Y, W\rangle W + ac\langle\varphi AY, X\rangle W$$
$$+ ac(\langle Y, W\rangle\varphi AX + \langle X, W\rangle\varphi AY) - c^2\langle Y, W\rangle\varphi X - c^2\langle\varphi X, Y\rangle W.$$

Finally, using the assumption that $4c \neq 0$, we obtain the desired conclusion.   □

In this final pair of lemmas we show that if $A\varphi + \varphi A = 0$, then $c = 0$. This is accomplished by computing

$$\sum (R(e_i, \varphi e_i) \cdot A)Z,$$

where $\{e_i\}$ is an orthonormal basis for $W^\perp$, in two different ways. The first way, described in Lemma 2.10, is to use Lemma 2.9, and compute the sum directly. The second way, described in Lemma 2.11, uses the Gauss equation.

LEMMA 2.10. *Let M^{2n-1}, where $n \geq 2$, be a real hypersurface in a complex space form of constant holomorphic sectional curvature $4c \neq 0$. Suppose $A\varphi + \varphi A = 0$ and $X \in W^\perp$. Then direct calculation of $(R(X, \varphi X) \cdot A)Z$ using Lemma 2.9 yields*

$$(R(X, \varphi X) \cdot A)Z = c(\langle X, Y \rangle \varphi AX - \langle X, \varphi Z \rangle AX + \langle X, AZ \rangle \varphi X + \langle X, \varphi AZ \rangle X).$$

Consequently, if $\{e_i\}$ is an orthonormal basis for $W^\perp$, we have

$$\sum (R(e_i, \varphi e_i) \cdot A)Z = 4c\varphi AZ$$

for all tangent vectors Z.

PROOF. We will first calculate $\nabla_X((\nabla_{\varphi X} A)Z)$, $(\nabla_{\varphi X} A)(\nabla_X Z)$ and $(\nabla_{\nabla_X \varphi X} A)Z$ separately, and then put the pieces together to get the desired results. We will make the first calculation in detail, and simply state the results for the others.

Direct calculation, using previous results, in particular Lemma 2.9, and repeated use of (1.7) and Proposition 1.3 allows us to conclude that

$$\nabla_X((\nabla_{\varphi X} A)Z)$$
$$= \nabla_X\big(a(\langle Z, W \rangle \varphi A \varphi X + \langle \varphi A \varphi X, Z \rangle W) + c(\langle \varphi Z, \varphi X \rangle W - \langle W, Z \rangle \varphi^2 X)\big)$$
$$= a\big(\langle \nabla_X Z, W \rangle AX + \langle Z, \nabla_X W \rangle AX + \langle Z, W \rangle \nabla_X(AX)$$
$$+ \langle \nabla_X(AX), Z \rangle W + \langle AX, \nabla_X Z \rangle W + \langle AX, Z \rangle \nabla_X W\big)$$
$$+ c\big(\langle \nabla_X X, Z \rangle W + \langle X, \nabla_X Z \rangle W + \langle X, Z \rangle \nabla_X W$$
$$+ \langle \nabla_X W, Z \rangle X + \langle W, \nabla_X Z \rangle X + \langle W, Z \rangle \nabla_X X\big), \quad (2.14)$$

where we have used (2.2) to dispose of Xa. Then, using the fact that

$$\nabla_X(AX) = A(\nabla_X X) + (\nabla_X A)X = A(\nabla_X X) + a\langle \varphi AX, X \rangle W + c\langle \varphi X, X \rangle W,$$

and noting that $\langle \varphi X, X \rangle = 0$, we rewrite (2.14) as

$$\nabla_X((\nabla_{\varphi X} A)Z)$$
$$= a\big(\langle \nabla_X Z, W \rangle AX + \langle Z, \varphi AX \rangle AX + \langle Z, W \rangle A(\nabla_X X) + a\langle \varphi AX, X \rangle \langle Z, W \rangle W$$
$$+ \langle A(\nabla_X X), Z \rangle W + a\langle \varphi AX, X \rangle \langle W, Z \rangle W + \langle AX, \nabla_X Z \rangle W + \langle AX, Z \rangle \varphi AX\big)$$
$$+ c\big(\langle \nabla_X X, Z \rangle W + \langle X, \nabla_X Z \rangle W + \langle X, Z \rangle \varphi AX$$
$$+ \langle \varphi AX, Z \rangle X + \langle W, \nabla_X Z \rangle X + \langle W, Z \rangle \nabla_X X\big). \quad (2.15)$$

Similar calculations can be performed to compute

$$(\nabla_{\varphi X} A)(\nabla_X Z) = a(\langle \nabla_X Z, W \rangle AX + \langle AX, \nabla_X Z \rangle W)$$
$$+ c(\langle \nabla_X Z, X \rangle W + \langle W, \nabla_X Z \rangle X), \quad (2.16)$$

and

$$(\nabla_{\nabla_X \varphi X} A)Z = -\beta \langle AX, X \rangle \langle Z, W \rangle W$$
$$+ a(\langle Z, W \rangle A\nabla_X X - \langle AX, X \rangle \varphi AZ - \langle Z, W \rangle \langle \nabla_X X, W \rangle aW$$
$$+ \langle A\nabla_X X, Z \rangle W - a\langle \nabla_X X, W \rangle \langle W, Z \rangle W)$$
$$+ c(\langle Z, \nabla_X X \rangle W - \langle Z, W \rangle \langle \nabla_X X, W \rangle W$$
$$+ \langle W, Z \rangle \nabla_X X - c\langle W, Z \rangle \langle \nabla_X X, W \rangle W). \quad (2.17)$$

Now define

$$N(X, Z) = (\nabla_X \nabla_{\varphi X} A - \nabla_{\nabla_X \varphi X} A)Z$$
$$= \nabla_X ((\nabla_{\varphi X} A)Z) - (\nabla_{\varphi X} A)(\nabla_X Z) - (\nabla_{\nabla_X \varphi X} A)Z.$$

Substituting the values from equations (2.15), (2.16), and (2.17), we get

$$N(X, Z) = \beta \langle AX, X \rangle \langle Z, W \rangle W$$
$$+ a(\langle Z, \varphi AX \rangle AX + \langle AX, Z \rangle \varphi AX + \langle AX, X \rangle \varphi AZ)$$
$$+ c(\langle X, Z \rangle \varphi AX + \langle \varphi AX, Z \rangle X - 2\langle Z, W \rangle \langle \varphi AX, X \rangle W). \quad (2.18)$$

Also, notice that, since we are assuming that X is orthogonal to W, we have

$$N(\varphi X, Z) = (\nabla_{\varphi X} \nabla_{\varphi^2 X} A - \nabla_{\nabla_{\varphi X} \varphi^2 X} A)Z - (\nabla_X \nabla_{\varphi X} A - \nabla_{\nabla_X \varphi X})Z,$$

from which we conclude that

$$(R(X, \varphi X) \cdot A)Z = N(X, Z) + N(\varphi X, Z).$$

Now, by direct calculation using (2.18), we see that

$$N(\varphi X, Z) = -\beta \langle AX, X \rangle \langle Z, W \rangle W$$
$$+ a(\langle Z, AX \rangle A\varphi X + \langle A\varphi X, Z \rangle AX - \langle AX, X \rangle \varphi AZ)$$
$$+ c(\langle \varphi X, Z \rangle AX + \langle AX, Z \rangle \varphi X - 2\langle Z, W \rangle \langle X, A\varphi X \rangle W).$$

Hence

$$(R(X, \varphi X) \cdot A)Z = c(\langle X, Z \rangle \varphi AX + \langle X, \varphi AZ \rangle X - \langle X, \varphi Z \rangle AX + \langle X, AZ \rangle \varphi X).$$

Now let $\{e_i\}$ be an orthonormal basis of $W^\perp$. Then

$$\sum (R(e_i, \varphi e_i) \cdot A)Z = c\Big(\varphi AZ + \varphi AZ - A\varphi Z + \varphi AZ \Big) = 4c\varphi AZ. \qquad \Box$$

LEMMA 2.11. *Under the conditions of Lemma 2.10, the Gauss equation yields*

$$\sum (R(e_i, \varphi e_i) \cdot A)Z = -4c(2n + 1)\varphi AZ.$$

PROOF. First look at the calculations for any $(1,1)$ tensor T, and evaluate

$$(TX \wedge T\varphi X)AZ - A(TX \wedge T\varphi X)Z.$$

Two of the terms needed in calculating the desired equality are of this form for various tensors T. Expand this to get

$$(TX \wedge T\varphi X)AZ - A(TX \wedge T\varphi X)Z$$
$$= \langle T\varphi X, AZ \rangle TX - \langle TX, AZ \rangle T\varphi X - \langle T\varphi X, Z \rangle ATX + \langle TX, Z \rangle AT\varphi X.$$

Summing this over $X = e_i$, the right side becomes

$$-T(\varphi T^* AZ) - T\varphi T^* AZ + AT\varphi T^* Z + AT\varphi T^* Z = -2T\varphi T^* AZ + 2AT\varphi T^* Z, \tag{2.19}$$

where T^* is the transpose of T. Now we can look at this for specific choices of T. In the case when $T = I$, (2.19) becomes

$$2(A\varphi - \varphi A)Z = -4\varphi AZ.$$

When $T = A$, (2.19) is

$$-2A\varphi A^2 Z + 2A^2\varphi AZ = -2(A\varphi A)AZ + 2A(A\varphi A)Z = -4c\varphi AZ.$$

Here we have substituted for $A\varphi A$ using (2.1). Now

$$R(e_i, \varphi e_i) = Ae_i \wedge A\varphi e_i + c(e_i \wedge \varphi e_i + \varphi e_i \wedge \varphi^2 e_i + 2\langle e_i, \varphi^2 e_i \rangle \varphi)$$
$$= Ae_i \wedge A\varphi e_i + 2c(e_i \wedge \varphi e_i) - 2c\varphi.$$

Summation of the last term in $(R(e_i, \varphi e_i) \cdot A)Z$ gives $-4c(2n - 2)\varphi AZ$. Using this and (2.19) with $T = I$ and $T = A$, we get

$$\sum (R(e_i, \varphi e_i) \cdot A)Z = -4c(2n + 1)\varphi AZ,$$

which is the claim of the lemma. □

COROLLARY 2.12. *Let M^{2n-1}, $n \geq 2$, be a real hypersurface in a complex space form of constant holomorphic sectional curvature $4c \neq 0$. Then $A\varphi + \varphi A$ cannot be identically zero.*

PROOF. Suppose $A\varphi + \varphi A = 0$. From Lemmas 2.10 and 2.11 we see that for all tangent vectors Z,

$$8c(n + 1)\varphi AZ = 0.$$

Therefore, $\varphi A = 0$. This is impossible by Proposition 1.4. □

Now all the ingredients for proving that a is constant have been assembled.

PROOF OF THEOREM 2.1. In view of Corollary 2.12 and Lemma 2.4, we must have $Wa = 0$. This implies that grad $a = 0$ by Lemma 2.2. Thus a is constant. □

Lemmas 2.7 and 2.8 are valid for $c = 0$, and, in fact, $\varphi A + A\varphi = 0$ leads to no contradiction in this case. However, the only new information that can be derived from these lemmas is that $A(\nabla_X A)Y = 0$ when X and Y are in $W^\perp$. Even this can be deduced more easily by differentiating the identity $AY = 0$.

We finish this section with two applications that make use of the preceding material. The first one will be useful for later classifications.

LEMMA 2.13. *Let M^{2n-1}, where $n \geq 2$, be a Hopf hypersurface in a complex space form of constant holomorphic sectional curvature $4c \neq 0$. Then $W\mathfrak{m} = 0$.*

PROOF. By the Codazzi equation (1.9), we have $(\nabla_W A)X = (\nabla_X A)W + c\varphi X$. Thus

$$W\mathfrak{m} = W(\text{trace } A) = \text{trace } \nabla_W A = \text{trace}\{X \mapsto a\nabla_X W - A\nabla_X W + c\varphi X\}$$
$$= \text{trace}(a\varphi A - A\varphi A + c\varphi) = 0. \qquad \square$$

The second application shows that there is a lower bound on the rank of A, even pointwise.

PROPOSITION 2.14. *Let M^{2n-1}, where $n \geq 2$, be a hypersurface in a complex space form of constant holomorphic sectional curvature $4c \neq 0$. Then the rank of the shape operator A is ≥ 2 at some point.*

PROOF. Suppose that the rank of A is ≤ 1 everywhere. Since M cannot be umbilic, by Theorem 1.5, there is an open connected set $\mathcal{U}$ where the rank is 1. We restrict our attention to $\mathcal{U}$. Let λ be the nonzero principal curvature with (one-dimensional) principal subspace T_λ. If X and Y are vector fields in the principal distribution $T_0 = \ker A = T_\lambda^\perp$, then Codazzi's equation gives

$$c(\langle X, W\rangle \varphi Y - \langle Y, W\rangle \varphi X + 2\langle X, \varphi Y\rangle W) = (\nabla_X A)Y - (\nabla_Y A)X$$
$$= -A(\nabla_X Y - \nabla_Y X). \qquad (2.20)$$

Clearly the right (and hence the left) side of this equation lies in $T_0^\perp = T_\lambda$. In particular, taking the inner product with W yields

$$3\langle X, W\rangle\langle \varphi Y, X\rangle = 0. \qquad (2.21)$$

If $W \notin T_\lambda$, we can choose $X \in T_0$ which is not orthogonal to W. For this X, we have $\langle \varphi Y, X\rangle = 0$ for all $Y \in T_0$ by (2.21). Thus $\varphi X \in T_\lambda$. At any point where $\varphi X = 0$, we have $X = \langle X, W\rangle W$ by (1.6) so that $W \in \text{span } \{X\} \subset T_0$. If $\varphi X \neq 0$, on the other hand, we have

$$0 = \langle \varphi W, X\rangle = -\langle W, \varphi X\rangle$$

so that $W \in T_\lambda^\perp = T_0$. In either case, we can conclude that $AW = 0$.

The net result is that either $AW \equiv \lambda W$ or that $AW \equiv 0$ on $\mathcal{U}$. Thus $\mathcal{U}$ is a Hopf hypersurface and we can apply Corollary 2.3(i). In the first case, we get

$$-\frac{\lambda}{2}A\varphi X = c\varphi X$$

for any $X \in T_0 = W^\perp$, which contradicts the fact that the rank is 1. If $AW = 0$, on the other hand, we have $\varphi X \in T_\lambda$ for any X orthogonal to W in T_0. This comes from setting $Y = W$ in (2.20). Corollary 2.3 then yields $c\varphi X = 0$, another contradiction. □

For other theorems involving lower bounds on the rank, see [Suh 1991].

3. Hypersurfaces in Complex Space Forms

In this section we will describe the standard examples of real hypersurfaces in spaces of constant holomorphic sectional curvature. We start with real hypersurfaces in complex hyperbolic space, and discuss them in detail.

3A. Examples in Complex Hyperbolic Space. We begin with a lemma that will be useful for discussing the first class of examples, the horospheres. The proof is a straightforward calculation.

LEMMA 3.1. *In* $\mathbb{C}^{n+1}$, *let* $p = (1, 1, 0, \ldots, 0)$ *and let* $\eta = (z_0 - z_1)p$. *Then treating* z *as the position vector field, we have, for any vector field* Z *in* $\mathbb{C}^{n+1}$,

(i) $\eta = (z_0 - z_1)p = -F(z, p)p$.
(ii) $\langle \eta, z \rangle = -|z_0 - z_1|^2 = -|F(z, p)|^2$.
(iii) $\langle \eta, Z \rangle = -\operatorname{Re}(F(z, p)\overline{F(Z, p)}) = \langle D_Z \eta, z \rangle$.
(iv) $D_Z \eta = -F(Z, p)p$.
(v) $Z \langle \eta, z \rangle = 2\langle \eta, Z \rangle$. ◁

We will use the notation introduced in Section 1. In particular, r is a positive number and the holomorphic curvature of $\mathbb{C}H^n$ is $4c = -4/r^2$. All of the examples, both in $\mathbb{C}H^n$ and in $\mathbb{C}P^n$, are tubes of some sort, and we will use these descriptive names for the purposes of identification. However, we will not attempt to justify these names, but refer the reader to [Cecil and Ryan 1982; Kimura 1986a; Berndt 1989a; 1990] for a full discussion.

The horospheres: Type A0. The first class of examples to be considered were called "self tubes" by Montiel [1985] and are now called horospheres. They form a one-parameter family, parametrized by $t > 0$. Let

$$M' = \{z \in \mathbb{C}^{n+1} : \langle z, z \rangle = -r^2, |z_0 - z_1|^2 = t\}.$$

Then the corresponding horosphere is $M = \pi M'$. We now investigate the geometry of M and its relation to that of M'.

LEMMA 3.2. *M' is a hypersurface in* $\mathbb{H}$. *For* $z \in M'$,

$$T_z M' = \{Z \in \mathbb{C}^{n+1} : \langle Z, z \rangle = 0, \langle Z, \eta \rangle = 0\}.$$

PROOF. First we show that z and η are linearly independent. Both are nonzero, since $\langle \eta, z \rangle = -t \neq 0$. However, $\langle \eta, \eta \rangle = \operatorname{Re}(F(z, p)\overline{F(z, p)}F(p, p)) = 0$ so z cannot be a multiple of η.

The map $\Phi : \mathbb{C}^{n+1} \to \mathbb{R}^2$, given by $\Phi(z) = (\langle z, z \rangle, \langle \eta, z \rangle)$ implicitly defines M'. That is

$$M' = \Phi^{-1}(-r^2, -t).$$

Its rank is 2 since for $Z \in \mathbb{C}^{n+1}$, we have $\Phi_* Z = 0$ if and only if $Z\langle z, z \rangle = 0 = Z\langle \eta, z \rangle$, if and only if $\langle z, Z \rangle = 0 = \langle \eta, Z \rangle$. Thus ker $\Phi_* = (\text{span}\{z, \eta\})^\perp$, which has dimension $2n$.

Let $\xi' = (r/t)\eta - (1/r)z$. Then ξ' is orthogonal to $T_z M'$ (since η and z are) and coefficients have been chosen so that $\langle \xi', z \rangle = 0$ and $\langle \xi', \xi' \rangle = 1$. Then $\{z/r, \xi'\}$ can be completed to an orthonormal basis for $\mathbb{C}^{n+1}$ by adding one timelike and $2n - 1$ spacelike vectors, so M' is a Lorentz hypersurface of $\mathbb{H}$.   □

The proof of the next lemma can be verified directly from the definitions, and the background developed in Section 1.

LEMMA 3.3. *Let $U = -J\xi'$ and $V = Jz$. Then $\{r^{-1}z, \xi', r^{-1}V, U = -J\xi'\}$ is an orthonormal set of vectors with $\{U, V\}$ in $T_z M'$ and $\{z, \xi'\}$ is orthogonal to $T_z M'$. The shape operator A' satisfies $A'V = U$, $A'U = -r^{-2}V + 2r^{-1}U$. For $Z \in T_z M' \cap \{U, V\}^\perp$, $A'Z = r^{-1}Z$.*   ◁

We now look at the geometry of $M = \pi M'$ which is a hypersurface in $\mathbb{C}H^n$. First note that ξ' is the horizontal lift of the unit normal ξ of M. For $W = -J\xi$, we have

$$\tilde{\nabla}_W \xi = \pi_*(\tilde{\nabla}_{W^L}\xi^L) = \pi_*\left(\frac{1}{r^2}V - \frac{2}{r}U\right) = -\frac{2}{r}W.$$

For X in $W^\perp$,

$$\tilde{\nabla}_X \xi = \pi_*(\tilde{\nabla}_{X^L}\xi^L) = -\frac{1}{r}X,$$

since $X^L \in \{U, V\}^\perp$. Thus we have established the following.

THEOREM 3.4. *The horospheres (Type A0) are hypersurfaces of complex hyperbolic space that have two distinct principal curvatures: $\lambda = 1/r$ of multiplicity $2n - 2$, and $a = 2/r$ of multiplicity 1.*   ◁

Tubes around complex hyperbolic spaces: Types A1, A2. Such tubes also form a one-parameter family, parametrized initially by $b > 0$ and later by u. Begin by writing $\mathbb{C}^{n+1} = \mathbb{C}^{p+1} \times \mathbb{C}^{q+1}$ where $p, q \geq 0$ and $p + q = n - 1 > 0$. Let

$$M' = \{z = (z_1, z_2) \in \mathbb{C}^{n+1} : F_1(z_1, z_1) = -(b^2 + r^2), F_2(z_2, z_2) = b^2\}$$

where F_1 and F_2 are the restrictions of F to $\mathbb{C}^{p+1}$ and $\mathbb{C}^{q+1}$ respectively. Then M' is the Cartesian product of an anti–de Sitter space and a sphere whose radii have been chosen so that M' lies in $\mathbb{H}$. Specifically,

$$M'^{2n} = H_1^{2p+1}((b^2 + r^2)^{1/2}) \times S^{2q+1}(b).$$

Write $b = r \sinh u$, $(b^2 + r^2)^{1/2} = r \cosh u$, $\lambda_1 = r^{-1}\tanh u$, $\lambda_2 = r^{-1}\coth u$, and $c = -r^{-2}$, so that $\lambda_1\lambda_2 + c = 0$. These turn out to be principal curvatures of M'.

Let $\xi' = -(\lambda_1 z_1 + \lambda_2 z_2)$. Note that we identify z_1 and $(z_1, 0)$ for convenience of notation. Similarly for z_2. It is easy to check that ξ' is a unit normal vector for M'.

Let $V = Jz$. Note that V is tangent to M' since $\langle V, z \rangle = \langle Jz, z \rangle = 0$ and $\langle V, \xi' \rangle = -\langle Jz_1 + Jz_2, \lambda_1 z_1 + \lambda_2 z_2 \rangle = -(\lambda_1 \langle Jz_1, z_1 \rangle + \lambda_2 \langle Jz_2, z_2 \rangle) = 0$. Both λ_1 and λ_2 satisfy the equation

$$\lambda^2 - \alpha\lambda + \frac{1}{r^2} = 0 \quad \text{with } \alpha = \frac{2}{r} \coth 2u.$$

A routine calculation yields the following two lemmas.

LEMMA 3.5. *For X tangent to H_1^{2p+1}, $A'X = \lambda_1 X$. For X tangent to S^{2q+1}, $A'X = \lambda_2 X$. If $U = -J\xi'$, then $A'V = U$ and $A'U = \alpha U - Vr^{-2}$.* ◁

LEMMA 3.6. *$\{z/r, \xi', V/r, U\}$ is an orthonormal set with ξ' and U spacelike. The other two vectors are timelike.* ◁

The hypersurface $\pi M'$ is denoted by $M_{2p+1,2q+1}$. At a typical point $z \in M'$, the horizontal subspace of $T_z M'$ is the orthogonal direct sum

$$(\text{span}\{U\}) \oplus T_1 \oplus T_2,$$

where

$$T_1 = \{Z \in T_z H_1^{2p+1} : \langle Z, U \rangle = 0, \langle Z, V \rangle = 0\},$$
$$T_2 = \{Z \in T_z S^{2q+1} : \langle Z, U \rangle = 0, \langle Z, V \rangle = 0\}.$$

Note that T_1 and T_2 are J-invariant. Thus

$$T_{\pi z} M_{2p+1,2q+1} = (\text{span}\{W\}) \oplus \pi_* T_1 \oplus \pi_* T_2.$$

We calculate the shape operator A using Lemma 1.7.

$$AW = \pi_*(A'U) - \frac{\varepsilon}{r^2}\pi_* V = \pi_*\left(A'U + \frac{1}{r^2}V\right)$$

$$= \pi_*\left(\alpha U - \frac{1}{r^2}V + \frac{1}{r^2}V\right) = \pi_*(\alpha U) = \alpha W.$$

For $X \in \pi_* T_1$,

$$AX = \pi_*\left(A'X^L - \langle X^L, U \rangle\frac{\varepsilon}{r^2}V\right) = \pi_*\lambda_1 X^L = \lambda_1 X.$$

Similarly, $AX = \lambda_2 X$ for $X \in \pi_* T_2$. When $p = 0$, M is a geodesic hypersphere with principal curvatures λ_2 (of multiplicity $2n - 2$) and α of multiplicity 1. The radius of the sphere is ru. When $q = 0$, M is a tube of radius ru over a complex hyperbolic hyperplane. There are only two principal curvatures λ_1 and α. These are the Type A1 hypersurfaces. The rest are Type A2 and have three distinct principal curvatures. Again, they are tubes of radius ru about complex hyperbolic spaces of codimension greater than 1. Summarizing this information, we have the following results.

THEOREM 3.7. *The tubes around complex hyperbolic hyperplanes (Type A1) in complex hyperbolic space have two distinct principal curvatures: $\lambda_1 = (1/r)\tanh u$ of multiplicity $2n - 2$ and $a = (2/r)\coth 2u$ of multiplicity 1.*   ◁

THEOREM 3.8. *The geodesic spheres (Type A1) in complex hyperbolic space have two distinct principal curvatures: $\lambda_2 = (1/r)\coth u$ of multiplicity $2n - 2$ and $a = (2/r)\coth 2u$ of multiplicity 1.*   ◁

THEOREM 3.9. *The Type A2 hypersurfaces in complex hyperbolic space have three distinct principal curvatures: $\lambda_1 = (1/r)\tanh u$ of multiplicity $2p$, $\lambda_2 = (1/r)\coth u$ of multiplicity $2q$, and $a = (2/r)\coth 2u$ of multiplicity 1, where $p > 0$, $q > 0$, and $p + q = n - 1$.*   ◁

Tubes around real hyperbolic space: Type B. Again, these examples form a one-parameter family. We begin with $t > r^4$ but later find a more convenient parameter u. Let

$$M' = \{z \in \mathbb{C}^{n+1} : \langle z, z \rangle = -r^2, |F(z, \bar{z})|^2 = t\}.$$

Write $Q(z) = F(z, \bar{z})$ and $\eta = Q(z)\bar{z}$.

LEMMA 3.10. *M' is a hypersurface in $\mathbb{H}$. For $z \in M'$,*

$$T_z M' = \{Z \in \mathbb{C}^{n+1} : \langle Z, z \rangle = 0, \langle Z, \eta \rangle = 0\}.$$

PROOF. First we note that z and η are nonzero. In fact, $\langle \eta, \eta \rangle = |Q(z)|^2 \langle \bar{z}, \bar{z} \rangle = -tr^2$. Further, they are linearly independent since if $\eta = \rho z$ then $-\rho r^2 = \langle \eta, z \rangle = \mathrm{Re}\, F(F(z, \bar{z})\bar{z}, z) = |F(z, \bar{z})|^2 = t$. On the other hand, by considering $\langle \eta, \eta \rangle$, we have $\rho^2 = t$. Thus $tr^4 = \rho^2 r^4 = (-\rho r^2)^2 = t^2$. In other words, $t(r^4 - t) = 0$, which is a contradiction.

The map $\Phi : \mathbb{C}^{n+1} \to R^2$, given by $\Phi(z) = (\langle z, z \rangle, |Q(z)|^2)$ implicitly defines M'. That is,

$$M' = \Phi^{-1}(-r^2, t).$$

Its rank is 2 since, for $Z \in \mathbb{C}^{n+1}$, we have $\Phi_* Z = 0$ if and only if $Z\langle z, z \rangle = 0 = Z\langle \eta, z \rangle$, if and only if $\langle z, Z \rangle = 0 = \langle \eta, Z \rangle$. To see this, note that

$$Z(F(z, \bar{z})F(\bar{z}, z)) = (F(Z, \bar{z}) + F(z, \bar{Z}))F(\bar{z}, z) + F(z, \bar{z})(F(\bar{Z}, z) + F(\bar{z}, Z))$$
$$= 2\,\mathrm{Re}(F(Z, \bar{z})F(\bar{z}, z)) + 2(\mathrm{Re}\, F(z, \bar{Z})F(\bar{z}, z))$$
$$= 2\,\mathrm{Re}\, F(Z, Q(z)\bar{z}) + 2\,\mathrm{Re}\, F(\overline{Q(z)}z, \bar{Z})$$
$$= 4\langle Z, Q(z)\bar{z} \rangle = 4\langle Z, \eta \rangle. \qquad \square$$

From the lemma we see that η is normal to M' but not tangent to $\mathbb{H}$. However, it is easy to verify that

$$\xi_0 = r^2 \eta + tz$$

satisfies $\langle \xi_0, z \rangle = 0$. Furthermore,

$$F(\xi_0, \xi_0) = r^4 \langle \eta, \eta \rangle + t^2 F(z, z) + 2r^2\,\mathrm{Re}\, F(\eta, tz) = -r^6 t - t^2 r^2 + 2t^2 r^2 = r^2 t(t - r^4),$$

so that we can define a unit normal ξ' by

$$\xi_0 = -rt^{1/2}(t - r^4)^{1/2}\xi'. \qquad (3.1)$$

We now describe the shape operator A' of M'. First let $V = Jz$ as in earlier examples. Note that V is tangent to M' since $F(Q(z)\bar{z}, V) = -iF(Q(z)\bar{z}, z) = -iQ(z)\overline{Q(z)}$ is purely imaginary. Also the vector field $U_0 = -J\xi_0 = -tV - r^2 J\eta$ is tangent to M' and orthogonal to V. In computing A', we first work with ξ_0 to avoid the normalization factor involved in (3.1).

LEMMA 3.11. (i) *For Z tangent to M', $D_Z\xi_0 = 2r^2 F(Z, \bar{z})\bar{z} + r^2 Q(z)\bar{Z} + tZ$.*
(ii) *In particular, $D_V\xi_0 = -U_0$.*
(iii) *$D_{U_0}\xi_0 = (t - r^4)(2U_0 + tV)$.*
(iv) *There are two $n - 1$ dimensional subspaces $\mathcal{V}^+$ and $\mathcal{V}^-$, orthogonal both to $\{U, V\}$ and to each other, such that $D_X\xi_0 = (t + r^2 t^{1/2})X$ for $X \in \mathcal{V}^+$, and $D_X\xi_0 = (t - r^2 t^{1/2})X$ for $X \in \mathcal{V}^-$.*

PROOF. We compute

$$D_Z\xi_0 = D_Z(r^2 F(z, \bar{z})\bar{z}) + D_Z(tz) = 2r^2 F(Z, \bar{z})\bar{z} + r^2 F(z, \bar{z})\bar{Z} + tZ$$
$$= 2r^2 F(Z, \bar{z})\bar{z} + r^2 Q(z)\bar{Z} + tZ,$$

which proves the general formula (i). Now, if $Z = V = iz$, we get

$$D_V\xi_0 = 2r^2 F(iz, \bar{z})\bar{z} + r^2 Q(z)(-i\bar{z}) + tiz = ir^2 Q(z)\bar{z} + tiz = i\xi_0 = -U_0.$$

For (iii) we compute

$$D_{U_0}\xi_0 = 2r^2 F(U_0, \bar{z})\bar{z} + r^2 Q(z)\bar{U}_0 + tU_0.$$

It is straightforward to verify that

$$F(U_0, \bar{z}) = -i(t - r^4)Q(z)$$

and that

$$Q(z)\bar{U}_0 = it(r^2 z + \eta),$$

from which it follows by substitution that

$$D_{U_0}\xi_0 = (t - r^4)(2U_0 + tV).$$

We now prove (iv). Choose α in the range $0 \le \alpha \le \pi/2$ and such that $Q(z) = e^{2i\alpha}|Q(z)|$. Let X be a totally real tangent vector (that is, $\bar{X} = X$). Then $Z = e^{i\alpha}X = \cos\alpha X + \sin\alpha JX$ satisfies

$$D_Z\xi_0 = r^2 Q(z)e^{-i\alpha}X + te^{i\alpha}X = r^2 Q(z)e^{-2i\alpha}Z + tZ$$
$$= (r^2|Q(z)| + t)Z = (t + r^2 t^{1/2})Z.$$

Similarly, if $Z = e^{i\alpha}JX = e^{i(\alpha+\pi/2)}X$, then

$$D_Z\xi_0 = r^2 Q(z)e^{-i\alpha}(-iX) + te^{i\alpha}(iX) = -r^2 Q(z)e^{-2i\alpha}Z + tZ = (t - r^2 t^{1/2})Z.$$

Now

$$\mathcal{V}^+ = \{e^{i\alpha}X : X \in T_zM', \ \bar{X} = X, \ \langle X, U_0 \rangle = \langle X, V \rangle = 0\}$$

and $\mathcal{V}^- = J\mathcal{V}^+$ have the properties stated in (iv). Note that $\mathcal{V}^-$ is obtained from "totally imaginary" tangent vectors as follows:

$$\mathcal{V}^- = \{e^{i\alpha}X : X \in T_zM', \ \bar{X} = -X, \ \langle X, U_0 \rangle = \langle X, V \rangle = 0\},$$

and that J interchanges the two principal spaces $\mathcal{V}^+$ and $\mathcal{V}^-$. $\qquad\square$

THEOREM 3.12. *The Type B hypersurfaces in complex hyperbolic space have three principal curvatures, namely, $\lambda_1 = (1/r)\coth u$ of multiplicity $n-1$, $\lambda_2 = (1/r)\tanh u$ of multiplicity $n-1$, and $a = (2/r)\tanh 2u$ of multiplicity 1. These curvatures are distinct unless $\coth u = \sqrt{3}$, in which case λ_1 and a coincide to make a principal curvature of multiplicity n.*

PROOF. Taking the normalization factor from (3.1) into account, we get

$$A'X = \frac{1}{r}\frac{(t^{1/2} + r^2)}{(t - r^4)^{1/2}}X = \lambda_1 X$$

for $X \in \mathcal{V}^+$, and

$$A'X = \frac{1}{r}\frac{(t^{1/2} - r^2)}{(t - r^4)^{1/2}}X = \lambda_2 X$$

for $X \in \mathcal{V}^-$. (These equations serve to define the principal curvatures λ_1 and λ_2.) Also, we get

$$D_{U_0}\xi' = -\frac{(t - r^4)^{1/2}}{rt^{1/2}}(2U_0 + tV).$$

Applying Lemma 1.7, $\mathcal{V}^+$ and $\mathcal{V}^-$ are horizontal subspaces projecting to $T_{\pi z}M$ and the principal curvatures are preserved. Also, we get

$$AW = \frac{(t - r^4)^{1/2}}{rt^{1/2}}(2W).$$

It is easy to verify that $\lambda_1\lambda_2 = 1/r^2 = -c$, so that there is a unique positive number u satisfying $\lambda_1 = (1/r)\coth u$ and $\lambda_2 = (1/r)\tanh u$. A direct calculation shows that $a = (2/r)\tanh 2u$. $\qquad\square$

The Type B hypersurfaces are tubes of radius ru around real hyperbolic space $\mathbb{R}H^n$.

3B. Examples in Complex Projective Space.

We now discuss the standard examples of real hypersurfaces in complex projective space. Since these examples may be more widely know than those in complex hyperbolic space, and since the constructions often resemble those in complex hyperbolic space, we will leave some of the details to the reader. Here r is a positive constant and $c = 1/r^2$.

Tubes around complex projective spaces: Types A1, A2. Begin by writing $\mathbb{C}^{n+1} = \mathbb{C}^{p+1} \times \mathbb{C}^{q+1}$, where $p, q \geq 0$ and $p + q = n - 1 > 0$. Choose b so that $0 < b < r$. Let

$$M' = \{z = (z_1, z_2) \in \mathbb{C}^{n+1} : F_1(z_1, z_1) = r^2 - b^2, \ F_2(z_2, z_2) = b^2\},$$

where F_1 and F_2 are the restrictions of F to $\mathbb{C}^{p+1}$ and $\mathbb{C}^{q+1}$ respectively. Then M' is the Cartesian product of spheres whose radii have been chosen so that M' lies in $S^{2n+1}(r)$. Specifically,

$$M'^{2n} = S^{2p+1}\left((r^2 - b^2)^{1/2}\right) \times S^{2q+1}(b).$$

Write $b = r \sin u$, so $(r^2 - b^2)^{1/2} = r \cos u$. We can choose u so that $0 < u < \pi/2$. Write $\lambda_1 = -(1/r)\tan u$ and $\lambda_2 = (1/r)\cot u$. Since $c = r^{-2}$ we have $\lambda_1 \lambda_2 + c = 0$. The numbers λ_1 and λ_2 turn out to be the principal curvatures of M'.

The principal curvatures λ_1, λ_2, and α project as in the hyperbolic case. There is only one kind of Type A1 hypersurface since tubes over complex projective hyperplanes are also geodesic spheres. The principal curvatures of one are related to those of the other by replacing the parameter u by $\frac{\pi}{2} - u$. We summarize as follows.

THEOREM 3.13. *The geodesic spheres (Type A1) in complex projective space have two distinct principal curvatures:* $\lambda_2 = (1/r)\cot u$ *of multiplicity* $2n - 2$ *and* $a = (2/r)\cot 2u$ *of multiplicity* 1. ◁

THEOREM 3.14. *The Type A2 hypersurfaces in complex projective space have three distinct principal curvatures:* $\lambda_1 = -(1/r)\tan u$ *of multiplicity* $2p$, $\lambda_2 = (1/r)\cot u$ *of multiplicity* $2q$, *and* $a = (2/r)\cot 2u$ *of multiplicity* 1, *where* $p > 0$, $q > 0$, *and* $p + q = n - 1$. ◁

Tubes around the complex quadric: Type B. Again, such tubes form a one-parameter family. We begin with $t < r^4$ but later find a more convenient parameter u. Let

$$M' = \{z \in \mathbb{C}^{n+1} : \langle z, z \rangle = r^2, \ |F(z, \bar{z})|^2 = t\}.$$

Write $Q(z) = F(z, \bar{z})$ and $\eta = Q(z)\bar{z}$. Then we have,

THEOREM 3.15. *The Type B hypersurfaces in complex projective space have three distinct principal curvatures:* $\lambda_1 = -(1/r)\cot u$ *of multiplicity* $n - 1$, $\lambda_2 = (1/r)\tan u$ *of multiplicity* $n - 1$, *and* $a = (2/r)\tan 2u$ *of multiplicity* 1. ◁

The Type B hypersurfaces are also tubes over totally geodesic real projective spaces $\mathbb{R}P^n$. The parameter u is chosen so that the tubes have radius ru. Then the tubes over the complex quadric have radius $r(\pi/4 - u)$. This is taken into account in the statement of Theorem 6.1. For more detail—in particular, the relationship to focal sets—see [Cecil and Ryan 1982].

Examples with Five Principal Curvatures. There are three additional standard types of hypersurfaces in $\mathbb{C}P^n$ and they have five distinct principal curvatures. They are listed at the end of Section 3 (page 261), but we will not describe their construction in detail. However we will list the principal curvatures for future reference.

THEOREM 3.16. *The Type C hypersurfaces in complex projective space have five distinct principal curvatures,*

(i) $\lambda_1 = -(1/r)\cot u$ *of multiplicity* $n - 3$,
(ii) $\lambda_2 = (1/r)\cot(\pi/4 - u)$ *of multiplicity 2,*
(iii) $\lambda_3 = (1/r)\cot(\pi/2 - u)$ *of multiplicity* $n - 3$,
(iv) $\lambda_4 = (1/r)\cot(3\pi/4 - u)$ *of multiplicity 2,*
(v) $a = -(2/r)\cot 2u$ *of multiplicity 1.*

These hypersurfaces occur for $n \geq 5$, *n odd.*   ◁

THEOREM 3.17. *The Type D hypersurfaces in complex projective space have five distinct principal curvatures,*

(i) $\lambda_1 = -(1/r)\cot u$ *of multiplicity 4,*
(ii) $\lambda_2 = (1/r)\cot(\pi/4 - u)$ *of multiplicity 4,*
(iii) $\lambda_3 = (1/r)\cot(\pi/2 - u)$ *of multiplicity 4,*
(iv) $\lambda_4 = (1/r)\cot(3\pi/4 - u)$ *of multiplicity 4,*
(v) $a = -(2/r)\cot 2u$ *of multiplicity 1.*

This hypersurface occurs only in $\mathbb{C}P^9$.   ◁

THEOREM 3.18. *The Type E hypersurfaces in complex projective space have five distinct principal curvatures,*

(i) $\lambda_1 = -(1/r)\cot u$ *of multiplicity 8,*
(ii) $\lambda_2 = (1/r)\cot(\pi/4 - u)$ *of multiplicity 6,*
(iii) $\lambda_3 = (1/r)\cot(\pi/2 - u)$ *of multiplicity 8,*
(iv) $\lambda_4 = (1/r)\cot(3\pi/4 - u)$ *of multiplicity 6,*
(v) $a = -(2/r)\cot 2u$ *of multiplicity 1.*

This hypersurface occurs only in $\mathbb{C}P^{15}$.   ◁

3C. Summary: Takagi's list and Montiel's list. In this section, we list, for reference purposes, standard examples of hypersurfaces in complex space forms. These examples are so prevalent in the subject that they have acquired a standard nomenclature. In $\mathbb{C}P^n$, they divide into five types, A–E, while $\mathbb{C}H^n$ has just two types. Types are further subdivided, e.g., A1, A2. The list is as follows. In complex projective space, $\mathbb{C}P^n$:

(A1) Geodesic spheres.
(A2) Tubes over totally geodesic complex projective spaces $\mathbb{C}P^k$, where $1 \leq k \leq n - 2$.

(B) Tubes over complex quadrics and $\mathbb{R}P^n$.

(C) Tubes over the Segre embedding of $\mathbb{C}P^1 \times \mathbb{C}P^m$ where $2m + 1 = n$ and $n \geq 5$.

(D) Tubes over the Plücker embedding of the complex Grassmann manifold $G_{2,5}$. Occur only for $n = 9$.

(E) Tubes over the canonical embedding of the Hermitian symmetric space $SO(10)/U(5)$. Occur only for $n = 15$.

This list consists precisely of the homogeneous real hypersurfaces in $\mathbb{C}P^n$ as determined by Takagi [1973], and is often referred to as "Takagi's list". The list itself with the type names is given in [Takagi 1975a]. In addition, every Hopf hypersurface with constant principal curvatures is an open subset of one of these. Many authors contributed to this result, which was completed by M. Kimura [Kimura 1986a]. In complex hyperbolic space $\mathbb{C}H^n$ the list is as follows:

(A0) Horospheres.

(A1) Geodesic spheres and tubes over totally geodesic complex hyperbolic hyperplanes.

(A2) Tubes over totally geodesic $\mathbb{C}H^k$, where $1 \leq k \leq n - 2$.

(B) Tubes over totally real hyperbolic space $\mathbb{R}H^n$.

These hypersurfaces are homogeneous, but there is yet no classification theorem for homogeneous hypersurfaces in $\mathbb{C}H^n$. However, every Hopf hypersurface with constant principal curvatures must be one of these. This classification was begun by S. Montiel [1985] (who also described the examples in detail) and completed by J. Berndt [1989a]. We refer to the list as "Montiel's list".

In subsequent sections, we will characterize certain subsets of these lists in terms of properties of the shape operator, Ricci tensor and other geometric objects.

4. Restrictions on the Shape Operator and the Number of Principal Curvatures

We recall that the principal spaces of the Type A hypersurfaces are invariant by the structure tensor φ. One of the first classification theorems in this subject is that this property is a characterization for Type A hypersurfaces.

THEOREM 4.1. *Let M^{2n-1}, where $n \geq 2$, be a real hypersurface in a complex space form of constant holomorphic sectional curvature $4c \neq 0$. Then $\varphi A = A\varphi$ if and only if M is an open subset of a Type A hypersurface.*

As a first step in proving Theorem 4.1, we show that this property is equivalent to parallelism of the shape operator of the lifted hypersurface M'.

LEMMA 4.2. *Under the hypothesis of Theorem 4.1, $\varphi A = A\varphi$ if and only if A' is parallel.*

PROOF. If A' is parallel, then $\varphi A = A\varphi$ by Proposition 1.9. Conversely, suppose $\varphi A = A\varphi$. Then $(A^2 - aA - cI)\varphi = 0$ by Lemma 2.2 and Corollary 2.5, so $A|_{W^\perp}$ satisfies the quadratic equation

$$t^2 - at - c = 0. \tag{4.1}$$

Lifting this condition to M' using Lemma 1.7, we see that A' satisfies (4.1) on $\{U, V\}^\perp$. Again using Lemma 1.7, we have

$$(A')^2 U = A'(aU + cV) = aA'U + cA'V = aA'U + cU$$

and

$$(A')^2 V = A'U = aU + cV = aA'V + cV.$$

Applying Lemma 1.10 to A', we see that A' is parallel provided that $a^2 + 4c \neq 0$. Suppose now that $a^2 + 4c = 0$. Without loss of generality we can assume that a is positive so that $a = 2/r$. Let

$$P = A' - \tfrac{1}{2}aI = A' - \frac{1}{r}I$$

so that (4.1) means that $P^2 = 0$. Note that $\nabla' A' = \nabla' P$ and that

$$\ker P = \{U, V\}^\perp \oplus \operatorname{span}\{rU - V\} = \{rU - V\}^\perp.$$

Writing $\bar{U} = rU - V$ and $\bar{V} = rU + V$, we have $P\bar{U} = 0$ and $P\bar{V} = (2/r)\bar{U}$. Take $Z \in \ker P$ and and differentiate $\langle Z, \bar{U} \rangle = 0$ with respect to an arbitrary tangent vector X. Using the fact that $\bar{U} = -J(r\xi + z)$, we arrive at

$$\langle \nabla'_X Z, \bar{U} \rangle = -r\langle Z, JPX \rangle.$$

On the other hand,

$$PX \in (\ker P) \cap (\ker P)^\perp = \operatorname{span} \bar{U}$$

so that $JPX \in \operatorname{span}\{r\xi + z\}$ which is normal to M'. Thus

$$\langle \nabla'_X Z, \bar{U} \rangle = 0. \tag{4.2}$$

Now take any tangent vectors X and Y and let Z be in $\ker P$. Then

$$\langle (\nabla'_X P)Y, Z \rangle = \langle \nabla'_X(PY), Z \rangle = -\langle PY, \nabla'_X Z \rangle = 0$$

since $\nabla'_X Z \in \ker P$ by (4.2). Noting that any expression of the form $\langle (\nabla'_X P)Y, Z \rangle$ is symmetric in its three arguments (by the Codazzi equation), we have shown that such an expression is zero if any of its three arguments lies in $\ker P$. To complete the proof that $\nabla' P = 0$, we need only show that $\langle (\nabla'_X P)Y, Z \rangle = 0$ when $X = Y = Z = \bar{V}$. For this Z we have $\langle PZ, Z \rangle = 2r^{-2}\langle \bar{U}, \bar{V} \rangle = 4r$, so that

$$\langle (\nabla'_Z P)Z, Z \rangle + 2\langle PZ, \nabla'_Z Z \rangle = 0.$$

On the other hand, if we differentiate $\langle Z, \bar{U} \rangle$, we get

$$\langle \nabla'_Z Z, \bar{U} \rangle = -\langle Z, \nabla'_Z \bar{U} \rangle = -\langle Z, r D_Z U - D_Z V \rangle.$$

Substituting $U = -J\xi$ and $V = Jz$, we show that the second argument of the inner product is $-J(r^{-1}\bar{V} + 2r^{-1}\bar{U})$ which is a normal vector to M'. Thus $\langle PZ, \nabla'_Z Z \rangle = 0$ and hence $\langle (\nabla'_Z P)Z, Z \rangle = 0$. This completes the proof that A' is parallel.   □

SKETCH OF PROOF OF THEOREM 4.1. Suppose that $\varphi A = A\varphi$. By Lemma 4.2, A' is parallel. It also follows from Lemma 1.10 that $a^2 + 4c \geq 0$. This is because $n \geq 2$ and the tangent space to M' cannot have a timelike subspace of dimension greater than one. If $a^2 + 4c > 0$, then (4.1) has two distinct roots. The classification is fairly straightforward in this case. If $a^2 + 4c = 0$, it is a little more difficult and leads to the horosphere. For details, see [Ryan 1971; Okumura 1975; Montiel and Romero 1986].   □

There are other characterizations of the Type A hypersurfaces. Theorem 1.8 together with Proposition 1.9 yield the following formula for ∇A.

THEOREM 4.3. *Let M^{2n-1}, where $n \geq 2$, be a real hypersurface in a complex space form of constant holomorphic sectional curvature $4c \neq 0$. Then $\varphi A = A\varphi$ if and only if*

$$(\nabla_X A)Y = -c(\langle \varphi X, Y \rangle W + \langle Y, W \rangle \varphi X). \qquad (4.3)$$

◁

COROLLARY 4.4. *Equality occurs in Theorem 1.11 if and only if M is an open subset of a Type A hypersurface.*   ◁

Theorem 4.1 is due to Okumura [1975] for $\mathbb{C}P^n$ and to Montiel and Romero [1986] for $\mathbb{C}H^n$. Theorem 1.11 and Corollary 4.4 were proven by Y. Maeda [1976] for $\mathbb{C}P^n$ and by B.-Y. Chen, G. D. Ludden, and Montiel [1984] for $\mathbb{C}H^n$. Also, a generalization of Theorem 1.11 and Corollary 4.4 to Type B hypersurfaces can be found in [Ki et al. 1990a]. As we saw in Theorem 1.5, there are no real hypersurfaces for which A is parallel. A $(1,1)$ tensor field T is said to be *cyclic parallel* if the cyclic sum

$$\langle (\nabla_X T)Y, Z \rangle + \langle (\nabla_Y T)Z, X \rangle + \langle (\nabla_Z T)X, Y \rangle$$

vanishes for all X, Y, and Z. This provides yet another characterization of the Type A hypersurfaces. Relevant references are [Chen and Vanhecke 1981; Chen et al. 1984; Ki 1988; Ki and Kim 1989].

THEOREM 4.5. *Let M^{2n-1}, where $n \geq 2$, be a real hypersurface in a complex space form of constant holomorphic sectional curvature $4c \neq 0$. Then the shape operator A is cyclic parallel if and only if (4.3) holds.*

PROOF. First assume (4.3). Then

$$\langle (\nabla_X A)Y, Z \rangle = -c(\langle \varphi X, Y \rangle \langle W, Z \rangle + \langle Y, W \rangle \langle \varphi X, Z \rangle),$$
$$\langle (\nabla_Y A)Z, X \rangle = -c(\langle \varphi Y, Z \rangle \langle W, X \rangle + \langle Z, W \rangle \langle \varphi Y, X \rangle),$$
$$\langle (\nabla_Z A)X, Y \rangle = -c(\langle \varphi Z, X \rangle \langle W, Y \rangle + \langle X, W \rangle \langle \varphi Z, Y \rangle).$$

The right sides sum to zero by skew-symmetry of φ. Thus A is cyclic parallel.

Conversely, suppose that A is cyclic parallel. Then we obtain (4.3) by applying the Codazzi equation twice, as follows:

$$
\begin{aligned}
-\langle(\nabla_X A)Y, Z\rangle &= \langle(\nabla_Y A)Z, X\rangle + \langle(\nabla_Z A)X, Y\rangle \\
&= \langle Z, (\nabla_Y A)X\rangle + \langle(\nabla_Z A)X, Y\rangle \\
&= \langle(\nabla_X A)Y, Z\rangle \\
&\quad - c(\langle X, W\rangle\langle\varphi Y, Z\rangle - \langle Y, W\rangle\langle\varphi X, Z\rangle + 2\langle X, \varphi Y\rangle\langle W, Z\rangle) \\
&\quad + \langle(\nabla_Z A)X, Y\rangle, \\
-2\langle(\nabla_X A)Y, Z\rangle &= \langle(\nabla_Z A)X, Y\rangle \\
&\quad - c(\langle X, W\rangle\langle\varphi Y, Z\rangle - \langle Y, W\rangle\langle\varphi X, Z\rangle + 2\langle X, \varphi Y\rangle\langle W, Z\rangle), \\
-2\langle Y, (\nabla_X A)Z\rangle &= \langle(\nabla_X A)Z, Y\rangle \\
&\quad - c(\langle X, W\rangle\langle\varphi Z, Y\rangle - \langle Z, W\rangle\langle\varphi X, Y\rangle + 2\langle X, \varphi Z\rangle\langle W, Y\rangle) \\
&\quad - c(\langle X, W\rangle\langle\varphi Y, Z\rangle - \langle Y, W\rangle\langle\varphi X, Z\rangle + 2\langle X, \varphi Y\rangle\langle W, Z\rangle), \\
-3\langle Y, (\nabla_X A)Z\rangle &= 3c(\langle Y, W\rangle\langle\varphi X, Z\rangle + \langle Z, W\rangle\langle\varphi X, Y\rangle).
\end{aligned}
$$

That is,

$$
(\nabla_X A)Y = -c(\langle\varphi X, Y\rangle W + \langle Y, W\rangle\varphi X). \qquad \square
$$

Theorem 4.5 is also trivially true when $c = 0$. In that case, A is cyclic parallel if and only if $\nabla A = 0$ because of the Codazzi equation. Also, when $c = 0$, (4.3) means that $\nabla A = 0$.

The "standard examples" listed in the summary at the end of Section 3 have constant principal curvatures, and there are classification theorems using this hypothesis. It is possible to get some results by merely imposing a limit on the *number* of distinct principal curvatures. As we noted in Theorem 1.5, there is no possibility of an umbilic hypersurface. The next two theorems show what can happen when the number of distinct principal curvatures is 2. For proofs, we refer to [Cecil and Ryan 1982; Montiel 1985].

THEOREM 4.6. *Let M^{2n-1}, where $n \geq 3$, be a real hypersurface in a complex space form of constant holomorphic sectional curvature $4c > 0$. Suppose that the number g of distinct principal curvatures is ≤ 2 at each point. Then M is an open subset of a geodesic hypersphere.* ◁

THEOREM 4.7. *Let M^{2n-1}, where $n \geq 3$, be a real hypersurface in a complex space form of constant holomorphic sectional curvature $4c < 0$. Suppose that the number g of distinct principal curvatures is ≤ 2 at each point. Then M is an open subset of one of the following:*

(i) *a geodesic sphere (Type A1);*
(ii) *a tube over a complex hyperbolic hyperplane (Type A1);*
(iii) *a horosphere (Type A0);*

(iv) *a tube of radius*

$$r \log\left(\frac{1+\sqrt{3}}{\sqrt{2}}\right) = \frac{r}{2} \log\left(2+\sqrt{3}\right)$$

over a totally real hyperbolic space (Type B).   ◁

Possibility (iv) occurs when u is chosen so that $\coth u = 2 \tanh 2u$. One principal subspace includes W and therefore has dimension n.

Since A cannot be parallel, we look for similar but weaker conditions that can be satisfied and can serve as characterizing properties. The condition $R \cdot A = 0$ is sometimes called "semi-parallel", and has been of interest for hypersurfaces in real space forms. However, here we find it is also too strong, as is shown by the next sequence of results.

LEMMA 4.8. *Let M^{2n-1}, where $n \geq 2$, be a real hypersurface in a complex space form of constant holomorphic sectional curvature $4c \neq 0$. Suppose that p is a point where $R \cdot A = 0$. If λ and μ are distinct principal curvatures at p, with associated principal orthonormal vectors X and Y, then*

$$\lambda\mu + c\left(1 + 3\langle \varphi X, Y \rangle^2\right) = 0.$$

PROOF. Since $A(R(X,Y)Y = R(X,Y)AY = \mu R(X,Y)Y$, we have $R(X,Y)Y \in T_\mu(p)$ so that $\langle R(X,Y)Y, X \rangle = 0$. On the other hand, by the Gauss equation,

$$\langle R(X,Y)Y, X \rangle$$
$$= (\lambda\mu + c) + c\langle \varphi Y, Y \rangle \langle \varphi X, X \rangle - c\langle \varphi X, Y \rangle \langle \varphi Y, X \rangle + 2c\langle X, \varphi Y \rangle \langle \varphi Y, X \rangle$$
$$= \lambda\mu + c(1 + 3\langle \varphi X, Y \rangle^2). \qquad \square$$

THEOREM 4.9. *Let M^{2n-1}, where $n \geq 2$, be a Hopf hypersurface in a complex space form of constant holomorphic sectional curvature $4c \neq 0$. Then $R \cdot A$ never vanishes.*

PROOF. Suppose $R \cdot A = 0$ at a point p. Take $Y = W$ in Lemma 4.8. This shows that $A = \lambda I$ on $W^\perp$ where $\lambda a + c = 0$. In particular, for any $X \in W^\perp$, $AX = \lambda X$ and $A\varphi X = \lambda \varphi X$. By Corollary 2.3, $0 \neq \lambda^2 = \lambda a + c = 0$, which is a contradiction. $\square$

THEOREM 4.10. *Let M^{2n-1}, where $n \geq 3$, be a real hypersurface in a complex space form of constant holomorphic sectional curvature $4c > 0$. Then $R \cdot A$ cannot be identically zero.*

PROOF. Suppose that $R \cdot A = 0$. As a result of Lemma 4.8, any two distinct principal curvatures have opposite signs. Therefore, there can be at most two of them at each point. Thus M must be an open subset of a geodesic hypersphere by Theorem 4.6. Since geodesic hyperspheres are Hopf hypersurfaces, this contradicts Theorem 4.9. $\square$

Theorem 4.10 is due to S. Maeda [1983]. Note that it deals only with the case $c > 0$. Also, because of the hypotheses required for Theorem 4.6, the proof only applies when $n \geq 3$. However, there is a direct proof for $n = 2$, both for positive and for negative c. For the proof, we refer to [Niebergall and Ryan 1996].

THEOREM 4.11. *Let M^3 be a real hypersurface in a complex space form $\tilde{M}$ of constant holomorphic sectional curvature $4c \neq 0$. Then $R \cdot A$ cannot be identically zero.*   ◁

In the next two theorems, we use the hypothesis of constant principal curvatures.

THEOREM 4.12. *Let M^{2n-1}, where $n \geq 2$, be a real hypersurface in a complex space form of constant holomorphic sectional curvature $4c > 0$. Suppose that the number of principal curvatures is 3 at each point and that these principal curvatures are constant. Then M is an open subset of a hypersurface of Type A2 or Type B.*   ◁

It is not known whether a similar theorem holds for complex hyperbolic space. In order to get further results along this line for either ambient space, we need to restrict our attention to Hopf hypersurfaces.

THEOREM 4.13. *Let M^{2n-1}, where $n \geq 2$, be a Hopf hypersurface in a complex space form of constant holomorphic sectional curvature $4c \neq 0$. If the principal curvatures are constants, then M is an open subset of a member of Takagi's list or Montiel's list.*   ◁

Theorem 4.12 is due to Takagi for $n \geq 3$ and to Q.-M. Wang for $n = 2$. Theorem 4.13 is due to Kimura for $c > 0$ and to Berndt for $c < 0$. Relevant references are [Takagi 1975a; 1975b; Li 1988; Wang 1983; Kimura 1986a; Berndt 1989a; 1990].

Böning [1995] obtained the following result, which we may regard as a generalization of Theorems 4.5 and 4.6 to the case $g = 3$. Note that he assumes that M is Hopf with $a^2 + 4c \neq 0$. He proves that under the stated hypotheses, the principal curvatures must be constants. Then Theorem 4.13 can be applied. It is not known whether the hypothesis that $a^2 + 4c \neq 0$ is necessary.

THEOREM 4.14. *Let M^{2n-1}, where $n \geq 3$, be a Hopf hypersurface in a complex space form of constant holomorphic sectional curvature $4c \neq 0$. Suppose that $a^2 + 4c \neq 0$ and the number of distinct principal curvatures is equal to 3 at each point. Then M is an open subset of a member of Takagi's list or Montiel's list.* ◁

Further results involving assumptions on the principal curvatures may be found in [Chen 1996; Deshmukh and Al-Gwaiz 1992; Ki and Takagi 1992; Kon 1980; Xu 1992]. Semi-parallelism and semi-symmetry are also discussed in [Kimura and Maeda 1993; Vernon 1991]. For extensions in other directions, including the indefinite case and the case of minimal hypersurfaces, see [Berndt et al. 1995; Bejancu and Duggal 1993; Garay and Romero 1990; Gotoh 1994; Kim and Pyo

1991; Maeda 1984; Martínez and Ros 1984; Miquel 1994; Nagai 1995; Shen 1985; Vernon 1989; Vernon 1987; Udagawa 1987]. The constructions in [Fornari et al. 1993] are also of interest, although their main result is incorrect, as noted in the errata to that paper.

5. The η-Parallel Condition

It is clear that the behavior of the structure vector W is crucial whenever we work with real hypersurfaces in complex space forms. The Hopf condition takes this into account. The next set of conditions we study will also do so. Specifically, we will take some familiar condition that is too strong to be useful for classification, and weaken it by only insisting that it apply on the holomorphic distribution $W^{\perp}$. We begin with η-parallelism which essentially restricts the $\nabla A = 0$ condition to $W^{\perp}$.

A $(1,1)$ tensor T on a hypersurface in a complex space form is said to be η-parallel if $\langle (\nabla_X T)Y, Z \rangle = 0$ for all X, Y, and $Z \in W^{\perp}$. Further, T is said to be cyclic η-parallel if the cyclic sum of this same expression vanishes. That is,

$$\langle (\nabla_X T)Y, Z \rangle + \langle (\nabla_Y T)Z, X \rangle + \langle (\nabla_Z T)X, Y \rangle = 0$$

for all X, Y, and $Z \in W^{\perp}$.

LEMMA 5.1. *Let* M^{2n-1}, *where* $n \geq 2$, *be a Hopf hypersurface in a complex space form of constant holomorphic sectional curvature* $4c \neq 0$. *Let* X *be a (smooth) principal vector field in* $W^{\perp}$ *with associated principal curvature* λ. *Then* $W\lambda = 0$. *Further, if* A *is* η-*parallel,* λ *must be constant.*

PROOF. The Codazzi equation gives

$$(\nabla_X A)W - (\nabla_W A)X = -c\varphi X,$$

which, using the fact that a is constant (Theorem 2.1), can be rewritten as

$$(aI - A)\varphi AX - (W\lambda)X - (\lambda I - A)\nabla_W X = -c\varphi X.$$

Taking the inner product with X yields $W\lambda = 0$.

Suppose now that A is η-parallel. Then if $Y \in W^{\perp}$, we have

$$0 = \langle (\nabla_Y A)X, X \rangle = (Y\lambda)\langle X, X \rangle + \langle (\lambda I - A)\nabla_Y X, X \rangle = (Y\lambda)\langle X, X \rangle,$$

so $Y\lambda = 0$. This proves that λ is constant. $\square$

LEMMA 5.2. *Let* M^{2n-1}, *where* $n \geq 2$, *be a Hopf hypersurface in a complex space form of constant holomorphic sectional curvature* $4c \neq 0$. *If* M *has* η-*parallel shape operator, then its principal curvatures are constant.*

PROOF. Suppose M has η-parallel shape operator. Let p be a point where the maximum number of principal curvatures are distinct. The $2n - 1$ principal curvature functions, numbered in nonincreasing order, are continuous. The set $\mathcal{U}$

of points where they assume the values taken at p is clearly closed by continuity. On the other hand, p must have a neighborhood where the distinct principal curvatures have constant multiplicities. Lemma 5.1 shows that these principal curvatures are constant in such a neighborhood. This shows that the set $\mathcal{U}$ is also open. We conclude that the principal curvatures are constant on M. (We have implicitly assumed that M is orientable. If not, apply the same argument to the twofold covering to reach the same conclusion.)   □

We make the observation that the expression $\langle (\nabla_X A)Y, Z \rangle$ restricted to $W^\perp$ is symmetric in its three arguments. This is immediate from the Codazzi equation and the symmetry of A.

THEOREM 5.3. *Let M^{2n-1}, where $n \geq 2$, be a Hopf hypersurface in a complex space form of constant holomorphic sectional curvature $4c \neq 0$. Then M has η-parallel shape operator if and only if it is an open subset of a Type A or Type B hypersurface from Takagi's list or Montiel's list.*

PROOF. In light of Theorem 4.13, there are two things to prove. First that every Type A or Type B hypersurface has η-parallel shape operator, and second, that hypersurfaces of type C, D, or E do not. The first part is easy. Since M is Hopf, $W^\perp$ is spanned by principal vector fields locally. Suppose M is a Type A or Type B hypersurface. Since the number of distinct principal curvatures on $W^\perp$ is 1 or 2, the arguments in any expression of the form $\langle (\nabla_X A)Y, Z \rangle$ with X, Y, and Z principal vectors in $W^\perp$, can be permuted so that Y and Z belong to the same principal distribution, say that of λ. Then

$$\langle (\nabla_X A)Y, Z \rangle = \langle (\lambda I - A)\nabla_X Y, Z \rangle = \langle \nabla_X Y, (\lambda I - A)Z \rangle = 0,$$

where we have used the fact that λ is constant. We write the second half of the proof as a separate lemma.   □

LEMMA 5.4. *Hypersurfaces of types C, D, and E in complex projective space are not η-parallel.*

PROOF. Denote the four distinct principal curvatures (other than a) by λ, μ, ρ, and σ, where T_λ and T_μ are φ-invariant, while φ interchanges T_ρ and T_σ. Assume that A is η-parallel. Choose $X \in T_\lambda$, $Z \in T_\rho$ (nonzero vectors) and compute $\langle R(X, \varphi X)Z, \varphi Z \rangle$ in two ways, once directly and once using the Gauss equation. This will lead to a contradiction. We first observe that since A is η-parallel, ∇_X takes any principal distribution $T_r \subseteq W^\perp$ into the span of T_r and W provided that $X \in W^\perp$. In the curvature calculation,

$$\langle \nabla_{\varphi X} Z, W \rangle = -\langle Z, \nabla_{\varphi X} W \rangle = -\langle Z, \varphi A \varphi X \rangle = \lambda \langle Z, X \rangle = 0.$$

Therefore, $\nabla_{\varphi X} Z \in T_\rho$ and $\nabla_X \nabla_{\varphi X} Z$ contributes nothing to the curvature since it has no T_σ component. Similarly $\nabla_{\varphi X} \nabla_X Z$ makes no contribution. Finally, the

$W^\perp$ component of $[X, \varphi X]$ makes no contribution since differentiation in a $W^\perp$ direction takes Z to T_ρ. However,

$$\langle [X, \varphi X], W \rangle = \langle \nabla_X(\varphi X), W \rangle - \langle \nabla_{\varphi X} X, W \rangle$$
$$= \langle \varphi X, \varphi A X \rangle + \langle X, \varphi A \varphi X \rangle = -2\lambda \langle X, X \rangle.$$

So far, we have

$$\langle R(X, \varphi X) Z, \varphi Z \rangle = 2\lambda \langle \nabla_W Z, \varphi Z \rangle \langle X, X \rangle.$$

We now simplify the right side further, as follows:

$$\langle (\nabla_W A) Z, \varphi Z \rangle = \langle (\rho I - A) \nabla_W Z, \varphi Z \rangle = (\rho - \sigma) \langle \nabla_W Z, \varphi Z \rangle.$$

Also,

$$\langle (\nabla_Z A) W, \varphi Z \rangle = \langle (aI - A) \nabla_Z W, \varphi Z \rangle = (a - \sigma) \langle \varphi A Z, \varphi Z \rangle = (a - \sigma) \rho \langle Z, Z \rangle.$$

Now apply the Codazzi equation to get

$$(\rho - \sigma) \langle \nabla_W Z, \varphi Z \rangle = ((a - \sigma) \rho - c) \langle Z, Z \rangle = \frac{a}{2}(\rho - \sigma) \langle Z, Z \rangle,$$

the last equality using the relationship between ρ and σ implicit in Corollary 2.3. Specifically, it is

$$\rho \sigma = \frac{\rho + \sigma}{2} a + c.$$

Thus we get

$$\langle R(X, \varphi X) Z, \varphi Z \rangle = \lambda a \langle X, X \rangle \langle Z, Z \rangle.$$

On the other hand, using the Gauss equation, we obtain

$$\langle R(X, \varphi X) Z, \varphi Z \rangle = -2c \langle X, X \rangle \langle Z, Z \rangle.$$

Since $\lambda a + 2c \neq 0$, as can be seen from Corollary 2.3, we have the desired contradiction.   □

Theorem 5.3 was proved by Kimura and S. Maeda [1989] for complex projective space. They also produced a class of examples which showed that the Hopf hypothesis is necessary (ruled real hypersurfaces). Suh [1990] extended it to complex hyperbolic space. In addition, the following characterization [Ki and Suh 1994] uses a condition that is stronger than η-parallelism. However, it does not assume that the hypersurface is Hopf, and establishing this is the heart of the proof.

THEOREM 5.5. *Let M^{2n-1}, where $n \geq 3$, be a real hypersurface in a complex space form of constant holomorphic sectional curvature $4c \neq 0$. Assume that*

$$(\nabla_X A) Y = -c \langle \varphi X, Y \rangle W$$

and

$$\langle (A\varphi - \varphi A) X, Y \rangle = 0$$

for all X and Y in $W^\perp$. Then M is an open subset of a Type A hypersurface from Takagi's list or Montiel's list.   ◁

For further results along these lines, see [Hamada 1995; Suh 1995].

6. Conditions on the Ricci Tensor

We recall from Section 1 that the $(1,1)$ Ricci tensor is denoted by S. A Riemannian manifold for which S is a constant multiple of the identity is called an *Einstein space*. A weaker condition is the *Ricci-parallel* condition which says that $\nabla S = 0$. As we shall see, both are too strong to be satisfied by a real hypersurface. A real hypersurface in a complex space form is said to be *pseudo-Einstein* if there are constants ρ and σ such that

$$SX = \rho X + \sigma \langle X, W \rangle W \tag{6.1}$$

for all tangent vectors X. (The terms *quasi-Einstein* and *η-Einstein* have also been used for this notion.) The following theorem classifies pseudo-Einstein hypersurfaces in $\mathbb{C}P^n$, and in fact proves a stronger result, namely that if a condition of the form (6.1) is satisfied, the coefficients are automatically constants. The proof can be found in [Cecil and Ryan 1982].

THEOREM 6.1. *Let M^{2n-1}, where $n \geq 3$, be a real hypersurface in a complex space form of constant holomorphic sectional curvature $4c > 0$. Suppose that there are smooth functions ρ and σ such that $SX = \rho X + \sigma \langle X, W \rangle W$ all tangent vectors X. Then ρ and σ must be constant and M is an open subset of one of*

(i) *a geodesic sphere (as in Theorem 3.13),*
(ii) *a tube of radius ur over a complex projective subspace $\mathbb{C}P^p$, with $1 \leq p \leq n-2$, $0 < u < \pi/2$, and $\cot^2 u = p/q$ (notation as in Theorem 3.14 with $\lambda_1^2 = qc/p$ and $\lambda_2^2 = pc/q$), or*
(iii) *a tube of radius ur over a complex quadric Q^{n-1} where $0 < u < \pi/4$ and $\cot^2 2u = n-2$ (as in Theorem 3.15).*   ◁

Theorem 6.1 was proved by M. Kon [1979] under the assumption that ρ and σ are constant. For complex hyperbolic space, the analogous theorem was proved by Montiel [1985].

THEOREM 6.2. *Let M^{2n-1}, where $n \geq 3$, be a real pseudo-Einstein hypersurface in a complex space form of constant holomorphic sectional curvature $4c < 0$. Then M is an open subset of one of*

(i) *a geodesic sphere,*
(ii) *a tube over a complex hyperbolic hyperplane, or*
(iii) *a horosphere.*   ◁

It is not trivial to prove directly that every pseudo-Einstein hypersurface is Hopf, even though we can observe it from Theorems 6.1 and 6.2, at least for $n \geq 3$.

When $\sigma + 3c \neq 0$, there is a straightforward proof, valid for $n = 2$ as well. However, if $\sigma + 3c = 0$, one essentially has to complete the classification. Since we have not included proofs of Theorems 6.1 and 6.2, it is worthwhile to present a few of the basic equations. Using (6.1) and (1.11), we observe that any pseudo-Einstein hypersurface satisfies

$$(A^2 - \mathfrak{m}A)X = -(\sigma + 3c)\langle X, W \rangle W + ((2n + 1)c - \rho)X.$$

Further, if the hypersurface is known to be Hopf, then

$$a^2 - \mathfrak{m}a = -(\sigma + \rho - 2(n - 1)c)$$

while on $W^{\perp}$, any principal curvature λ must satisfy

$$\lambda^2 - \mathfrak{m}\lambda - ((2n + 1)c - \rho) = 0.$$

In fact, a Hopf hypersurface will be pseudo-Einstein if and only if any two principal curvatures λ_1 and λ_2 on $W^{\perp}$ satisfy

$$(\lambda_1 - \lambda_2)(\lambda_1 + \lambda_2 - \mathfrak{m}) = 0. \tag{6.2}$$

The reason that hypersurfaces of types A2 and B in $\mathbb{C}H^n$ cannot be pseudo-Einstein is that the necessary condition $\lambda_1 + \lambda_2 = \mathfrak{m}$ cannot hold when all the principal curvatures have the same sign, which is the case for $\mathbb{C}H^n$. In $\mathbb{C}P^n$, on the other hand, the signs of λ_1 and λ_2 differ, so in each family there is a one choice of radius for which the hypersurface will be pseudo-Einstein.

In the rest of this section we study several conditions that represent ways of weakening the Ricci-parallel condition $\nabla S = 0$. Surprisingly, many of these turn out to be equivalent to, or to imply, the pseudo-Einstein condition.

When a hypersurface is pseudo-Einstein, it is easy to check that ∇S satisfies the identity

$$(\nabla_X S)Y = \sigma(\langle \varphi AX, Y \rangle W + \langle Y, W \rangle \varphi AX), \tag{6.3}$$

where the constant σ is as in (6.1). We shall now investigate how a condition of the form of (6.3) restricts a hypersurface. A routine calculation yields the following information on ∇S.

PROPOSITION 6.3. *Let* M^{2n-1}, *where* $n \geq 2$, *be a real hypersurface in a complex space form of constant holomorphic sectional curvature* $4c \neq 0$. *If there is a function* κ *such that*

$$(\nabla_X S)Y = \kappa(\langle \varphi AX, Y \rangle W + \langle Y, W \rangle \varphi AX), \tag{6.4}$$

then

$$|\nabla S|^2 = 2\kappa^2(\text{trace } A^2 - |AW|^2). \qquad \triangleleft$$

We now look at the standard examples in light of the condition (6.4).

THEOREM 6.4. *Let M^{2n-1}, where where $n \geq 2$, be a member of Takagi's list or Montiel's list satisfying the hypothesis of Proposition 6.3. Then M is pseudo-Einstein. Specifically it is a Type A0 or A1 hypersurface or one of the Type A2 or B hypersurfaces occurring in case (ii) or (iii) of Theorem 6.1. The latter two occur only when $c > 0$ and $n \geq 3$.*

PROOF. We consider first a Type A hypersurface and derive an expression for ∇S. From Theorem 4.1 and formulas (1.10) and (4.3), we calculate

$$(\nabla_X S)Y = -2c(\langle \varphi AX, Y \rangle W + \langle Y, W \rangle \varphi AX) + c(a - m)(\langle \varphi X, Y \rangle W + \langle Y, W \rangle \varphi X).$$
(6.5)

This implies

$$(\nabla_W S)Y = 0,$$

and hence both sides of (6.4) vanish when $X = W$. Now suppose that $X \in W^{\perp}$. If the hypersurface is of type A0 or A1, then $m - a = (2n - 2)\lambda$ where λ is the principal curvature for $W^{\perp}$ so that (6.4) holds with $\kappa = -2nc$. However, a Type A2 hypersurface will have linearly independent values of X corresponding to distinct principal curvatures. The only way that (6.4) can be satisfied is if $m - a = 0$. This occurs for just one choice of radius for each value of p (see Theorem 3.14), namely the one that makes the hypersurface pseudo-Einstein. For a Type B hypersurface, we note that the principal curvatures satisfy $\lambda_1 \lambda_2 = -c$ so that neither is zero. If (6.4) holds, then (as we shall see later) so does (6.7). Hence $m\lambda_1 - \lambda_1^2 = m\lambda_2 - \lambda_2^2$ which is precisely the condition for M to be pseudo-Einstein. It will become clear in the proofs of the next few theorems that hypersurfaces of types C, D, or E cannot satisfy an equation of the form (6.4).   □

Using Proposition 6.3 and the information on the standard examples in Section 3, we can compute the following information.

COROLLARY 6.5. *For the hypersurfaces of Theorem 6.4, $|\nabla S|^2$ is equal to*

(i) $16n^2(n-1)|c|^3$ *for Type A0,*

(ii) $16n^2(n-1)|c|^3 \tanh^2 u$ *for Type A1 with $c < 0$,*

(iii) $16n^2(n-1)|c|^3 \cot^2 u$ *for Type A1 with $c > 0$,*

(iv) $16(n-1)|c|^3$ *for Type A2,*

(v) $16n(n-1)(2n-1)^2|c|^3/(n-2)$ *for Type B.*   ◁

In particular, the value of κ occurring in any equation of the form (6.4) is nonzero when M is one of the standard examples. Thus:

COROLLARY 6.6. *For the hypersurfaces in Takagi's and Montiel's lists, the Ricci tensor is never parallel, that is, ∇S never vanishes. In particular, none of these hypersurfaces are Einstein spaces.*   ◁

This also proves, for $n \geq 3$, the remark made in the introduction that there are no Einstein hypersurfaces in $\mathbb{C}P^n$ or $\mathbb{C}H^n$. This statement is also true for $n = 2$. A proof can be found in [Niebergall and Ryan 1996]. As a consequence of the calculations performed in proving Theorem 6.4, we can make the following further observation.

THEOREM 6.7. *The Type A0 and Type A1 hypersurfaces in* $\mathbb{C}P^n$ *and* $\mathbb{C}H^n$, *where* $n \geq 2$, *satisfy*

$$(\nabla_X S)Y = -2nc\lambda(\langle \varphi X, Y \rangle W + \langle Y, W \rangle \varphi X)$$

where λ *is the principal curvature for the principal space* $W^\perp$.   ◁

We now look at the converse of formula (6.3). As a first step we show that (6.4) implies the Hopf condition if $\kappa \neq 0$.

LEMMA 6.8. *Let* M^{2n-1}, *where* $n \geq 2$, *be a real hypersurface in a complex space form of constant holomorphic sectional curvature* $4c \neq 0$. *If there is a nonvanishing function* κ *satisfying* (6.4), *then* M *is a Hopf hypersurface, and* $\mathrm{m}^2 - \operatorname{trace} A^2$ *is constant.*

PROOF. We first differentiate (1.11) to find the general expression for ∇S in terms of ∇A to get

$$(\nabla_X S)Y = -3c(\langle Y, \varphi AX \rangle W + \langle Y, W \rangle \varphi AX) + (X\mathrm{m})AY + \mathrm{m}(\nabla_X A)Y - (\nabla_X(A^2))Y. \tag{6.6}$$

Taking the trace of (6.6) and of (6.4) with respect to Y, we get

$$\mathrm{m}(X\mathrm{m}) + \mathrm{m}\operatorname{trace}(\nabla_X A) - \operatorname{trace}(\nabla_X A^2) = 2\mathrm{m}(X\mathrm{m}) - X(\operatorname{trace} A^2)$$
$$= X(\mathrm{m}^2 - \operatorname{trace} A^2),$$

and we conclude that $\mathrm{m}^2 - \operatorname{trace} A^2$ is constant.

On the other hand, using the fact that $\nabla_X A^2 = (\nabla_X A)A + A\nabla_X A$ in (6.6), substituting for $(\nabla_X A)Y$ and $(\nabla_X A)AY$ from the Codazzi equation, and taking the trace with respect to X gives

$$-3c\langle Y, \varphi AW \rangle + \langle \operatorname{grad} \mathrm{m}, AY \rangle + \mathrm{m}(\operatorname{trace}(\nabla_Y A))$$
$$- \operatorname{trace}(\nabla_{AY} A) - (\operatorname{trace} A\nabla_Y A + c\langle A\varphi Y, W \rangle + 2c\langle AW, \varphi Y \rangle)$$
$$= \kappa\langle \varphi AW, Y \rangle.$$

This equation simplifies to

$$\kappa\langle \varphi AW, Y \rangle = \tfrac{1}{2}\langle \operatorname{grad}(\mathrm{m}^2 - \operatorname{trace} A^2), Y \rangle.$$

Since the right side vanishes, we must have $\varphi AW = 0$ so that M is a Hopf hypersurface.   □

LEMMA 6.9. *Let* M^{2n-1}, *where* $n \geq 2$, *be a Hopf hypersurface in a complex space form of constant holomorphic sectional curvature* $4c \neq 0$. *Suppose that there is a function* κ *satisfying* (6.4). *Then either* a *is zero or* $\mathfrak{m}$ *is constant. Furthermore,*

$$(\kappa + 3c)\varphi AX = ((\mathfrak{m}a - a^2)I - \mathfrak{m}A + A^2)\varphi AX \qquad (6.7)$$

for all $X \in W^\perp$.

PROOF. Combine (6.4) and (6.6) and take the inner product with W. The resulting equation gives $X(a\mathfrak{m}) = 0$ and (6.7). $\square$

LEMMA 6.10. *Let* M^{2n-1}, *where* $n \geq 2$, *be a Hopf hypersurface in a complex space form of constant holomorphic sectional curvature* $4c \neq 0$. *Suppose that there is a function* κ *such that*

$$(\nabla_X S)Y = \kappa(\langle \varphi AX, Y \rangle W + \langle Y, W \rangle \varphi AX).$$

If $a/2$ *occurs as a principal curvature on* $W^\perp$ *at every point of* M, *then* $c < 0$ *and* M *is an open subset of a horosphere.*

PROOF. To make the argument more general, we first assume only that there is a point where $a/2$ occurs as a principal curvature on $W^\perp$ and work at this point. Then $a^2 + 4c = 0$ by Corollary 2.3 and hence a is not zero. If, in addition, 0 is a principal curvature, $X \in T_0$ implies that $A\varphi X = \frac{1}{2}a\varphi X$, again by Corollary 2.3. Substituting φX for X in (6.7), we get

$$\kappa + 3c = \mathfrak{m}a - a^2.$$

This reduces (6.7) to

$$(\mathfrak{m}A - A^2)\varphi AX = 0, \qquad (6.8)$$

which holds for all $X \in W^\perp$.

Now let $\mathcal{V} = T_0 \oplus \varphi T_0$ and note that $A\mathcal{V} \subseteq \mathcal{V}$. Then $\tilde{\mathcal{V}} = (\mathcal{V} \oplus \mathrm{span}\{W\})^\perp$ is A-invariant and φ-invariant. From (6.8) we see that the only possible principal curvature on $\tilde{\mathcal{V}}$ has the value $\mathfrak{m} = a/2$. Whether $\tilde{\mathcal{V}}$ is the zero subspace or not, we are led to the absurd conclusion that $\mathfrak{m} = \mathrm{trace}\, A = a + (k-1)\mathfrak{m} = (k+1)\mathfrak{m}$ where k is the rank of A. The result is that if $\frac{a}{2}$ occurs as a principal curvature (with principal vector in $W^\perp$) at some point, then 0 is not a principal curvature at that point.

We now continue to work at a point where $a/2$ is a principal curvature on $W^\perp$. Since 0 is not a principal curvature,

$$(\kappa + 3c)I = (\mathfrak{m}a - a^2)I - \mathfrak{m}A + A^2.$$

holds on all of $W^\perp$. The fact that $a/2$ is a principal curvature allows us to compute that $\kappa = \mathfrak{m}a/2$. Further, the equation two lines above shows that any principal curvature not equal to $a/2$ must be equal to $\mathfrak{m} - a/2$.

Now assume the full hypothesis of the lemma. Recall that $a \neq 0$ and so $\mathfrak{m}$ is constant by Lemma 6.9. If three principal curvatures are distinct at any point, then there is a neighborhood where they are all distinct, hence constant by the arguments of the preceding paragraph. Since none of the standard examples with three distinct principal curvature satisfy $a^2 + 4c = 0$, we have a contradiction in view of Theorem 4.13. We therefore must have exactly two principal curvatures at each point, a and $a/2$. Again by Theorem 4.13, M must be an open subset of a horosphere. $\square$

LEMMA 6.11. *Let M^{2n-1}, where $n \geq 2$, be a Hopf hypersurface in a complex space form of constant holomorphic sectional curvature $4c \neq 0$. Suppose that there is a constant κ such that*

$$(\nabla_X S)Y = \kappa(\langle \varphi AX, Y \rangle W + \langle Y, W \rangle \varphi AX).$$

If $a/2$ does not occur as a principal curvature on $W^\perp$, M is an open subset of a Type A or Type B hypersurface provided $a \neq 0$. If $n \geq 3$, the same conclusion holds without the assumption that $a \neq 0$.

PROOF. Choose a point p where the maximum number of principal curvatures are distinct and work in a neighborhood of that point. Suppose first that 0 occurs as a principal curvature at p so that $AX = 0$ for a nonzero $X \in W^\perp$. Again invoking Corollary 2.3, we have that $a \neq 0$ and $-2c/a$ is also a principal curvature with principal vector φX. Using the fact that $\frac{1}{2}$ is not a principal curvature, we can verify that 0 and $-2c/a$ have equal multiplicities. Formula (6.8) still holds and arguing as in Lemma 6.10, we find that $\kappa + 3c - \mathfrak{m}a + a^2 = 0$ and hence that any further principal curvature λ must satisfy $\lambda = \mathfrak{m}$. In addition, T_λ must be φ-invariant and so $\mathfrak{m}^2 = a\mathfrak{m} + c$ by Corollary 2.3. If all four principal curvatures we have identified are distinct, then there are also four distinct principal curvatures at nearby points. The zero principal curvature must remain zero since (6.8) does not allow for more than two distinct principal curvatures on $W^\perp$. As a consequence, M has four distinct constant principal curvatures in a neighborhood of p. This is impossible by Theorem 4.13. There is still the possibility that $a = \mathfrak{m}$ or that $a = -2c/a$ so that only three principal curvatures are distinct. In the first case, $\mathfrak{m} = a$ must be a principal curvature of the same multiplicity nearby. If the zero principal curvature were to become nonzero nearby, then (6.8) would imply that this nonzero value must be equal to $\mathfrak{m}$, which will not be true for sufficiently nearby points. If $a = -2c/a$, the same argument shows that the zero principal curvature remains zero nearby. In either case, M has three distinct principal curvatures, all constant, in a neighborhood of p. Observing that none of the standard examples has this particular configuration of principal curvatures, we conclude that 0 cannot occur as a principal curvature on $W^\perp$ at p.

Now that we know that 0 is excluded, there are two possibilities arising from (6.7). The first is that $a \neq 0$ in which case $\mathfrak{m}$ is constant and any principal curvature on $W^\perp$ satisfies a quadratic equation with constant coefficients in a neighborhood of p. Thus, M is locally a Type A or Type B hypersurface. If $a = 0$, then $\mathfrak{m}$ is not automatically constant. However, the quadratic equation is still satisfied. If only two principal curvatures are distinct at p, then the same holds in a neighborhood and locally M is one of the standard examples as in Theorems 4.6 and 4.7 provided that $n \geq 3$. On the other hand, if three principal curvatures are distinct, say 0, λ, and μ, then λ and μ satisfy the quadratic equation

$$\kappa + 3c + \mathfrak{m}t - t^2 = 0.$$

Noting that $a = 0$, we observe that $\lambda\mu = c$ by Corollary 2.3. Thus $\kappa = -4c$. If the two eigenspaces are φ-invariant, then each principal curvature is constant, again by Corollary 2.3. If not, then $\lambda + \mu = \mathfrak{m}$ and we get $(n - 2)\mathfrak{m} = 0$. Thus $\mathfrak{m} = 0$ and the principal curvatures are constants provided that $n \geq 3$. □

We can now state the result that we have been working towards. It determines the hypersurfaces that can satisfy a condition of the form of (6.4).

THEOREM 6.12. *Let M^{2n-1}, where $n \geq 3$, be a real hypersurface in a complex space form of constant holomorphic sectional curvature $4c \neq 0$. If there is a nonzero constant κ such that*

$$(\nabla_X S)Y = \kappa(\langle \varphi AX, Y \rangle W + \langle Y, W \rangle \varphi AX),$$

then M is an open subset of a pseudo-Einstein hypersurface as listed in Theorems 6.1 and 6.2.

PROOF. From Lemma 6.8, M is a Hopf hypersurface. Suppose there is a point p where $a/2$ is not a principal curvature on $W^\perp$. Then Lemma 6.11 shows that the set where all principal curvatures have the same values as they have at p is open. Since this set is also closed, M is an open subset of a Type A or Type B hypersurface. However, as we have seen in Theorem 6.4, a Type A or Type B hypersurface satisfying the given condition must be pseudo-Einstein. On the other hand, if no such p exists, the desired conclusion follows from Lemma 6.10. □

Theorem 6.12 was proved by Kimura [1986b] for the case of $\mathbb{C}P^n$. Although the assumption that $n \geq 3$ is implicit in his proof (which relies on [Cecil and Ryan 1982] to handle the possibility that $a = 0$), it is omitted from the stated hypothesis.

The above argument also holds for $n = 2$, provided that we assume $a \neq 0$. We state the result separately as follows.

THEOREM 6.13. *Let M^3 be a real hypersurface in $\mathbb{C}P^2$ or $\mathbb{C}H^2$. Suppose that $a = \langle AW, W \rangle \neq 0$. If there is a nonzero constant κ such that*

$$(\nabla_X S)Y = \kappa(\langle \varphi AX, Y \rangle W + \langle Y, W \rangle \varphi AX),$$

M is an open subset of a hypersurface of type A0 or type A1. ◁

The Type A0 and A1 hypersurfaces are also characterized by a refinement of this condition as follows [Kimura and Maeda 1992; Taniguchi 1994; Choe 1995].

THEOREM 6.14. *Let M^{2n-1}, where $n \geq 2$, be a real hypersurface in a complex space form of constant holomorphic sectional curvature $4c \neq 0$. Then there is a nonzero constant κ such that*

$$(\nabla_X S)Y = \kappa(\langle \varphi X, Y \rangle W + \langle Y, W \rangle \varphi X)$$

if and only if M is an open subset of a hypersurface of type A0 or type A1. ◁

If $n \geq 3$, one need not specify that κ is constant in Theorem 6.14. If κ is assumed to be a function, it turns out to be constant. However, $\kappa \neq 0$ is essential, although this hypothesis is missing from [Choe 1995, Theorem 3.2].

We remarked in Corollary 6.6 that none of the standard examples have parallel Ricci tensor. It follows from Lemmas 6.9 through 6.11 that Hopf hypersurfaces cannot have parallel Ricci tensor, at least when $n \geq 3$. In fact, the following stronger result is known [Ki 1989]. We will not give the proof here, since it will be a consequence of Theorem 6.29.

THEOREM 6.15. *Let M^{2n-1}, where $n \geq 3$, be a real hypersurface in a complex space form of constant holomorphic sectional curvature $4c \neq 0$. Then the Ricci tensor of M cannot be parallel everywhere.* ◁

Ki and Suh [1992] proved yet another characterization of the Type A hypersurfaces. Recalling (6.5), which is an expression for ∇S, the following could be regarded a kind of converse.

THEOREM 6.16. *Let M^{2n-1}, where $n \geq 3$, be a real hypersurface in a complex space form of constant holomorphic sectional curvature $4c > 0$. Suppose that the mean curvature $\mathfrak{m} = \text{trace } A$ and $a = \langle AW, W \rangle$ are constants. Then*

$$\begin{aligned}(\nabla_X S)Y = &-c\mathfrak{m}(\langle Y, \varphi X \rangle W + \langle Y, W \rangle \varphi X) \\ &+ c(\langle Y, \varphi X \rangle AW + \langle AW, Y \rangle \varphi X) - 2c(\langle Y, W \rangle \varphi AX + \langle \varphi AX, Y \rangle W)\end{aligned}$$

if and only if M is a Type A hypersurface from Takagi's list. ◁

Kimura and Maeda [1991] investigated the consequences of assuming only that S is parallel in the direction of the structure vector W. Although we will go into some detail concerning a similar condition on the shape operator in the next section, we will merely state their theorem concerning for $\nabla_W S$. See [Maeda 1993] for a related result.

THEOREM 6.17. *Let M^{2n-1}, where $n \geq 2$, be a Hopf hypersurface in a complex space form of constant holomorphic sectional curvature $4c > 0$. Assume that the mean curvature is constant. If $\nabla_W S = 0$, then M is an open subset of a hypersurface from Takagi's list. There is a restriction on the radii that can occur.* ◁

For further results along these lines, see [Cho et al. 1991; Kim 1988b; Maeda 1994].

We now look at a condition analogous to the one characterizing the Type A hypersurfaces in Theorem 4.1. However, we ask only that φ commute with S, not with A. This condition is significantly weaker as it allows at least some of each type of homogeneous hypersurface as well as as certain nonhomogeneous ones. Relevant references are [Aiyama et al. 1990; Ki and Suh 1990; Kimura 1987b]. The original statements have been modified to take certain corrections into account.

THEOREM 6.18. *Let M^{2n-1}, where $n \geq 3$, be a Hopf hypersurface in a complex space form of constant holomorphic sectional curvature $4c > 0$. If $\varphi S = S\varphi$, then M is an open subset of a hypersurface from Takagi's list or a certain non-homogeneous hypersurface. Although all types A–E occur, there is a restriction on the radii of the tubes.* ◁

THEOREM 6.19. *Let M^{2n-1}, where $n \geq 3$, be a Hopf hypersurface in a complex space form of constant holomorphic sectional curvature $4c < 0$. If $\varphi S = S\varphi$, then M is an open subset of a hypersurface of type A from Montiel's list.* ◁

A Riemannian manifold is said to have *harmonic curvature* if its Ricci tensor S is a Codazzi tensor, i.e., $(\nabla_X S)Y = (\nabla_Y S)X$. Concerning this condition, we can state the following [Ki 1989; Kwon and Nakagawa 1989a; Kim 1988a].

THEOREM 6.20. *Let M^{2n-1}, where $n \geq 3$, be a Hopf hypersurface in a complex space form of constant holomorphic sectional curvature $4c \neq 0$. Then M cannot have harmonic curvature, that is, $(\nabla_X S)Y - (\nabla_Y S)X$ cannot vanish identically.*

◁

Other versions of the harmonicity condition are pursued in [Ki and Nakagawa 1991; Ki et al. 1989; Ki et al. 1990b].

We now proceed to look at further conditions on the ∇S. First of all, it is possible for S to be cyclic-parallel, but the condition is still a rather strong one [Kwon and Nakagawa 1988].

THEOREM 6.21. *Let M^{2n-1}, where $n \geq 2$, be a real hypersurface in a complex space form of constant holomorphic sectional curvature $4c > 0$. If the Ricci tensor S is cyclic parallel, then M is an open subset of a hypersurface from Takagi's list. There is a restriction on the radii that can occur.* ◁

As long as we assume the Hopf condition, η-parallelism of S turns out to be strong enough to characterize Type A and Type B hypersurfaces. Suh [1990] proved the following theorem.

THEOREM 6.22. *Let M^{2n-1}, where $n \geq 2$, be a Hopf hypersurface in a complex space form of constant holomorphic sectional curvature $4c \neq 0$. Then M has η-parallel Ricci tensor if and only if it is an open subset of a Type A or Type B hypersurface from Takagi's list or Montiel's list.* ◁

For $n \geq 3$, the same conclusion was obtained by J.-H. Kwon and H. Nakagawa [1989b] under a weaker assumption. We will present a full account of their result, organized as a sequence of lemmas preceded by a statement of the theorem and the core of the proof.

THEOREM 6.23. *Let M^{2n-1}, where $n \geq 3$, be a Hopf hypersurface in a complex space form of constant holomorphic sectional curvature $4c \neq 0$. Then the Ricci tensor of M is cyclic η-parallel if and only if M is an open subset of a Type A or Type B hypersurface from Takagi's list or Montiel's list.*

PROOF. Suppose that S is cyclic η-parallel. Let p be a point where the maximum number of principal curvatures are distinct. Let $\mathcal{V}$ be a neighborhood of p where we can find an orthonormal basis of principal vectors. Then Lemma 6.27 (to be proved below) shows that m is constant. For any two distinct principal curvatures λ and μ, we have from Lemma 6.26 below that

$$(2\lambda - \mathrm{m})X\lambda = (2\lambda + 4\mu - 3\mathrm{m})X\mu = 0,$$

where X is the principal vector corresponding to λ. Let $\mathcal{U} = \{x \in \mathcal{V} : 2\lambda \neq \mathrm{m}\}$. Then $X\lambda = 0$ on $\bar{\mathcal{U}}$. For any point in the complement of $\bar{\mathcal{U}}$, if $X\lambda \neq 0$, then $2\lambda = m$ in a neighborhood of this point, which contradicts the fact that $X\mathrm{m} = 0$. We conclude that $X\lambda = 0$ on all of $\mathcal{V}$. Similarly, we see that $X\mu = 0$ on the closure of the set where $2\lambda + 4\mu - 3\mathrm{m} \neq 0$. On the complement, however, we have $4\mu = 3\mathrm{m} - 2\lambda$. By the same argument as before, we see that $X\mu = 0$ there as well. We have shown that every principal curvature function is constant along every direction in $W^{\perp}$. By Lemma 5.1, it is also constant in the W direction. We conclude that $\mathcal{V}$ has constant principal curvatures. By Theorem 4.13, it is an open subset of a member of Takagi's list or Montiel's list.

It remains to check which of the standard examples actually have cyclic η-parallel Ricci tensor. This is covered in the next theorem, Theorem 6.24. To complete the proof, observe that the set of points where the principal curvatures match their values at p is open since any such point will be a point where the maximum number of principal curvatures are distinct. Also, it will have a neighborhood where the principal curvatures are constant. On the other hand, such a set is closed by continuity.   □

THEOREM 6.24. *Let M^{2n-1}, where $n \geq 2$, be a member of Takagi's list or Montiel's list. Then the Ricci tensor is cyclic η-parallel if and only if the shape operator is η-parallel; that is if M is of type A or type B.*

PROOF. Lemma 6.25 below shows that in the case of the standard examples, the cyclic sum of the expression $\langle (\nabla_X S)Y, Z \rangle$ over principal vectors X, Y, and Z in $W^{\perp}$ is equal to

$$(2(\lambda + \mu + \sigma) - 3\mathrm{m})\langle (\nabla_X A)Y, Z \rangle,$$

where λ, μ, and σ are the respective principal curvatures. Thus a hypersurface with η-parallel shape operator will have an η-parallel Ricci tensor. In view of Theorem 5.3, it remains to show that hypersurfaces of types C, D, and E in $\mathbb{C}P^n$ do not have η-parallel Ricci tensor.

Let M be a hypersurface of type C, D, or E. Assume that the Ricci tensor is cyclic η-parallel. Pick any three of the four principal curvatures whose principal spaces lie in $W^\perp$. Call them λ, μ, and σ. See Theorems 3.16 to 3.18 for the values. For a particular type (say type C), note that the expression $2(\lambda+\mu+\sigma) - 3m$ varies continuously with u so that among all the hypersurfaces in the one-parameter family, only a finite number will give a value of zero for this expression. Thus, except for a finite number of values of u, the corresponding expression involving ∇A will vanish. This holds for any combination of three distinct principal curvatures and corresponding principal vectors. By continuity, it holds for all values of u as well. On the other hand, the argument given in the proof of Theorem 5.3 takes care of the case of vectors which do not belong to distinct principal spaces. The net result is that $\langle(\nabla_X A)Y, Z\rangle = 0$ on $W^\perp$. Since this contradicts Theorem 5.3, we must conclude that no hypersurface of type C, D, or E can have cyclic η-parallel Ricci tensor. $\qquad\square$

LEMMA 6.25. *Let M^{2n-1}, where $n \geq 2$, be a Hopf hypersurface in a complex space form of constant holomorphic sectional curvature $4c \neq 0$. Then the cyclic sum of the expression $\langle(\nabla_X S)Y, Z\rangle$ over principal vectors X, Y, and Z in $W^\perp$ is equal to*

$$(2(\lambda+\mu+\sigma)-3m)(\langle(\nabla_X A)Y, Z\rangle-(\lambda(Zm)\langle X, Y\rangle+\mu(Xm)\langle Y, Z\rangle+\sigma(Ym)\langle Z, X\rangle))),$$

where λ, μ, and σ are the respective principal curvatures. $\qquad\triangleleft$

LEMMA 6.26. *Let M^{2n-1}, where $n \geq 2$, be a Hopf hypersurface in a complex space form of constant holomorphic sectional curvature $4c \neq 0$. Suppose that the Ricci tensor S is cyclic η-parallel. Suppose that X and Y are smooth orthonormal principal vector fields in $W^\perp$ with corresponding principal curvatures λ and μ. Then*

$$(2\lambda - m)X\lambda = \lambda Xm,$$
$$(2\lambda + 4\mu - 3m)X\mu = \mu Xm.$$ $\qquad\triangleleft$

LEMMA 6.27. *Let M^{2n-1}, where $n \geq 2$, be a Hopf hypersurface in a complex space form of constant holomorphic sectional curvature $4c \neq 0$. Suppose that the Ricci tensor S is cyclic η-parallel. Suppose further that the principal curvatures have constant multiplicities. Then the mean curvature m is constant.*

PROOF. First note that $Wm = 0$ by Lemma 2.13. Because the principal curvatures have constant multiplicities, we can always find a local orthonormal principal frame when desired. We begin with the case $a^2 + 4c = 0$. Assume that there is a principal curvature $\lambda \neq a/2$ with corresponding unit principal vector $X \in W^\perp$. Set $Y = \varphi X$ so that in the notation of Theorem 6.26, $\mu = a/2$. Then

$X\mathfrak{m} = 0$ by Theorem 6.26. Also $Y\mathfrak{m} = 0$ for any $Y \in T_\mu$, by the first part of Theorem 6.26. We conclude that $\mathfrak{m}$ is constant.

Now consider the case $a^2 + 4c \neq 0$. Then $a/2$ does not occur as a principal curvature. Suppose that there is a unit principal vector field X such that $X\mathfrak{m} \neq 0$. Take $Y = \varphi X$ and use the setup of Lemma 6.26. A straightforward but tedious calculation using Corollary 2.3 and Lemma 6.26 yields

$$f(\lambda)X\mathfrak{m} = 0, \tag{6.9}$$

where f is the polynomial $a_0 t^4 + a_1 t^3 + a_2 t^2 + a_3 t + a_4$ whose coefficients are

$$
\begin{aligned}
a_0 &= 8a, \\
a_1 &= -4(a^2 + ma - 8c), \\
a_2 &= 2(2a^3 - ma^2 - 4ca - 16mc), \\
a_3 &= 2(ma^3 + 6ca^2 + 10mca + 16c^2), \\
a_4 &= -2mca^2.
\end{aligned}
$$

If $a = 0$, then $\lambda\mu = c$ by Corollary 2.3 and equation (6.9) reduces to

$$32c\lambda(\lambda^2 - m\lambda + c) = 0.$$

Direct substitution shows that this equation is also satisfied by μ. Thus $\lambda + \mu = m$. Suppose now that ν is a nonzero principal curvature distinct from λ and μ. Applying Lemma 6.26 again, we get

$$
\begin{aligned}
(2\lambda + 4\nu - 3m)X\nu &= \nu X\mathfrak{m}, \\
(2\lambda + 4\sigma - 3m)X\sigma &= \sigma X\mathfrak{m},
\end{aligned}
$$

where σ is the principal curvature for $\varphi(T_\nu)$. Noting that $\nu\sigma = c$, we have

$$
\begin{aligned}
((2\lambda - 3m)\sigma + 4c)X\nu &= cX\mathfrak{m}, \\
((2\lambda - 3m)\nu + 4c)X\sigma &= cX\mathfrak{m};
\end{aligned}
$$

these two equations can be added to yield $X(\nu + \sigma) = \frac{1}{2}X\mathfrak{m}$. Recalling that $a = 0$, we know that $\mathfrak{m}$ is the sum of n terms, each of which is either $\lambda + \mu$ or a term of the form $\nu + \sigma$. Thus

$$X\mathfrak{m} = kX\mathfrak{m} + \frac{n-k}{2}X\mathfrak{m},$$

where k is the multiplicity of λ. This implies that $n = k = 1$, a contradiction. We conclude that $a \neq 0$.

Now differentiate $f(\lambda) = 0$ with respect to X and use the first equality in the conclusion of Lemma 6.26 to get

$$g(\lambda)X\mathfrak{m} = 0,$$

where $g = b_0 t^4 + b_1 t^3 + b_2 t^2 + b_3 t + b_4$ with

$$
\begin{aligned}
b_0 &= 24a, \\
b_1 &= -8(2a^2 + ma - 4c), \\
b_2 &= 2(6a^3 - ma^2 + 12ca - 16mc), \\
b_3 &= 8c(a^2 + 4c), \\
b_4 &= 2mca^2.
\end{aligned}
$$

Now $f(\lambda) = 0$ and $g(\lambda) = 0$ may be regarded as equations in m of degree 1 from which m can be eliminated to yield a degree 7 polynomial equation in λ with constant coefficients. Thus λ is constant. In particular, $X\lambda = 0$ and $X\mu = 0$ so that $Xm = 0$ by Lemma 6.26. Again, this is a contradiction. We must conclude that m is constant. $\qquad\square$

Lemma 6.27 is the final ingredient required for the proof of Theorem 6.23. In view of the argument in the proof of Theorem 6.23, we now see that Lemma 6.27 is true even without the hypothesis of constant multiplicities.

A Riemannian manifold is said to be a *Ryan space* if $R \cdot S = 0$. As far as we can determine, this term (and variations of it) were first used by Ki, Nakagawa, and Suh [Ki et al. 1990b]. The spaces themselves were introduced in [Ryan 1971; 1972] and independently by R. L. Bishop and S. I. Goldberg [1972], and have been studied by many authors in the intervening years. Subsequently, the same spaces have been called *Ricci-semisymmetric*. See [Deszcz 1992], for example, where the term is part of a comprehensive naming scheme for a number of conditions, all related to the notion of *pseudosymmetry*. We will adopt the terminology used in the literature being surveyed.

A Riemannian manifold is a *cyclic-Ryan space* if the cyclic sum over tangent vectors X, Y, and Z of $(R(X,Y) \cdot S)Z$ vanishes. A real hypersurface in a complex space form is said to be *pseudo-Ryan* if $\langle (R(X_1, X_2) \cdot S)X_3, X_4 \rangle = 0$ provided all X_i lie in $W^\perp$. The Ryan condition is too strong to be satisfied by a real hypersurface, as we shall see in Theorem 6.29. We discuss the weaker cyclic-Ryan condition first. The following result was proved in [Ki et al. 1990b].

THEOREM 6.28. *Let M^{2n-1}, where $n \geq 3$, be a real hypersurface in a complex space form of constant holomorphic sectional curvature $4c \neq 0$. If M satisfies the cyclic-Ryan condition, then M is a Hopf hypersurface.*

PROOF. Our initial discussion will be valid for any hypersurface in a complex space form. We begin by applying the Gauss equation and (1.11) to get

$$
R(X,Y)(SZ) = (AX \wedge AY + c(X \wedge Y + \varphi X \wedge \varphi Y + 2\langle X, \varphi Y \rangle \varphi))
$$
$$
\times ((2n+1)cZ - 3c\langle Z, W \rangle W - PZ),
$$

where $P = A^2 - (\text{trace } A)A = A^2 - \mathfrak{m}A$. The right side is the sum of the following terms:

(1) $\quad (2n+1)c^2(X \wedge Y)Z$,

(2) $\quad -3c^2(\langle Z, W \rangle \langle Y, W \rangle X - \langle Z, W \rangle \langle X, W \rangle Y)$,

(3) $\quad -c(X \wedge Y)PZ - c(\varphi X \wedge \varphi Y)PZ$,

(4) $\quad (2n+1)c^2(\varphi X \wedge \varphi Y)Z$,

(5) $\quad 2(2n+1)c^2\langle X, \varphi Y \rangle \varphi Z$,

(6) $\quad -2c\langle X, \varphi Y \rangle \varphi PZ$,

(7) $\quad (2n+1)c(AX \wedge AY)Z$,

(8) $\quad -3c\langle Z, W \rangle (AX \wedge AY)W$,

(9) $\quad (AX \wedge AY)PZ$.

Because of the first Bianchi identity, the cyclic sum of $(R(X,Y) \cdot S)Z$ is equal to the cyclic sum of $R(X,Y)(SZ)$. We look at the cyclic sums of each of the terms (1)–(9) above and conclude by straightforward calculation the following facts. Terms (1) and (2) and the first part of (3) all sum to 0, while cyclic sum of the remainder of (3) is equal to the cyclic sum of

$$-c\langle (P\varphi + \varphi P)Y, Z \rangle \varphi X.$$

The cyclic sum of (4) and (5) taken together is zero, as is the cyclic sum of (7). The cyclic sum of (8) is equal to that of

$$-3c(\langle Z, W \rangle \langle AY, W \rangle - \langle Y, W \rangle \langle AZ, W \rangle)AX.$$

The cyclic sum of (9) is zero. Here we need to use the fact the P commutes with A. The results of these observations is that the cyclic sum of $(R(X,Y) \cdot S)Z$ is equal to the cyclic sum of

$$\langle (P\varphi + \varphi P)Y, Z \rangle \varphi X + 2\langle X, \varphi Y \rangle \varphi PZ + 3(\langle Z, W \rangle \langle AY, W \rangle - \langle Y, W \rangle \langle AZ, W \rangle)AX, \tag{6.10}$$

multiplied by $-c$. Now suppose that the cyclic-Ryan condition holds. Take $X = W$ and compute the cyclic sum of (6.10) to obtain

$$\begin{aligned}
0 = {} & \langle \varphi PW, Y \rangle \varphi Z + \langle P\varphi Z, W \rangle \varphi Y + 2\langle Y, \varphi Z \rangle \varphi PW \\
& + 3(\langle Z, W \rangle \langle AY, W \rangle - \langle Y, W \rangle \langle AZ, W \rangle)AW \\
& + 3(\langle W, W \rangle \langle AZ, W \rangle - \langle Z, W \rangle \langle AW, W \rangle)AY \\
& + 3(\langle Y, W \rangle \langle AW, W \rangle - \langle W, W \rangle \langle AY, W \rangle)AZ.
\end{aligned} \tag{6.11}$$

Apply φ to the right side of (6.11) and take the trace (as a linear map in Z) to get

$$0 = \langle \varphi PW, Y \rangle (-2n + 3) - 2\langle \varphi PW, Y \rangle$$
$$+ 3(\langle \varphi AW, W \rangle \langle AY, W \rangle - \langle Y, W \rangle \langle A\varphi AW, W \rangle)$$
$$+ 3(\langle A\varphi AY, W \rangle - \langle \varphi AY, W \rangle \langle AW, W \rangle)$$
$$+ 3(\langle Y, W \rangle \langle AW, W \rangle - \langle AY, W \rangle). \tag{6.12}$$

Note that $\langle A\varphi AW, W \rangle = 0$ and $\text{trace}(\varphi A) = 0$, so the equation reduces to

$$(2n - 1)\varphi PW + 3A\varphi AW = 0. \tag{6.13}$$

Let $U = \varphi AW$ and rewrite (6.12) as

$$\langle AU, Y \rangle \varphi Z - \langle AU, Z \rangle \varphi Y + 2\langle Y, \varphi Z \rangle AU$$
$$= (2n - 1)(\langle Z, W \rangle \langle AY, W \rangle - \langle Y, W \rangle \langle AZ, W \rangle)AW$$
$$+ (2n - 1)(\langle \varphi Z, U \rangle AY - \langle \varphi Y, U \rangle AZ), \tag{6.14}$$

where we have used the fact that $(A - aI)W = -\varphi U$. Upon taking the inner product with U and using the fact that $\langle AW, U \rangle = 0$, we get

$$(n - 1)(\langle \varphi Z, U \rangle \langle AY, U \rangle - \langle \varphi Y, U \rangle \langle AZ, U \rangle) = \langle Y, \varphi Z \rangle \langle AU, U \rangle. \tag{6.15}$$

Taking $Y = U$ gives

$$(n - 2)\langle AU, U \rangle \varphi U = 0$$

and hence $\langle AU, U \rangle = 0$ provided that $n \geq 3$. If we put $Z = U$ in (6.14), we get

$$(2n - 3)\langle \varphi Y, U \rangle AU - \langle AY, U \rangle \varphi U = 0.$$

Combining this with (6.15) yields

$$2(n - 2)\langle \varphi Y, U \rangle AU = 0,$$

so that AU must be zero. (Again, we have used the hypothesis that $n \geq 3$.) Then $\varphi PW = 0$ by (6.13) and we can simplify (6.11). Put $Z = \varphi U$ in (6.11) to get

$$-\langle Y, W \rangle \langle A\varphi U, W \rangle AW + \langle A\varphi U, W \rangle AY + \langle (aI - A)W, Y \rangle A\varphi U = 0.$$

In other words,

$$\langle U, U \rangle (\langle Y, W \rangle AW - AY) = \langle \varphi U, Y \rangle A\varphi U. \tag{6.16}$$

We intend to show that $AW = aW$, that is, that $U = 0$. Suppose then, that $U \neq 0$ at some point of M. For the rest of this proof, we work in a neighborhood where U is nonzero. Let $\mathcal{V}$ be the two-dimensional subspace spanned by $\{W, \varphi U\}$ and let $\bar{U}$ be a unit vector in the direction of φU. We see immediately from (6.16) that $AY = 0$ for Y orthogonal to $\mathcal{V}$. Thus $\mathcal{V}$ is A-invariant, and if we write $AW = aW + b\bar{U}$ and $A\bar{U} = bW + d\bar{U}$, a short calculation reveals that

$PX = (b^2 - ad)X = pX$ (say) for $X \in \mathcal{V}$. In the cyclic sum of (6.10), substitute $X = \varphi U$ and $Y = U$, and take Z orthogonal to X, Y, and W to get

$$p\,|U|^2\varphi Z = 0.$$

So $p = 0$, and we conclude that $P = 0$. Now, let X and Y be principal vectors in $\mathcal{V}$ with corresponding principal curvatures λ and μ. Since $P = 0$, we have $\lambda^2 - m\lambda = \lambda^2 - (\lambda + \mu)\lambda = 0$ so that $\lambda\mu = 0$. Take μ to be the zero principal curvature. To see that this leads to a contradiction, take $Z \in \mathcal{V}^\perp$ and use Codazzi's equation to get

$$-A[Y, Z] = c\langle Y, W\rangle\varphi Z + 2\langle Y, \varphi Z\rangle W,$$

which reduces to

$$c\langle Y, W\rangle\,|Z|^2 = 0$$

upon taking the inner product with φZ. We have to use the fact that $A(\varphi Z) = 0$. We conclude that $\langle Y, W\rangle = 0$. This makes $\bar{U}$ and hence W principal, which is a contradiction. Thus M is a Hopf hypersurface. $\qquad\square$

We can use Theorem 6.28 to get a direct proof that Ryan spaces cannot occur as hypersurfaces when $n \geq 3$.

THEOREM 6.29. *In a complex space form of constant holomorphic sectional curvature $4c \neq 0$, there exists no real hypersurface M^{2n-1}, $n \geq 3$, satisfying $R \cdot S = 0$. For $n = 2$, there are no Hopf hypersurfaces satisfying $R \cdot S = 0$.*

PROOF. Suppose that M^{2n-1}, where $n \geq 2$, satisfies $R \cdot S = 0$. The cyclic-Ryan condition is satisfied so M is Hopf by Theorem 6.28 if $n \geq 3$. Otherwise, it is Hopf by hypothesis. Let X be any principal vector orthogonal to W with associated principal curvature λ. Evaluating $(R(X, W) \cdot S)W$ using the Gauss equation and (1.11), we see that $R \cdot S = 0$ if and only if

$$(\lambda a + c)(3c + (\lambda - a)(m - \lambda - a)) = 0 \tag{6.17}$$

for all principal curvatures λ whose principal spaces are in $W^\perp$. If λ and μ are principal curvatures corresponding to X and φX respectively, we also must have

$$\lambda\mu = \frac{\lambda + \mu}{2}a + c, \tag{6.18}$$

by Corollary 2.3. Of the various ways in which these two conditions might be satisfied, $\lambda a + c = \mu a + c = 0$ is impossible since it implies $a \neq 0$, $\lambda \neq 0$, $\mu \neq 0$, $\lambda = \mu$, and $\lambda^2 = \lambda a + c$. On the other hand, if $\lambda a + c$ and $\mu a + c$ are both nonzero, then both λ and μ satisfy the quadratic equation

$$t^2 - mt + ma - a^2 - 3c = 0. \tag{6.19}$$

If λ and μ are distinct, any other principal curvature ρ (on $W^\perp$) must satisfy $\rho a + c = 0$, since it cannot satisfy (6.19). However, given the φ-invariance of

$W^\perp$, and the first remark, there can be no such ρ. Thus λ, μ, and a are the only principal curvatures.

Suppose that there is a point where two distinct principal curvatures exist as in (6.18). First check that neither λ nor μ is equal to $a/2$. With this possibility eliminated, there is a neighborhood in which $\mathfrak{m} = (n-1)(\lambda+\mu) + a$, $\mathfrak{m} = \lambda + \mu$, and

$$\lambda\mu = \mathfrak{m}a - a^2 - 3c = \frac{\mathfrak{m}a}{2} + c.$$

We now calculate that

$$(2n-3)a^2 + 8(n-2)c = 0.$$

If $c > 0$, this is already a contradiction. If $c < 0$, we continue, obtaining

$$\lambda\mu = \frac{2n+1}{2n-3}c.$$

Since the relevant examples of Hopf hypersurfaces with constant principal curvatures have $\lambda\mu > 0$, the situation in this paragraph cannot occur. (The coefficients of (6.19) are constant since $(n-2)\mathfrak{m} = -a$ and $a \neq 0$ by (6.18).)

A third possibility is that for all points and every principal $\lambda a + c = 0$ but $\mu a + c \neq 0$. Then μ satisfies (6.19) as does any further principal curvature ν. In addition, T_ν must be φ-invariant and $\nu^2 = a\nu + c$. Using (6.18) and (6.19), respectively, we can compute μ and ν in terms of a and c. If this situation holds at a point, then it also holds in some neighborhood, which is therefore a Hopf hypersurface with constant principal curvatures. This is clearly impossible since none of the standard examples have three distinct principal curvatures on $W^\perp$. We conclude that no such ν can exist.

The only remaining possibility is that there are two distinct constant principal curvatures on $W^\perp$,

$$\lambda = -\frac{c}{a} \quad\text{and}\quad \mu = -\frac{ca}{a^2 + 2c},$$

the latter having been calculated from (6.18). Because φ interchanges the principal subspaces, the only possible candidates among the standard examples are the Type B hypersurfaces. However, one can check that none of the Type B hypersurfaces in fact satisfy (6.17). This concludes the proof that a hypersurface satisfying the hypothesis cannot exist. □

Theorem 6.29 was proved in [Kimura and Maeda 1989] for $c > 0$. For any $c \neq 0$ it can be deduced from the (ii) $\Rightarrow$ (i) implication of the next theorem (Theorem 6.30) by checking that none of the pseudo-Einstein hypersurfaces satisfy $R \cdot S = 0$. We do not prove that part of Theorem 6.30 here but refer to [Ki et al. 1990b]. Weakening the condition $R \cdot S = 0$ in either of two directions, we get two additional characterizations of the pseudo-Einstein hypersurfaces discussed in Theorems 6.1 and 6.2. In the first case, we look at the cyclic sum. In the second case, we restrict the condition to $W^\perp$.

THEOREM 6.30. *For a real hypersurface M^{2n-1}, where $n \geq 3$, in a complex space form of constant holomorphic sectional curvature $4c \neq 0$, the following conditions are equivalent:*

(i) *M satisfies the pseudo-Einstein condition.*
(ii) *M satisfies the cyclic-Ryan condition.*
(iii) *M satisfies the pseudo-Ryan condition and is a Hopf hypersurface.*

Thus, if M satisfies any of these three conditions, it is one of the hypersurfaces listed in Theorems 6.1 and 6.2.

The equivalence of (i) and (ii) was established by Ki, Nakagawa, and Suh [Ki et al. 1990b]. Condition (iii) was studied by S.-B. Lee, N.-G. Kim, and S.-S. Ahn [Lee et al. 1990] who proved that condition (iii) implies that M is of type A or B. However, one can check that among these hypersurfaces only pseudo-Einstein ones actually satisfy (iii). Thus, the three conditions are equivalent.

We will give the proof of the equivalence of (i) and (iii). To simplify the proof, we set $TZ = \langle Z, W \rangle W$ and $P = A^2 - mA$. Then the Ricci tensor S can be expressed as $S = c_1 I + c_2 T - P$ for suitable constants c_1 and c_2; see (1.11). It is easy to check that $\langle (R(X,Y)T)Z_1, Z_2 \rangle = 0$ for any X and Y, and for Z_1 and Z_2 in $W^\perp$. Consequently, for such arguments,

$$\langle (R(X,Y)S)Z_1, Z_2 \rangle = -\langle (R(X,Y)P)Z_1, Z_2 \rangle.$$

We now give an alternate characterization of pseudo-Ryan hypersurfaces, which follows immediately from this discussion.

PROPOSITION 6.31. *A real hypersurface in a complex space form is pseudo-Ryan if and only if*

$$\langle R(X,Y)PZ_1, Z_2 \rangle + \langle R(X,Y)PZ_2, Z_1 \rangle = 0 \qquad (6.20)$$

for all $X, Y, Z_1, Z_2 \in W^\perp$.   ◁

LEMMA 6.32. *Let M^{2n-1}, where $n \geq 2$, be a Hopf hypersurface in a complex space form of constant holomorphic sectional curvature $4c$. Suppose that M satisfies the pseudo-Ryan condition at a point p. Let X and Y be unit principal vectors in $W^\perp$ at p with $AX = \lambda X$, $AY = \mu Y$, $PX = \alpha X$, and $PY = \beta Y$. Then*

$$(\alpha - \beta)(\lambda \mu + 2nc)\langle \varphi X, Y \rangle = 0$$

and

$$(\lambda - \mu)(\lambda + \mu - m)(\lambda \mu + 2nc)\langle \varphi X, Y \rangle = 0.$$

PROOF. From the definition of P, we see that $\alpha = \lambda^2 - m\lambda$ and $\beta = \mu^2 - m\mu$. Consequently,

$$\alpha - \beta = (\lambda - \mu)(\lambda + \mu - m). \qquad (6.21)$$

By Proposition 6.31,

$$0 = (\alpha - \beta)\langle R(Z_1, Z_2)X, Y \rangle, \qquad (6.22)$$

for all $Z_1, Z_2 \in W^\perp$. In particular, using the Gauss equation, we compute

$$
\begin{aligned}
0 &= (\alpha - \beta)\langle R(X,Y)Y, X\rangle \\
&= (\alpha - \beta)\big(c + \lambda\mu + c\langle(\varphi X \wedge \varphi Y)Y, X\rangle + 2c\langle X, \varphi Y\rangle\langle\varphi Y, X\rangle\big) \\
&= (\alpha - \beta)\big(c + \lambda\mu + c\langle X, \varphi Y\rangle^2 + 2c\langle X, \varphi Y\rangle^2\big) \\
&= (\alpha - \beta)\big(c + \lambda\mu + 3c\langle X, \varphi Y\rangle^2\big).
\end{aligned}
\tag{6.23}
$$

We also conclude that, for any unit vector $Z \in W^\perp$,

$$
\begin{aligned}
\langle R(\varphi Z, Z)&X, Y\rangle \\
&= c\langle(\varphi Z \wedge Z)X, Y\rangle + c\langle(\varphi^2 Z \wedge \varphi Z)X, Y\rangle \\
&\quad + 2c\langle\varphi Z, \varphi Z\rangle\langle\varphi X, Y\rangle + \langle(A\varphi Z \wedge AZ)X, Y\rangle \\
&= (2c + \lambda\mu)(\langle Z, X\rangle\langle\varphi Z, Y\rangle - \langle\varphi Z, X\rangle\langle Z, Y\rangle) + 2c\langle\varphi X, Y\rangle.
\end{aligned}
\tag{6.24}
$$

Unless $\alpha = \beta$ (in which case, we are finished), the left hand side of (6.24) must be zero by (6.22). Let Z run through an orthonormal basis of $W^\perp$ and take the sum, to get

$$
0 = -2(\lambda\mu + 2nc)\langle X, \varphi Y\rangle,
$$

and the conclusion of the lemma follows by using (6.21). $\qquad\square$

Lemma 6.32 does not require that $c \neq 0$. However, we will have no further use for the result in the case $c = 0$.

COROLLARY 6.33. *In Lemma 6.32, suppose that $Y = \varphi X$. If $n \geq 3$ and $c \neq 0$, then $\alpha = \beta$.*

PROOF. Taking $Z = X$ and $Y = \varphi X$ in (6.24) yields $\lambda\mu + 4c = 0$. On the other hand, Lemma 6.32 gives $(\alpha - \beta)(\lambda\mu + 2nc) = 0$. Subtracting these equations, we obtain $2c(n-2)(\alpha - \beta) = 0$, as required. $\qquad\square$

PROOF THAT (i) $\Longleftrightarrow$ (iii) IN THEOREM 6.30. Assuming (iii), we will prove that the principal curvatures must be constant. Then Theorem 4.13 limits the possibilities to those on Takagi's and Montiel's lists. Then we can pick from these lists the examples that actually satisfy the pseudo-Ryan condition.

First suppose that there is a point where P has two or more distinct eigenvalues on $W^\perp$. Then, there are orthonormal principal vectors X and Y as in the hypothesis of Lemma 6.32 with $\alpha \neq \beta$. Assuming that neither λ nor μ is equal to $a/2$, let $A\varphi X = \nu\varphi X$. By Corollary 6.33, $\lambda^2 - m\lambda = \nu^2 - m\nu$. Also, by (6.23), $\lambda\mu + c = \nu\mu + c = 0$ since span$\{T_\lambda, \varphi T_\lambda\}$ and span$\{T_\nu, \varphi T_\nu\}$ are orthogonal. Thus $\lambda = \nu$. This shows that λ (and similarly μ) satisfies the quadratic equation

$$
t^2 - at - c = 0.
\tag{6.25}
$$

Note that $a/2$ cannot occur as a principal curvature since an application of Corollary 2.3 would imply that $a^2 + 4c = 0$ and hence, by (6.25), that $\lambda = a/2$. On the other hand, the argument just completed shows that no further principal curvature is possible since it would have to satisfy (6.25). Thus, $W^\perp$ decomposes into two φ-invariant principal subspaces and the associated principal curvatures are constant. This shows that M is (at least locally) a Type A hypersurface.

The alternative possibility is that P is a multiple of the identity on $W^\perp$. In other words, for any two principal curvatures λ and μ (corresponding to principal vectors in $W^\perp$) at any point, we must have

$$(\lambda - \mu)(\lambda + \mu - \mathfrak{m}) = 0.$$

In particular, the number of distinct eigenvalues of A on $W^\perp$ is at most 2. Suppose that at some point there is a principal curvature $\lambda \neq a/2$ with associated unit principal vector $X \in W^\perp$ such that $A\varphi X = \mu\varphi X$, with $\mu \neq \lambda$. Then $\lambda^2 - \mathfrak{m}\lambda = \mu^2 - \mathfrak{m}\mu$ so that $\lambda + \mu = \mathfrak{m}$. This shows that there can be no other principal curvature on $W^\perp$. Each of λ and μ has multiplicity $n - 1$. Thus, we have

$$(n - 1)\lambda + (n - 1)\mu + a = \mathfrak{m} = \lambda + \mu$$

and hence,

$$(n - 2)(\lambda + \mu) + a = 0.$$

Again by Corollary 2.3,

$$\lambda\mu - (\lambda + \mu)a/2 - c = 0,$$

so that λ and μ satisfy the quadratic equation

$$t^2 + \frac{a}{n - 2}t + c - \frac{a^2}{2(n - 2)} = 0.$$

Again, A has two distinct constant curvatures on $W^\perp$. Because the principal spaces are interchanged by φ, we have locally a Type B hypersurface.

The existence of one point satisfying either of the conditions discussed in the preceding two paragraphs implies that the condition holds globally. The remaining possibility is that for all points and all principal curvatures $\lambda \neq a/2$ on $W^\perp$, the principal space T_λ is φ-invariant. If there is no such λ, then $A = (a/2)I$ on $W^\perp$ and our hypersurface is a horosphere (Type A0). Now suppose that there is at least one such λ. Then there can be at most one other principal curvature μ. Further, arguing as in the first part of this proof, any such μ satisfies $\mu \neq a/2$. We can ignore this case since it was covered in the first step of the proof (Type A2). If there is no such μ, then $A = \lambda I$ on $W^\perp$ and we have a geodesic sphere or tube over a complex hyperbolic hyperplane (Type A1).

It remains to determine which of the manifolds above actually satisfy the pseudo-Ryan condition. In this discussion, we look at (6.20) and assume that the vectors are taken from a principal orthonormal basis that is φ-invariant up to sign. We first note that if M is pseudo-Einstein, then P is a multiple of the identity on $W^\perp$, so that (iii) is satisfied. Thus all hypersurfaces of types A0 and A1 are pseudo-Ryan. Now look at the Type A2 hypersurfaces as described in

Section 3. In (6.20), we may choose Z_1 and Z_2 in distinct principal distributions (since the formula clearly holds otherwise). Then (6.20) reduces to

$$(\lambda_1 - \lambda_2)(\lambda_1 + \lambda_2 - \mathfrak{m})\langle R(X,Y)Z_1, Z_2\rangle = 0. \tag{6.26}$$

If X and Y are in the same principal distribution, we calculate the curvature term using the Gauss equation (1.8) and find, noting φ-invariance, that it vanishes. Thus, we take X and Z_2 to be in the λ_1-distribution and Y and Z_1 to be principal for λ_2. Now, using the fact that $\lambda_1\lambda_2 + c = 0$, the Gauss equation for (X,Y) becomes

$$R(X,Y) = c(\varphi X \wedge \varphi Y) + 2c\langle X, \varphi Y\rangle\varphi,$$

so that the curvature term in (6.26) is $c(\langle \varphi Y, Z_1\rangle\langle \varphi X, Z_2\rangle)$, which vanishes unless $Z_1 = \pm\varphi Y$ and $Z_2 = \pm\varphi X$. In this case, however, the curvature term is nonzero and it is necessary that $(\lambda_1 - \lambda_2)(\lambda_1 + \lambda_2 - \mathfrak{m}) = 0$ in order for M to be pseudo-Ryan. This is precisely the condition for M to be pseudo-Einstein. We examine the Type B hypersurfaces in the same way, using the fact that the principal spaces are interchanged by φ and conclude that the only Type B hypersurfaces satisfying (iii) are the pseudo-Einstein ones.

Conversely, if (i) is assumed, P is a multiple of the identity on $W^\perp$. We can use (6.21) which is a valid expression for comparing the eigenvalues of P and does not depend on the pseudo-Ryan condition in the hypothesis of Lemma 6.32. Because of (6.2), we get $\alpha = \beta$ in (6.21) as required.   □

7. A is W-Parallel

As we have seen, $\nabla A = 0$ is too strong a condition to be satisfied by a hypersurface in $\mathbb{C}P^n$ or $\mathbb{C}H^n$. However, if we merely ask that A be parallel in the direction of the structure vector W, interesting results are possible.

THEOREM 7.1. *Let* M^{2n-1}, *where* $n \geq 3$, *be a real hypersurface in a complex space form of constant holomorphic sectional curvature* $4c > 0$. *Assume that the principal curvatures of* M *have constant multiplicities. If* $\nabla_W A = 0$, *then* M *is an open subset of a Type A hypersurface from Takagi's list or is a non-homogeneous hypersurface which is a tube of radius* $r\pi/4$ *over a certain Kähler submanifold of* $\mathbb{C}P^n$.

This classification was performed by Kimura and Maeda [1991]. The key to the theorem is the following proposition, which we generalize to the case $4c \neq 0$.

PROPOSITION 7.2. *Let* M^{2n-1}, *where* $n \geq 2$, *be a real hypersurface in a complex space form of constant holomorphic sectional curvature* $4c \neq 0$. *If* $\nabla_W A = 0$, *then* M *is a Hopf hypersurface.*

The first step in the proof of Proposition 7.2 is to show that under the conditions of the proposition, AW must be principal.

LEMMA 7.3. *Under the assumptions of Proposition 7.2, at any point $p \in M$, either $AW = 0$ or AW is principal. If $AW \neq 0$ then $\langle AW, W \rangle \neq 0$.*

PROOF. Let X be any tangent vector field. Applying the Codazzi equation (1.9) to X and W yields

$$(\nabla_X A)W = -c\varphi X,$$

and consequently

$$\nabla_X(AW) = -c\varphi X + A\varphi AX. \tag{7.1}$$

On the other hand, applying the Codazzi equation to any tangent X and Y gives

$$A\varphi((\nabla_X A)Y - (\nabla_Y A)X) = -c(\langle X, W \rangle AY - \langle Y, W \rangle AX),$$

and thus

$$\nabla_X \nabla_Y(AW) - \nabla_{\nabla_X Y}(AW) = -c(\nabla_X \varphi)Y + (\nabla_X A)\varphi AY + A\varphi(\nabla_X A)Y$$
$$+ \langle Y, AW \rangle A^2 X - \langle AX, AY \rangle AW.$$

The curvature tensor R now satisfies

$$R(X, Y)(AW) = (\nabla_X A)\varphi AY - (\nabla_Y A)\varphi AX + \langle Y, AW \rangle A^2 X - \langle X, AW \rangle A^2 Y.$$

On the other hand, using the Gauss equation, we can calculate that

$$R(X, Y)(AW) = c(\langle Y, AW \rangle X - \langle X, AW \rangle Y) + (\langle AY, AW \rangle AX - \langle AX, AW \rangle AY)$$
$$+ c(\langle \varphi Y, AW \rangle \varphi X - \langle \varphi X, AW \rangle \varphi Y) + 2c\langle X, \varphi Y \rangle \varphi AW.$$

The last two equations give us two expressions for $\langle R(X, Y)AW, W \rangle$, which we equate to get

$$0 = 2(\langle Y, AW \rangle \langle X, A^2 W \rangle - \langle X, AW \rangle \langle Y, A^2 W \rangle)$$
$$+ \langle W, c\langle \varphi Y, AW \rangle \varphi X - c\langle \varphi X, AW \rangle \varphi Y + 2c\langle X, \varphi Y \rangle \varphi AW \rangle.$$

Since Y was arbitrary, we conclude that

$$\langle A^2 W, X \rangle AW = \langle X, AW \rangle A^2 W.$$

Setting $X = W$ this becomes

$$|AW|^2 AW = \langle AW, W \rangle A^2 W. \tag{7.2}$$

If AW is orthogonal to W, then $AW = 0$. On the other hand, if $AW \neq 0$, we can rewrite (7.2) as

$$A(AW) = \alpha AW \tag{7.3}$$

for some nonzero α, and the conclusion of the lemma then follows. $\square$

PROOF OF PROPOSITION 7.2. Suppose that there is a point p where W is not principal, that is, where $\varphi AW \neq 0$. Work in a neighborhood where this condition holds. By Lemma 7.3, we know that AW is a principal vector, with principal curvature

$$\alpha = \frac{|AW|^2}{\langle AW, W \rangle}.$$

Let $a = \langle AW, W \rangle$, which is nonzero by Lemma 7.3. Then we compute

$$\nabla_X(A^2 W) = (\nabla_X A)AW + A(\nabla_X A)W + A^2 \varphi AX,$$
$$\nabla_X(aAW) = (Xa)AW - ca\varphi X + aA\varphi AX.$$

By (7.3) the left sides of these two equations are equal. Equating the right sides, taking inner product with Y, and subtracting the same expression with X and Y interchanged, we get

$$(X\alpha)\langle AW, Y \rangle - (Y\alpha)\langle AW, X \rangle = c(\langle X, W \rangle \langle \varphi Y, AW \rangle$$
$$- \langle Y, W \rangle \langle \varphi X, AW \rangle + 2\langle X, \varphi Y \rangle a) + \langle (A^2 \varphi A + A\varphi A^2)X, Y \rangle$$
$$- c\langle (A\varphi + \varphi A)X, Y \rangle + 2c\alpha\langle \varphi X, Y \rangle - 2\alpha\langle A\varphi AX, Y \rangle. \quad (7.4)$$

If we set $Y = W$ in this equality, we get

$$a(X\alpha) = W\alpha\langle AW, X \rangle + c\langle X, \varphi AW \rangle + c\langle \varphi AW, X \rangle$$
$$- \langle A\varphi A^2 W, X \rangle - \langle A^2 \varphi AW, X \rangle + 2\alpha\langle A\varphi AW, X \rangle.$$

Consequently,

$$a(\operatorname{grad}\alpha) = (W\alpha)AW + \alpha A\varphi AW + 2c\varphi AW - A^2 \varphi AW. \quad (7.5)$$

Using (7.5), we can rewrite the left-hand side of (7.4) and let $X = AW$ in (7.4) to get

$$\frac{1}{a}(-\alpha^2 aA\varphi AW - 2c\alpha a\varphi AW + \alpha a A^2 \varphi AW)$$
$$= -ca\varphi AW - 2ca\varphi AW - cA\varphi AW - c\alpha\varphi AW$$
$$+ \alpha A^2 \varphi AW + \alpha^2 A\varphi AW + 2c\alpha\varphi AW - 2\alpha^2 A\varphi AW.$$

Simplifying and noting that $c \neq 0$, we get

$$A\varphi AW = 3(\alpha - a)\varphi AW.$$

This further simplifies (7.5) and we obtain

$$a \operatorname{grad}\alpha = (W\alpha)AW + (2c - 3(3a - 2\alpha)(a - \alpha))\varphi AW. \quad (7.6)$$

We can also calculate directly that

$$Xa = 2\langle A\varphi AX, W \rangle = 6(a - \alpha)\langle \varphi AW, X \rangle.$$

Consequently, if we let $\sigma = 6(a - \alpha)$, we get

$$\operatorname{grad} a = \sigma\varphi AW \quad (7.7)$$

and

$$\operatorname{grad} \sigma = \frac{6}{a}(\rho \varphi AW - (W\alpha)AW),$$

where ρ is a scalar. Next, compute

$$\nabla_X(\operatorname{grad} a) = \frac{6}{a}(\rho\langle \varphi AW, X\rangle \varphi AW - (W\alpha)\langle AW, X\rangle \varphi AW)$$
$$+ a\sigma AX - \sigma\langle A^2W, X\rangle W - c\sigma(-X + \langle X, W\rangle W) + \sigma\varphi A\varphi AX,$$

and then

$$\langle \nabla_X(\operatorname{grad} a), Y\rangle = -\frac{6}{a}((W\alpha)\langle AW, X\rangle\langle \varphi AW\, Y\rangle) - \sigma\alpha\langle AW, X\rangle\langle W, Y\rangle$$
$$+ \sigma\langle(\varphi A)^2 X, Y\rangle + \text{terms symmetric in } X \text{ and } Y.$$

Using the fact that

$$\langle \nabla_X(\operatorname{grad} a), Y\rangle = \langle \nabla_Y(\operatorname{grad} a), X\rangle$$

we see that

$$0 = -\frac{6(W\alpha)}{a}(\langle AW, X\rangle\langle \varphi A\dot{W}, Y\rangle - \langle AW, Y\rangle\langle \varphi AW, X\rangle)$$
$$- \sigma\alpha(\langle AW, X\rangle\langle W, Y\rangle - \langle AW, Y\rangle\langle W, X\rangle) + \sigma\langle((\varphi A)^2 + (A\varphi)^2)X, Y\rangle = 0.$$

Let $Y = W$ in this equality to get

$$(W\alpha)\varphi AW = (a - \alpha)(3a - 2\alpha)(AW - aW).$$

Taking the inner product of this with φAW and AW in turn yields

$$(W\alpha)\,|\varphi AW| = 0$$

and

$$(3a - 2\alpha)(a - \alpha)^2 a = 0.$$

This again allows us to simplify (7.6) to

$$a \operatorname{grad} \alpha = 2c\varphi AW, \tag{7.8}$$

while

$$\operatorname{grad} a = -6(a - \alpha)\varphi AW.$$

Now $a \neq \alpha$, since $0 \neq |AW - aW|^2 = \alpha a - 2a^2 + a^2 = (\alpha - a)a$, and hence $3a = 2\alpha$. Comparing (7.4) and (7.8) now gives

$$9a^2 + 4c = 0.$$

Thus a is constant and $\operatorname{grad} a = 0$, a contradiction. This completes the proof of Proposition 7.2 since, by (7.7), W is again principal.   □

PROOF OF THEOREM 7.1. By Proposition 7.2, if $\nabla_W A = 0$, then M is a Hopf hypersurface and we can write $AW = aW$. By Theorem 3.1, a is constant. We can then divide the discussion into two cases, $a = 0$ and $a \neq 0$.

Suppose that $a = 0$. Then $A\varphi A = -c\varphi$, by (7.1). If no principal curvature is equal to $(1/r)\cot(\pi/4) = 1/r$, the focal map associated to the principal curvature a has constant rank and M lies on a tube of radius $r\pi/4$ over a complex submanifold. This is an application of [Cecil and Ryan 1982, Theorem 1, p. 489]. Even if $\lambda = 1/r$ is a principal curvature of constant multiplicity, we can make the same claim. In this case, T_λ can be easily seen to be φ-invariant. Whether or not M is a Type A hypersurface (e.g., a geodesic sphere of radius $r\pi/4$) or a nonhomogeneous hypersurface would depend on whether there were additional principal curvatures.

In the case $a \neq 0$, the following theorem of S. Maeda and S. Udagawa [1990] completes the proof.   □

THEOREM 7.4. *Let M^{2n-1}, where $n \geq 3$, be a Hopf hypersurface in a complex space form of constant holomorphic sectional curvature $4c > 0$. Suppose that $a \neq 0$. If $\nabla_W A = 0$, then M is an open subset of a Type A hypersurface from Takagi's list.*

PROOF. Let X be any principal vector in $W^\perp$ with corresponding principal curvature λ. By (7.1),

$$0 = a\varphi AX + c\varphi X - A\varphi AX = a\lambda\varphi X + c\varphi X - \lambda A\varphi X. \qquad (7.9)$$

Thus $\lambda A\varphi X = (\lambda a + c)\varphi X$. Comparing this with the formula in Corollary 2.3 yields $a(\lambda^2 - a\lambda - c) = 0$. This equation has two distinct roots, hence there are at most two distinct principal curvatures and they are locally constant. Each principal space is φ-invariant, as can be seen from (7.9). This rules out the Type B hypersurfaces as possibilities and completes the proof.   □

Related results are found in [Pyo 1994a; 1994b]. In addition, some authors have studied similar conditions using Lie derivatives instead of covariant derivatives; see, for example, [Ki and Suh 1995; Ki et al. 1991; Ki et al. 1992; Ki et al. 1994; Ki et al. 1996; Kim et al. 1992a; Kimura and Maeda 1995; Pyo and Suh 1995].

8. Additional Topics

In this section we briefly discuss some topics related to the preceding material that we have not been able to include in this article.

Ruled Real Hypersurfaces. This class of hypersurfaces that does not belong to our list of "standard examples" but have occurred in some recent classification results. We introduce their definition and state a few of their properties.

Take a regular curve γ in $\tilde{M}$ ($\mathbb{C}P^n$ or $\mathbb{C}H^n$) with tangent vector field X. At each point of γ there is a unique complex projective or hyperbolic hyperplane cutting γ so as to be orthogonal not only to X but to JX. The union of these hyperplanes is called a *ruled real hypersurface*. It will be an embedded hypersurface locally although globally it will in general have self-intersections and singularities.

THEOREM 8.1. *Ruled real hypersurfaces have the following properties.*

(i) *The holomorphic distribution* $W^\perp$ *is integrable.*

(ii) *The structure vector* W *is not principal.*

(iii) *The principal curvatures are not all constant.*

(iv) *The shape operator has rank 2 and is* η-*parallel.*

(v) *The principal space for the zero principal curvature lies in* $W^\perp$.   ◁

We will not discuss the related classification results here. Relevant references are [Kimura 1987a; Kimura and Maeda 1989; Maeda and Udagawa 1990; Suh 1992; 1995; Ahn et al. 1993; Taniguchi 1994; Pyo 1994c; Ki and Kim 1994; Ki and Suh 1995; 1996].

Isoparametric Hypersurfaces. In real space forms, isoparametric hypersurfaces are characterized by the fact that all their principal curvatures are constant. However, there are other equivalent characterizations. In the case of complex space forms, the analogous properties turn out not to be equivalent.

The following conditions can be considered.

(i) M has constant principal curvatures.

(ii) M is one of a parallel family of hypersurfaces of constant mean curvature.

(iii) M is one of a transnormal system, a system of parallel hypersurfaces with common normal geodesics.

(iv) The lifted hypersurface M' is an isoparametric hypersurface of $\tilde{M}'$.

Relevant references are [D'Atri 1979; Bolton 1973; Carter and West 1985; Li 1988; Park 1989; Wang 1982; 1983; 1987]. The new examples in $\mathbb{C}P^n$ are hypersurfaces M whose lifts M' are isoparametric but have more than two principal spaces that are not horizontal. Then the principal curvatures of M are need not be constant.

Real Hypersurfaces in Quaternionic Projective Space. Several authors have studied such hypersurfaces. Instead of the structure vector W, there is a three-dimensional distinguished subspace of the tangent space to be considered. Many basic questions analogous to those treated in the complex space forms have been asked and answered. Relevant references are [Berndt 1991; Berndt and Vanhecke 1992; 1993; Dong 1993; Hamada 1993; Ki et al. 1997; Martínez 1988; Martínez and Pérez 1986; Pak 1977; Pérez 1991; 1992; 1993a; 1993b; 1994; 1996a; 1996b; Pérez and Santos 1985; 1991; 1993; Pérez and Suh 1996a; 1996b].

9. Conclusion and Open Problems

In the preceding sections we have tried to present in an orderly fashion the central results concerning real hypersurfaces in $\mathbb{C}P^n$ and $\mathbb{C}H^n$. Because of the limitations of time and space, we have had to forego presenting the details of all the stated theorems, though we have done so for a good proportion of them. Two facts emerge from our study. First, some results hold for all dimensions $n \geq 2$, while many others require $n \geq 3$. Second, many results require the hypersurface to be a Hopf hypersurface ($AW = aW$) while some hold more generally. Some papers in the literature are vague about which of these conditions are being assumed, or rely on results that may require stronger hypotheses than those explicitly presented. We have attempted to be as explicit as possible in our presentation and in case of ambiguity in the literature have tried to err on the side of caution in those cases where we did not work through the complete proofs.

The following are questions and problems that appear to us to be open.

QUESTION 9.1. (See Proposition 1.4.) Although φA cannot vanish on an open set, are there examples for which φA vanishes at isolated points or on sets of lower dimension? The same question can be asked about umbilics (see Theorem 1.5) and the vanishing of $\varphi A + A\varphi$ (see Corollary 2.12).

QUESTION 9.2. Do Theorems 4.6 and 4.7 extend to $n = 2$? Are there hypersurfaces in $\mathbb{C}P^2$ or $\mathbb{C}H^2$ that have ≤ 2 principal curvatures, other than the standard examples?

QUESTION 9.3. (See Theorems 4.9 and 4.10.) $R \cdot A$ never vanishes for a Hopf hypersurface. Are there non-Hopf examples for which it does? $R \cdot A$ cannot vanish on an open set if $c > 0$. What about $c < 0$? Any counterexamples would have to satisfy $n \geq 3$ in view of Theorem 4.11.

QUESTION 9.4. (See Theorem 4.12.) A hypersurface in $\mathbb{C}P^n$ with three principal curvatures, all constants, must be a Hopf hypersurface, and hence one of the standard examples. Is the same true for $\mathbb{C}H^n$?

QUESTION 9.5. (See Theorems 6.1 and 6.2.) Classify the pseudo-Einstein hypersurfaces in $\mathbb{C}P^2$ and $\mathbb{C}H^2$.

QUESTION 9.6. Does Theorem 6.2 still hold if the pseudo-Einstein hypothesis is relaxed to allow σ and ρ to be functions as is the case in Theorem 6.1?

QUESTION 9.7. (See Corollary 6.5.) Although there are some estimates involving $|\nabla S|$ in the literature, there do not seem to be any that are as simple as seen in Theorem 1.11 for $|\nabla A|$. Are there converses for any of the statements in Corollary 6.5?

QUESTION 9.8. (See Theorem 6.15.) Are there any Ricci-parallel hypersurfaces in $\mathbb{C}P^2$ or $\mathbb{C}H^2$? As we have seen, these could not be Hopf hypersurfaces.

QUESTION 9.9. Is Theorem 6.17 true for $c < 0$? (This theorem is concerned with the condition $\nabla_W S = 0$).

QUESTION 9.10. Many results have been proved for $n \geq 3$ but questions remain concerning the case $n = 2$. For example, Theorems 5.5, 6.18, 6.19, 6.20, 6.21, 6.23, and 6.30 can be considered from this point of view.

QUESTION 9.11. The question of classifying homogeneous real hypersurfaces in $\mathbb{C}H^n$ remains an outstanding open question. See [Berndt 1990].

Acknowledgements

We thank J. Berndt, A. Romero, and L. Verstraelen for helpful suggestions.

References

[Ahn et al. 1993] S.-S. Ahn, S.-B. Lee, and Y. J. Suh, "On ruled real hypersurfaces in a complex space form", *Tsukuba J. Math.* **17**:2 (1993), 311–322.

[Aiyama et al. 1990] R. Aiyama, H. Nakagawa, and Y. J. Suh, "A characterization of real hypersurfaces of type C, D and E of a complex projective space", *J. Korean Math. Soc.* **27**:1 (1990), 47–67.

[Bejancu and Deshmukh 1996] A. Bejancu and S. Deshmukh, "Real hypersurfaces of $\mathbb{C}P^n$ with non-negative Ricci curvature", *Proc. Amer. Math. Soc.* **124**:1 (1996), 269–274.

[Bejancu and Duggal 1993] A. Bejancu and K. L. Duggal, "Real hypersurfaces of indefinite Kaehler manifolds", *Internat. J. Math. and Math. Sci.* **16** (1993), 545–556.

[Berndt 1989a] J. Berndt, "Real hypersurfaces with constant principal curvatures in complex hyperbolic space", *J. Reine Angew. Math.* **395** (1989), 132–141.

[Berndt 1989b] J. Berndt, *Über untermannigfaltigkeiten von komplexen Raumformen*, Dissertation, Universität zu Köln, 1989.

[Berndt 1990] J. Berndt, "Real hypersurfaces with constant principal curvatures in complex space forms", pp. 10–19 in *Geometry and topology of submanifolds, II* (Avignon, 1988), edited by M. Boyom et al., World Sci. Publishing, Teaneck, NJ, 1990.

[Berndt 1991] J. Berndt, "Real hypersurfaces in quaternionic space forms", *J. Reine Angew. Math.* **419** (1991), 9–26.

[Berndt and Vanhecke 1992] J. Berndt and L. Vanhecke, "Curvature-adapted submanifolds", *Nihonkai Math. J.* **3**:2 (1992), 177–185.

[Berndt and Vanhecke 1993] J. Berndt and L. Vanhecke, "Naturally reductive Riemannian homogeneous spaces and real hypersurfaces in complex and quaternionic space forms", pp. 353–364 in *Differential geometry and its applications* (Opava, 1992), edited by O. Kowalski and D. Krupka, Math. Publ. 1, Silesian Univ. Opava, Opava, 1993.

[Berndt et al. 1995] J. Berndt, J. Bolton, and L. M. Woodward, "Almost complex curves and Hopf hypersurfaces in the nearly Kähler 6-sphere", *Geom. Dedicata* **56**:3 (1995), 237–247.

[Bishop and Goldberg 1972] R. L. Bishop and S. I. Goldberg, "On conformally flat spaces with commuting curvature and Ricci transformations", *Canad. J. Math.* **24** (1972), 799–804.

[Blair 1976] D. E. Blair, *Contact manifolds in Riemannian geometry*, Lecture Notes in Math. **509**, Springer, Berlin, 1976.

[Bolton 1973] J. Bolton, "Transnormal systems", *Quart. J. Math. Oxford* (2) **24** (1973), 385–395.

[Böning 1992] R. Böning, "Curvature surfaces of real hypersurfaces in a complex space form", pp. 43–50 in *Geometry and topology of submanifolds, IV* (Leuven, 1991), edited by F. Dillen and L. Verstraelen, World Scientific, River Edge, NJ, 1992.

[Böning 1995] R. Böning, "Curvature surfaces of Hopf hypersurfaces in complex space forms", *Manuscripta Math.* **87**:4 (1995), 449–458.

[Carter and West 1985] S. Carter and A. West, "Isoparametric systems and transnormality", *Proc. London Math. Soc.* (3) **51**:3 (1985), 520–542.

[Cecil and Ryan 1982] T. E. Cecil and P. J. Ryan, "Focal sets and real hypersurfaces in complex projective space", *Trans. Amer. Math. Soc.* **269** (1982), 481–499.

[Cecil and Ryan 1985] T. E. Cecil and P. J. Ryan, *Tight and taut immersions of manifolds*, Research Notes in Mathematics **107**, Pitman, Boston, 1985.

[Chen 1996] B.-Y. Chen, "A general inequality for submanifolds in complex-space-forms and its applications", *Arch. Math.* (*Basel*) **67**:6 (1996), 519–528.

[Chen and Vanhecke 1981] B.-Y. Chen and L. Vanhecke, "Differential geometry of geodesic spheres", *J. Reine Angew. Math.* **325** (1981), 28–67.

[Chen et al. 1984] B.-Y. Chen, G. D. Ludden, and S. Montiel, "Real submanifolds of a Kaehlerian manifold", *Algebras, Groups, and Geometries* **1** (1984), 174–216.

[Cho et al. 1991] J.-K. Cho, S.-B. Lee, and N.-G. Kim, "Real hypersurfaces with infinitesimal invariant Ricci tensor of a complex projective space", *Honam Math. J.* **13**:1 (1991), 15–20.

[Choe 1995] Y.-W. Choe, "Characterizations of certain real hypersurfaces of a complex space form", *Nihonkai Math. J.* **6**:1 (1995), 97–114.

[D'Atri 1979] J. E. D'Atri, "Certain isoparametric families of hypersurfaces in symmetric spaces", *J. Differential Geom.* **14**:1 (1979), 21–40.

[Deshmukh and Al-Gwaiz 1992] S. Deshmukh and M. A. Al-Gwaiz, "Hypersurfaces of the two-dimensional complex projective space", *Nihonkai Math. J.* **3**:1 (1992), 1–7.

[Deszcz 1992] R. Deszcz, "On pseudosymmetric spaces", *Bull. Soc. Math. Belg. Sér. A* **44**:1 (1992), 1–34.

[Dong 1993] Y. X. Dong, "Real hypersurfaces of a quaternionic projective space", *Chinese Ann. Math. Ser. A* **14**:1 (1993), 75–80.

[Fornari et al. 1993] S. Fornari, K. Frensel, and J. Ripoll, "Hypersurfaces with constant mean curvature in the complex hyperbolic space", *Trans. Amer. Math. Soc.* **339**:2 (1993), 685–702. Errata in *Trans. Amer. Math. Soc.* **347**:8 (1995), 3177.

[Garay and Romero 1990] O. J. Garay and A. Romero, "An isometric embedding of the complex hyperbolic space in a pseudo-Euclidean space and its application to the study of real hypersurfaces", *Tsukuba J. Math.* **14**:2 (1990), 293–313.

[Gotoh 1994] T. Gotoh, "The nullity of compact minimal real hypersurfaces in a complex projective space", *Tokyo J. Math.* **17**:1 (1994), 201–209.

[Gray 1967] A. Gray, "Pseudo-Riemannian almost product manifolds and submersions", *J. Math. Mech.* **16** (1967), 715–737.

[Hamada 1993] T. Hamada, "On real hypersurfaces of a quaternionic projective space", *Saitama Math. J.* **11** (1993), 29–39.

[Hamada 1995] T. Hamada, "On real hypersurfaces of a complex projective space with η-recurrent second fundamental tensor", *Nihonkai Math. J.* **6**:2 (1995), 153–163.

[Ki 1988] U-H. Ki, "Cyclic-parallel real hypersurfaces of a complex space form", *Tsukuba J. Math.* **12**:1 (1988), 259–268.

[Ki 1989] U-H. Ki, "Real hypersurfaces with parallel Ricci tensor of a complex space form", *Tsukuba J. Math.* **13**:1 (1989), 73–81.

[Ki and Kim 1989] U-H. Ki and H.-J. Kim, "A note on real hypersurfaces of a complex space form", *Bull. Korean Math. Soc.* **26**:1 (1989), 69–74.

[Ki and Kim 1994] U-H. Ki and N.-G. Kim, "Ruled real hypersurfaces of a complex space form", *Acta Math. Sinica* (N.S.) **10**:4 (1994), 401–409.

[Ki and Nakagawa 1991] U-H. Ki and H. Nakagawa, "Hypersurfaces with harmonic Weyl tensor of a real space form", *J. Korean Math. Soc.* **28**:2 (1991), 229–244.

[Ki and Suh 1990] U-H. Ki and Y. J. Suh, "On real hypersurfaces of a complex space form", *Math. J. Okayama Univ.* **32** (1990), 207–221.

[Ki and Suh 1992] U-H. Ki and Y. J. Suh, "Characterizations of some real hypersurfaces in $P_n(\mathbb{C})$ in terms of Ricci tensor", *Nihonkai Math. J.* **3**:2 (1992), 133–162.

[Ki and Suh 1994] U-H. Ki and Y. J. Suh, "On a characterization of real hypersurfaces of type A in a complex space form", *Canad. Math. Bull.* **37**:2 (1994), 238–244.

[Ki and Suh 1995] U-H. Ki and Y. J. Suh, "Characterizations of some real hypersurfaces in a complex space form in terms of Lie derivative", *J. Korean Math. Soc.* **32**:2 (1995), 161–170.

[Ki and Suh 1996] U-H. Ki and Y. J. Suh, "Some characterizations of ruled real hypersurfaces in a complex space form", *J. Korean Math. Soc.* **33**:1 (1996), 101–119.

[Ki and Takagi 1992] U-H. Ki and R. Takagi, "Real hypersurfaces in $P_n(\mathbb{C})$ with constant principal curvatures", *Math. J. Okayama Univ.* **34** (1992), 233–240.

[Ki et al. 1989] U-H. Ki, H.-J. Kim, and H. Nakagawa, "Real hypersurfaces with η-harmonic Weyl tensor of a complex space form", *J. Korean Math. Soc.* **26**:2 (1989), 311–322.

[Ki et al. 1990a] U-H. Ki, S. Kim, and H. Nakagawa, "A characterization of a real hypersurface of type B", *Tsukuba J. Math.* **14**:1 (1990), 9–26.

[Ki et al. 1990b] U-H. Ki, H. Nakagawa, and Y. J. Suh, "Real hypersurfaces with harmonic Weyl tensor of a complex space form", *Hiroshima Math. J.* **20**:1 (1990), 93–102.

[Ki et al. 1991] U-H. Ki, S. J. Kim, and S.-B. Lee, "Some characterizations of a real hypersurface of type A", *Kyungpook Math. J.* **31**:1 (1991), 73–82.

[Ki et al. 1992] U-H. Ki, N.-G. Kim, and S.-B. Lee, "On certain real hypersurfaces of a complex space form", *J. Korean Math. Soc.* **29**:1 (1992), 63–77.

[Ki et al. 1994] U-H. Ki, S. Maeda, and Y. J. Suh, "Lie derivatives on homogeneous real hypersurfaces of type B in a complex projective space", *Yokohama Math. J.* **42**:2 (1994), 121–131.

[Ki et al. 1996] U-H. Ki, Y. S. Pyo, and Y. J. Suh, "Classification of real hypersurfaces in complex space forms in terms of curvature tensors", *Kyungpook Math. J.* **36**:1 (1996), 201–218.

[Ki et al. 1997] U-H. Ki, J. D. Pérez, and Y. J. Suh, "Real hypersurfaces of type A in quaternionic projective space", *Internat. J. Math. Sci.* **20**:1 (1997), 115–122.

[Kim 1988a] H.-J. Kim, "A note on real hypersurfaces of a complex hyperbolic space", *Tsukuba J. Math.* **12**:2 (1988), 451–457.

[Kim 1988b] S. J. Kim, "Semi-invariant submanifolds with parallel Ricci tensor of a complex hyperbolic space", *J. Korean Math. Soc.* **25**:2 (1988), 227–243.

[Kim and Pyo 1991] J. J. Kim and Y.-S. Pyo, "Real hypersurfaces with parallelly cyclic condition of a complex space form", *Bull. Korean Math. Soc.* **28**:1 (1991), 1–10.

[Kim et al. 1992a] N.-G. Kim, S.-S. Ahn, S.-B. Lee, and I.-T. Lim, "On characterizations of a real hypersurface of type A in a complex space form", *Honam Math. J.* **14**:1 (1992), 37–44.

[Kim et al. 1992b] N.-G. Kim, S.-B. Lee, and I.-T. Lim, "On certain real hypersurfaces of a complex space form", *Comm. Korean Math. Soc.* **7** (1992), 99–110.

[Kimura 1986a] M. Kimura, "Real hypersurfaces and complex submanifolds in complex projective space", *Trans. Amer. Math. Soc.* **296**:1 (1986), 137–149.

[Kimura 1986b] M. Kimura, "Real hypersurfaces of a complex projective space", *Bull. Austral. Math. Soc.* **33**:3 (1986), 383–387.

[Kimura 1987a] M. Kimura, "Sectional curvatures of holomorphic planes on a real hypersurface in $P^n(\mathbb{C})$", *Math. Ann.* **276**:3 (1987), 487–497.

[Kimura 1987b] M. Kimura, "Some real hypersurfaces of a complex projective space", *Saitama Math. J.* **5** (1987), 1–5. Correction in **10** (1992), 33–34.

[Kimura and Maeda 1989] M. Kimura and S. Maeda, "On real hypersurfaces of a complex projective space", *Math. Z.* **202**:3 (1989), 299–311.

[Kimura and Maeda 1991] M. Kimura and S. Maeda, "On real hypersurfaces of a complex projective space II", *Tsukuba J. Math.* **15**:2 (1991), 547–561.

[Kimura and Maeda 1992] M. Kimura and S. Maeda, "Characterizations of geodesic hyperspheres in a complex projective space in terms of Ricci tensors", *Yokohama Math. J.* **40**:1 (1992), 35–43.

[Kimura and Maeda 1993] M. Kimura and S. Maeda, "On real hypersurfaces of a complex projective space III", *Hokkaido Math. J.* **22**:1 (1993), 63–78.

[Kimura and Maeda 1995] M. Kimura and S. Maeda, "Lie derivatives on real hypersurfaces in a complex projective space", *Czechoslovak Math. J.* **45**:1 (1995), 135–148.

[Kobayashi and Nomizu 1969] S. Kobayashi and K. Nomizu, *Foundations of differential geometry II*, Wiley, New York, 1969.

[Kon 1979] M. Kon, "Pseudo-Einstein real hypersurfaces in complex space forms", *J. Diff. Geom.* **14** (1979), 339–354.

[Kon 1980] M. Kon, "Real minimal hypersurfaces in a complex projective space", *Proc. Amer. Math. Soc.* **79** (1980), 285–288.

[Kwon and Nakagawa 1988] J.-H. Kwon and H. Nakagawa, "Real hypersurfaces with cyclic-parallel Ricci tensor of a complex projective space", *Hokkaido Math. J.* **17**:3 (1988), 355–371.

[Kwon and Nakagawa 1989a] J.-H. Kwon and H. Nakagawa, "A note on real hypersurfaces of a complex projective space", *J. Austral. Math. Soc. Ser. A* **47**:1 (1989), 108–113.

[Kwon and Nakagawa 1989b] J.-H. Kwon and H. Nakagawa, "Real hypersurfaces with cyclic η-parallel Ricci tensor of a complex space form", *Yokohama Math. J.* **37**:1 (1989), 45–55.

[Lawson 1970] H. B. Lawson, Jr., "Rigidity theorems in rank-1 symmetric spaces", *J. Differential Geometry* **4** (1970), 349–357.

[Lee et al. 1990] S.-B. Lee, N.-G. Kim, and S.-S. Ahn, "Pseudo-Ryan real hypersurfaces of a complex space form", *Kyungpook Math. J.* **30**:2 (1990), 127–135.

[Li 1988] Z. Q. Li, "Isoparametric hypersurfaces in $\mathbb{CP}^n$ with constant principal curvatures", *Chinese Ann. Math. Ser. B* **9**:4 (1988), 485–493.

[Maeda 1976] Y. Maeda, "On real hypersurfaces of a complex projective space", *J. Math. Soc. Japan* **28**:3 (1976), 529–540.

[Maeda 1983] S. Maeda, "Real hypersurfaces of complex projective spaces", *Math. Ann.* **263**:4 (1983), 473–478.

[Maeda 1984] S. Maeda, "Real hypersurfaces of a complex projective space II", *Bull. Austral. Math. Soc.* **30**:1 (1984), 123–127.

[Maeda 1991] S. Maeda, "Second fundamental form of a real hypersurface in a complex projective space", pp. 139–153 in *Nonassociative algebras and related topics* (Hiroshima, 1990), edited by K. Yamaguti and N. Kawamoto, World Sci. Publishing, River Edge, NJ, 1991.

[Maeda 1993] S. Maeda, "Geometry of submanifolds which are neither Kaehler nor totally real in a complex projective space", *Bull. Nagoya Inst. Tech.* **45** (1993), 1–50.

[Maeda 1994] S. Maeda, "Ricci tensors of real hypersurfaces in a complex projective space", *Proc. Amer. Math. Soc.* **122**:4 (1994), 1229–1235.

[Maeda and Udagawa 1990] S. Maeda and S. Udagawa, "Real hypersurfaces of a complex projective space in terms of holomorphic distribution", *Tsukuba J. Math.* **14**:1 (1990), 39–52.

[Martínez 1988] A. Martínez, "Ruled real hypersurfaces in quaternionic projective space", *An. Ştiinţ. Univ. "Al. I. Cuza" Iaşi Secţ. I a Mat.* (N.S.) **34**:1 (1988), 73–78.

[Martínez and Pérez 1986] A. Martínez and J. D. Pérez, "Real hypersurfaces in quaternionic projective space", *Ann. Mat. Pura Appl.* (4) **145** (1986), 355–384.

[Martínez and Ros 1984] A. Martínez and A. Ros, "On real hypersurfaces of finite type of $\mathbb{CP}^m$", *Kodai Math. J.* **7**:3 (1984), 304–316.

[Miquel 1994] V. Miquel, "Compact Hopf hypersurfaces of constant mean curvature in complex space forms", *Ann. Global Anal. Geom.* **12**:3 (1994), 211–218.

[Montiel 1985] S. Montiel, "Real hypersurfaces of a complex hyperbolic space", *J. Math. Soc. Japan* **37**:3 (1985), 515–535.

[Montiel and Romero 1986] S. Montiel and A. Romero, "On some real hypersurfaces of a complex hyperbolic space", *Geom. Dedicata* **20**:2 (1986), 245–261.

[Nagai 1995] S. Nagai, "Naturally reductive Riemannian homogeneous structure on a homogeneous real hypersurface in a complex space form", *Boll. Un. Mat. Ital. A* (7) **9**:2 (1995), 391–400.

[Niebergall and Ryan 1996] R. Niebergall and P. J. Ryan, "Semi-parallel and semi-symmetric real hypersurfaces in complex space forms", preprint, McMaster University, 1996.

[Okumura 1975] M. Okumura, "On some real hypersurfaces of a complex projective space", *Trans. Amer. Math. Soc.* **212** (1975), 355–364.

[Okumura 1978] M. Okumura, "Compact real hypersurface with constant mean curvature of a complex projective space", *J. Diff. Geom.* **13** (1978), 43–50.

[O'Neill 1966] B. O'Neill, "The fundamental equations of a submersion", *Michigan Math. J.* **13** (1966), 459–469.

[Oproiu 1990] V. Oproiu, "The C-projective curvature of some real hypersurfaces in complex space forms", *An. Ştiinţ. Univ. "Al. I. Cuza" Iaşi Secţ. I a Mat.* (N.S.) **36**:4 (1990), 385–391.

[Pak 1977] J. S. Pak, "Real hypersurfaces in quaternionic Kaehlerian manifolds with constant Q-sectional curvature", *Kōdai Math. Sem. Rep.* **29**:1-2 (1977), 22–61.

[Park 1989] K. S. Park, "Isoparametric families on projective spaces", *Math. Ann.* **284**:3 (1989), 503–513.

[Pérez 1991] J. D. Pérez, "A characterization of real hypersurfaces of quaternionic projective space", *Tsukuba J. Math.* **15**:2 (1991), 315–323.

[Pérez 1992] J. D. Pérez, "Some conditions on real hypersurfaces of quaternionic projective space", *An. Ştiinţ. Univ. "Al. I. Cuza" Iaşi Secţ. Ia Mat.* **38**:1 (1992), 103–110.

[Pérez 1993a] J. D. Pérez, "Cyclic-parallel real hypersurfaces of quaternionic projective space", *Tsukuba J. Math.* **17**:1 (1993), 189–191.

[Pérez 1993b] J. D. Pérez, "On certain real hypersurfaces of quaternionic projective space II", *Algebras Groups Geom.* **10**:1 (1993), 13–24.

[Pérez 1994] J. D. Pérez, "Real hypersurfaces of quaternionic projective space satisfying $\nabla_{U_i} A = 0$", *J. Geom.* **49**:1-2 (1994), 166–177.

[Pérez 1996a] J. D. Pérez, "A characterization of almost-Einstein real hypersurfaces of quaternionic projective space", preprint, Universidad de Granada, 1996.

[Pérez 1996b] J. D. Pérez, "On the Ricci tensor of real hypersurfaces of quaternionic projective space", *Internat. J. Math. Math. Sci.* **19**:1 (1996), 193–197.

[Pérez and Santos 1985]   J. D. Pérez and F. G. Santos, "On pseudo-Einstein real hypersurfaces of the quaternionic projective space", *Kyungpook Math. J.* **25**:1 (1985), 15–28. Erratum in **25**:2 (1985), 185.

[Pérez and Santos 1991]   J. D. Pérez and F. G. Santos, "On real hypersurfaces with harmonic curvature of a quaternionic projective space", *J. Geom.* **40**:1-2 (1991), 165–169.

[Pérez and Santos 1993]   J. D. Pérez and F. G. Santos, "Cyclic-parallel Ricci tensor on real hypersurfaces of a quaternionic projective space", *Rev. Roumaine Math. Pures Appl.* **38**:1 (1993), 37–40.

[Pérez and Suh 1996a]   J. D. Pérez and Y. J. Suh, "On real hypersurfaces in quaternionic projective space with $D^\perp$-parallel second fundamental form", *Nihonkai Math. J.* **7**:2 (1996), 185–195.

[Pérez and Suh 1996b]   J. D. Pérez and Y. J. Suh, "Real hypersurfaces of quaternionic projective space satisfying $\nabla_{U_i} R = 0$", preprint, Kyungpook National University, 1996.

[Pyo 1994a]   Y.-S. Pyo, "On real hypersurfaces of type A in a complex space form I", *Tsukuba J. Math.* **18**:2 (1994), 483–492.

[Pyo 1994b]   Y.-S. Pyo, "On real hypersurfaces of type A in a complex space form II", *Comm. Korean Math. Soc.* **9** (1994), 369–383.

[Pyo 1994c]   Y.-S. Pyo, "On ruled real hypersurfaces in a complex space form", *Math. J. Toyama Univ.* **17** (1994), 55–71.

[Pyo and Suh 1995]   Y.-S. Pyo and Y. J. Suh, "Characterizations of real hypersurfaces in complex space forms in terms of curvature tensors", *Tsukuba J. Math.* **19**:1 (1995), 163–172.

[Ryan 1971]   P. J. Ryan, "Hypersurfaces with parallel Ricci tensor", *Osaka J. Math.* **8** (1971), 251–259.

[Ryan 1972]   P. J. Ryan, "A class of complex hypersurfaces", *Colloq. Math.* **26** (1972), 175–182, 385.

[Shen 1985]   Y. B. Shen, "On real minimal hypersurfaces in a complex projective space", *Acta Math. Sinica* **28**:1 (1985), 85–90.

[Suh 1990]   Y. J. Suh, "On real hypersurfaces of a complex space form with η-parallel Ricci tensor", *Tsukuba J. Math.* **14**:1 (1990), 27–37.

[Suh 1991]   Y. J. Suh, "On type number of real hypersurfaces in $P^n(\mathbb{C})$", *Tsukuba J. Math.* **15**:1 (1991), 99–104.

[Suh 1992]   Y. J. Suh, "A characterization of ruled real hypersurfaces in $P_n(\mathbb{C})$", *J. Korean Math. Soc.* **29**:2 (1992), 351–359.

[Suh 1995]   Y. J. Suh, "Characterizations of real hypersurfaces in complex space forms in terms of Weingarten map", *Nihonkai Math. J.* **6**:1 (1995), 63–79.

[Suh and Takagi 1991]   Y. J. Suh and R. Takagi, "A rigidity for real hypersurfaces in a complex projective space", *Tôhoku Math. J.* (2) **43**:4 (1991), 501–507.

[Takagi 1973]   R. Takagi, "On homogeneous real hypersurfaces in a complex projective space", *Osaka J. Math.* **10** (1973), 495–506.

[Takagi 1975a]   R. Takagi, "Real hypersurfaces in a complex projective space with constant principal curvatures", *J. Math. Soc. Japan* **27** (1975), 43–53.

[Takagi 1975b] R. Takagi, "Real hypersurfaces in a complex projective space with constant principal curvatures II", *J. Math. Soc. Japan* **27**:4 (1975), 507–516.

[Takagi 1990] R. Takagi, "Real hypersurfaces in a complex projective space", *Proc. Topology and Geometry Research Center (Taegu)* **1** (1990), 1–17.

[Takagi and Takahashi 1972] R. Takagi and T. Takahashi, "On the principal curvatures of homogeneous hypersurfaces in a sphere", pp. 469–481 in *Differential geometry (in honor of Kentaro Yano)*, edited by S. Kobayashi et al., Kinokuniya, Tokyo, 1972.

[Taniguchi 1994] T. Taniguchi, "Characterizations of real hypersurfaces of a complex hyperbolic space in terms of Ricci tensor and holomorphic distribution", *Tsukuba J. Math.* **18**:2 (1994), 469–482.

[Tashiro and Tachibana 1963] Y. Tashiro and S. Tachibana, "On Fubinian and *C*-Fubinian manifolds", *Kōdai Math. Sem. Rep.* **15** (1963), 176–183.

[Udagawa 1987] S. Udagawa, "Bi-order real hypersurfaces in a complex projective space", *Kodai Math. J.* **10**:2 (1987), 182–196.

[Vernon 1987] M. H. Vernon, "Contact hypersurfaces of a complex hyperbolic space", *Tôhoku Math. J.* (2) **39**:2 (1987), 215–222.

[Vernon 1989] M. H. Vernon, "Some families of isoparametric hypersurfaces and rigidity in a complex hyperbolic space", *Trans. Amer. Math. Soc.* **312**:1 (1989), 237–256.

[Vernon 1991] M. H. Vernon, "Semi-symmetric hypersurfaces of anti-de Sitter spacetime that are S^1-invariant", *Tensor* (N.S.) **50**:2 (1991), 99–105.

[Wang 1982] Q. M. Wang, "Isoparametric hypersurfaces in complex projective spaces", pp. 1509–1523 in *Symposium on Differential Geometry and Differential Equations* (Beijing, 1980), vol. 3, Science Press, Beijing, 1982.

[Wang 1983] Q. M. Wang, "Real hypersurfaces with constant principal curvatures in complex projective spaces I", *Sci. Sinica Ser. A* **26**:10 (1983), 1017–1024.

[Wang 1987] Q. M. Wang, "Isoparametric functions on Riemannian manifolds I", *Math. Ann.* **277**:4 (1987), 639–646.

[Xu 1992] Z. C. Xu, "A geometric characteristic of real hypersurface in $\mathbb{C}H^n$ with constant principal curvatures", *Northeast. Math. J.* **8**:3 (1992), 283–286.

Ross Niebergall
Department of Mathematics and Computer Science
University of Northern British Columbia
Prince George, BC, Canada V2L 5P2
rossn@unbc.edu

Patrick J. Ryan
Department of Mathematics and Statistics
McMaster University
Hamilton, Ontario, Canada L8S 4K1
pjr@maccs.dcss.mcmaster.ca

Bibliography on Tight, Taut
and Isoparametric Submanifolds

COMPILED BY

WOLFGANG KÜHNEL

AND

THOMAS E. CECIL

Note. An author index for this bibliography can be found starting starting on page 337, and can be consulted by the reader who wishes to find all listed papers having a given person as a coauthor.

UWE ABRESCH

[1982] *Notwendige Bedingungen für isoparametrische Hyperflächen in Sphären mit mehr als drei verschiedenen Hauptkrümmungen*, Bonner Math. Schriften **146**, Mathematisches Institut Universität Bonn, Bonn, 1982.

[1983] "Isoparametric hypersurfaces with four or six distinct principal curvatures. Necessary conditions on the multiplicities", *Math. Ann.* **264**:3 (1983), 283–302.

REIKO AIYAMA, HISAO NAKAGAWA, AND YOUNG JIN SUH

[1990] "A characterization of real hypersurfaces of type C, D and E of a complex projective space", *J. Korean Math. Soc.* **27**:1 (1990), 47–67.

A. D. ALEKSANDROV

[1938] "On a class of closed surfaces", *Mat. Sbornik* **4** (1938), 69–77. In Russian.

CARL B. ALLENDOERFER AND ANDRÉ WEIL

[1943] "The Gauss-Bonnet theorem for Riemannian polyhedra", *Trans. Amer. Math. Soc.* **53** (1943), 101–129.

Y. A. AMINOV

[1984] "Embedding problems: Geometric and topological aspects", *J. Sov. Math.* **25** (1984), 1308–1331.

THOMAS F. BANCHOFF

[1965] "Tightly embedded 2-dimensional polyhedral manifolds", *Amer. J. Math.* **87** (1965), 462–472.

[1967] "Critical points and curvature for embedded polyhedra", *J. Diff. Geom.* **1** (1967), 245–256.

[1970a] "Critical points and curvature for embedded polyhedral surfaces", *Amer. Math. Monthly* **77** (1970), 475–485.

[1970b] "Non-rigidity theorems for tight polyhedra", *Arch. Math. (Basel)* **21** (1970), 416–423.

[1970c] "The spherical two-piece property and tight surfaces in spheres", *J. Differential Geometry* **4** (1970), 193–205.

[1970d] "Total central curvature of curves", *Duke Math. J.* **37** (1970), 281–289.

[1971a] "High codimensional 0-tight maps on spheres", *Proc. Amer. Math. Soc.* **29** (1971), 133–137.

[1971b] "The two-piece property and tight n-manifolds-with-boundary in E^n", *Trans. Amer. Math. Soc.* **161** (1971), 259–267.

[1974] "Tight polyhedral Klein bottles, projective planes, and Möbius bands", *Math. Ann.* **207** (1974), 233–243.

[1982] "Global geometry of polygons, III: Frenet frames and theorems of Jacobi and Milnor for space polygons", *Rad Jugoslav. Akad. Znan. Umjet.* **396** (1982), 101–108.

[1983] "Critical points and curvature for embedded polyhedra II", pp. 34–55 in *Differential geometry* (College Park, MD, 1981/1982), Progr. Math. **32**, Birkhäuser, Boston, 1983.

[1986] "Computer graphics applications in geometry: 'because the light is better over here'", pp. 1–14 in *The merging of disciplines: new directions in pure, applied, and computational mathematics* (Laramie, WY, 1985), Springer, New York, 1986.

THOMAS F. BANCHOFF AND WOLFGANG KÜHNEL

[1992] "Equilibrium triangulations of the complex projective plane", *Geom. Dedicata* **44**:3 (1992), 313–333.

[1997] "Tight submanifolds, smooth and polyhedral", pp. 51–118 in *Tight and Taut Submanifolds*, edited by T. E. Cecil and S.-s. Chern, Cambridge U. Press, 1997.

THOMAS F. BANCHOFF AND NICOLAAS H. KUIPER

[1981] "Geometrical class and degree for surfaces in three-space", *J. Diff. Geom.* **16** (1981), 559–576.

THOMAS F. BANCHOFF AND FLORIS TAKENS

[1975] "Height functions on surfaces with three critical points", *Illinois J. Math.* **19** (1975), 325–335.

M. A. BELTAGY

[1986] "Two-piece property in manifolds without focal points", *Indian J. Pure Appl. Math.* **17**:7 (1986), 883–889.

RICCARDO BENEDETTI AND MARIA DEDÒ

[1987] "A geometric inequality for the total curvature of plane curves", *Geom. Dedicata* **22**:1 (1987), 105–115.

JÜRGEN BERNDT

[1989] "Real hypersurfaces with constant principal curvatures in complex hyperbolic space", *J. Reine Angew. Math.* **395** (1989), 132–141.

[1990] "Real hypersurfaces with constant principal curvatures in complex space forms", pp. 10–19 in *Geometry and topology of submanifolds, II* (Avignon, 1988), edited by M. Boyom et al., World Sci. Publishing, Teaneck, NJ, 1990.

[1991] "Real hypersurfaces in quaternionic space forms", *J. Reine Angew. Math.* **419** (1991), 9–26.

JÜRGEN BERNDT, JOHN BOLTON, AND LYNDON M. WOODWARD

[1995] "Almost complex curves and Hopf hypersurfaces in the nearly Kähler 6-sphere", *Geom. Dedicata* **56**:3 (1995), 237–247.

JÜRGEN BERNDT AND LIEVEN VANHECKE

[1992] "Curvature-adapted submanifolds", *Nihonkai Math. J.* **3**:2 (1992), 177–185.

[1993] "Naturally reductive Riemannian homogeneous spaces and real hypersurfaces in complex and quaternionic space forms", pp. 353–364 in *Differential geometry and its applications* (Opava, 1992), edited by O. Kowalski and D. Krupka, Math. Publ. **1**, Silesian Univ. Opava, Opava, 1993.

GRACIELA SILVIA BIRMAN

[1993] "Total absolute curvature of curves in Lorentz manifolds", *Rev. Colombiana Mat.* **27** (1993), 9–13. Proceedings of the Second Latin American Colloquium on Analysis (Spanish) (Santafé de Bogotá, 1992).

CAROL BLOMSTROM

[1985] "Symmetric immersions in pseudo-Riemannian space forms", pp. 30–45 in *Global differential geometry and global analysis* (Berlin, 1984), edited by D. Ferus et al., Lecture Notes in Math. **1156**, Springer, Berlin, 1985.

NICOLAE BOJA

[1994] "The total absolute curvature for immersed *p*-adic compact submanifolds", *An. Univ. Timişoara Ser. Mat.-Inform.* **32**:1 (1994), 9–20.

JOHN BOLTON

[1973] "Transnormal systems", *Quart. J. Math. Oxford* (2) **24** (1973), 385–395.

[1982] "Tight immersions into manifolds without conjugate points", *Quart. J. Math. Oxford* (2) **33** (1982), 159–167.

ROLF BÖNING

[1992] "Curvature surfaces of real hypersurfaces in a complex space form", pp. 43–50 in *Geometry and topology of submanifolds, IV* (Leuven, 1991), edited by F. Dillen and L. Verstraelen, World Scientific, River Edge, NJ, 1992.

[1995] "Curvature surfaces of Hopf hypersurfaces in complex space forms", *Manuscripta Math.* **87**:4 (1995), 449–458.

KAROL BORSUK

[1947] "Sur la courbure totale des courbes fermées", *Ann. Polon. Math.* **20** (1947), 251–265.

RAOUL BOTT

[1956] "An application of Morse theory to symmetric spaces", *Bull. Soc. Math. France* **4** (1956), 251–281.

RAOUL BOTT AND HANS SAMELSON

[1958] "Applications of the theory of Morse to symmetric spaces", *Amer. J. Math.* **80** (1958), 964–1029. Corrections in **83** (1961), pp. 207–208.

DIETRICH BRAESS

[1974] "Morse-Theorie für berandete Mannigfaltigkeiten", *Math. Ann.* **208** (1974), 133–148.

ULRICH BREHM AND WOLFGANG KÜHNEL

[1981] "Smooth approximation of polyhedral surfaces with respect to curvature measures", pp. 64–68 in *Global Differential Geometry and Global Analysis* (Berlin, 1979), edited by D. Ferus et al., Lecture Notes in Math. **838**, Springer, Berlin, 1981.

[1982] "Smooth approximation of polyhedral surfaces regarding curvatures", *Geom. Dedicata* **12** (1982), 61–85.

[1992] "15-vertex triangulations of an 8-manifold", *Math. Ann.* **294**:1 (1992), 167–193.

PETER BREUER AND WOLFGANG KÜHNEL

[1997] "The tightness of tubes", 1997. To appear in *Forum Math.*

FREDERICK BRICKELL AND CHUAN-CHIH HSIUNG

[1974] "The total absolute curvature of closed curves in Riemannian manifolds", *J. Differential Geometry* **9** (1974), 177–193.

STEVEN G. BUYSKE

[1989] "Lie sphere transformations and the focal sets of certain taut immersions", *Trans. Amer. Math. Soc.* **311**:1 (1989), 117–133.

ÉLIE CARTAN

[1938] "Familles de surfaces isoparamétriques dans les espaces à courbure constante", *Annali di Mat.* **17** (1938), 177–191.

[1939a] "Sur des familles remarquables d'hypersurfaces isoparamétriques dans les espaces sphériques", *Math. Z.* **45** (1939), 335–367.

[1939b] "Sur quelques familles remarquables d'hypersurfaces", pp. 30–41 in *Comptes rendus du congrès des sciences mathématiques de Liège*, Georges Thone, Liège, 1939.

[1940] "Sur des familles d'hypersurfaces isoparamétriques des espaces sphériques à 5 et à 9 dimensions", *Revista Univ. Tucumán Ser. A* **1** (1940), 5–22.

SHEILA CARTER

[1973] "The focal set of a hypersurface with abelian fundamental group", *Topology* **12** (1973), 1–4.

[1991] "Focal sets and covering spaces", pp. 81–89 in *Geometry and topology of submanifolds, III* (Leeds, 1990), edited by L. Verstraelen and A. West, World Sci. Publishing, River Edge, NJ, 1991.

[1993] "Immersions parallel to a given immersion", pp. 68–76 in *Geometry and topology of submanifolds, V* (Leuven/Brussels, 1992), edited by F. Dillen et al., World Sci. Publishing, River Edge, NJ, 1993.

SHEILA CARTER AND RIDVAN EZENTAS

[1992] "Immersions with totally reducible focal set", *J. Geom.* **45**:1-2 (1992), 1–7.

[1994] "Embeddings of nonorientable surfaces with totally reducible focal set", *Glasgow Math. J.* **36**:1 (1994), 11–16.

SHEILA CARTER, NADIR G. MANSOUR, AND ALAN WEST

[1982] "Cylindrically taut immersions", *Math. Ann.* **261** (1982), 133–139.

SHEILA CARTER AND STEWART A. ROBERTSON

[1967] "Relations between a manifold and its focal set", *Invent. Math.* **3** (1967), 300–307.

SHEILA CARTER AND ZERRIN ŞENTÜRK

[1994a] "On the index of parallel immersions", pp. 43–49 in *Geometry and topology of submanifolds, VI* (Leuven/Brussels, 1993), edited by F. Dillen et al., World Sci. Publishing, River Edge, NJ, 1994.

[1994b] "The space of immersions parallel to a given immersion", *J. London Math. Soc.* (2) **50**:2 (1994), 404–416.

SHEILA CARTER AND ALAN WEST

[1972] "Tight and taut immersions", *Proc. London Math. Soc.* (3) **25** (1972), 701–720.

[1978] "Totally focal embeddings", *J. Diff. Geom.* **13** (1978), 251–261.

[1981] "Totally focal embeddings: special cases", *J. Diff. Geom.* **16** (1981), 685–697.

[1982] "A characterisation of isoparametric hypersurfaces in spheres", *J. London Math. Soc.* (2) **26** (1982), 183–192.

[1985a] "Convexity and cylindrical two-piece properties", *Illinois J. Math.* **29**:1 (1985), 39–50.

[1985b] "Generalised Cartan polynomials", *J. London Math. Soc.* (2) **32**:2 (1985), 305–316.

[1985c] "Isoparametric systems and transnormality", *Proc. London Math. Soc.* (3) **51**:3 (1985), 520–542.

[1990] "Isoparametric and totally focal submanifolds", *Proc. London Math. Soc.* (3) **60**:3 (1990), 609–624.

[1995] "Partial tubes about immersed manifolds", *Geom. Dedicata* **54**:2 (1995), 145–169.

MARIO CASELLA AND WOLFGANG KÜHNEL

[1996]   "A triangulated K3 surface with the minimum number of vertices", preprint,
1996.

ARTHUR CAYLEY

[1873]   "On the cyclide", *Quart. J. of Pure and Appl. Math.* **12** (1873), 148–165.
Reprinted as pp. 64–78 in *Collected mathematical papers*, vol. 9, Cambridge U. Press,
1896.

THOMAS E. CECIL

[1974a]   "A characterization of metric spheres in hyperbolic space by Morse theory",
Tôhoku Math. J. (2) **26** (1974), 341–351.

[1974b]   "Geometric applications of critical point theory to submanifolds of complex
projective space", *Nagoya Math. J.* **55** (1974), 5–31.

[1975]   "Geometric applications of critical point theory to submanifolds of complex
projective space and hyperbolic space", pp. 115–117 in *Differential geometry*
(Stanford, 1973), edited by S. S. Chern and R. Osserman, Proceedings of Symposia
in Pure Mathematics **27** (part 1), Amer. Math. Soc., Providence, 1975.

[1976]   "Taut immersions of noncompact surfaces into a Euclidean 3-space", *J.
Differential Geometry* **11**:3 (1976), 451–459.

[1989]   "Reducible Dupin submanifolds", *Geom. Dedicata* **32**:3 (1989), 281–300.

[1990]   "On the Lie curvature of Dupin hypersurfaces", *Kodai Math. J.* **13**:1 (1990),
143–153.

[1991]   "Lie sphere geometry and Dupin submanifolds", pp. 90–107 in *Geometry and
topology of submanifolds, III* (Leeds, 1990), edited by L. Verstraelen and A. West,
World Sci. Publishing, River Edge, NJ, 1991.

[1992]   *Lie sphere geometry, with applications to submanifolds*, Universitext, Springer,
New York, 1992.

[1993]   "Dupin submanifolds", pp. 77–102 in *Geometry and topology of submanifolds,
V* (Leuven/Brussels, 1992), edited by F. Dillen et al., World Sci. Publishing, River
Edge, NJ, 1993.

[1997]   "Taut and Dupin submanifolds", pp. 135–180 in *Tight and Taut Submanifolds*,
edited by T. E. Cecil and S.-s. Chern, Cambridge U. Press, 1997.

THOMAS E. CECIL AND SHIING-SHEN CHERN

[1987]   "Tautness and Lie sphere geometry", *Math. Ann.* **278**:1-4 (1987), 381–399.

[1989]   "Dupin submanifolds in Lie sphere geometry", pp. 1–48 in *Differential geometry
and topology* (Tianjin, 1986–87), edited by B. Jiang et al., Lecture Notes in Math.
1369, Springer, Berlin, 1989.

THOMAS E. CECIL AND GARY R. JENSEN

[1995]   "Dupin hypersurfaces", pp. 100–105 in *Geometry and topology of submanifolds,
VII* (Leuven/Brussels, 1994), edited by F. Dillen et al., World Scientific, River Edge,
NJ, 1995.

[1997]   "Dupin hypersurfaces with three principal curvatures", 1997. To appear in
Invent. Math.

THOMAS E. CECIL AND PATRICK J. RYAN

[1978a] "Focal sets of submanifolds", *Pacific J. Math.* **78** (1978), 27–39.

[1978b] "Focal sets, taut embeddings and the cyclides of Dupin", *Math. Ann.* **236** (1978), 177–190.

[1979a] "Distance functions and umbilic submanifolds of hyperbolic space", *Nagoya Math. J.* **74** (1979), 67–75.

[1979b] "Tight and taut immersions into hyperbolic space", *J. London Math. Soc.* (2) **19** (1979), 561–572.

[1980] "Conformal geometry and the cyclides of Dupin", *Can. J. Math.* **32** (1980), 767–782.

[1981] "Tight spherical embeddings", pp. 94–104 in *Global Differential Geometry and Global Analysis* (Berlin, 1979), edited by D. Ferus et al., Lecture Notes in Math. **838**, Springer, Berlin, 1981.

[1982] "Focal sets and real hypersurfaces in complex projective space", *Trans. Amer. Math. Soc.* **269** (1982), 481–499.

[1984] "On the number of top-cycles of a tight surface in 3-space", *J. London Math. Soc.* (2) **30**:2 (1984), 335–341.

[1985] *Tight and taut immersions of manifolds*, Research Notes in Mathematics **107**, Pitman, Boston, 1985.

DAVIDE P. CERVONE

[1994] "A tight polyhedral immersion of the real projective plane with one handle", 1994. Available at http://www.geom.umn.edu/locate/rp2-handle.

[1996] "Tight immersions of simplicial surfaces in three space", *Topology* **35**:4 (1996), 863–873.

[1997] "On tight immersions of the real projective plane with one handle", pp. 119–133 in *Tight and Taut Submanifolds*, edited by T. E. Cecil and S.-s. Chern, Cambridge U. Press, 1997.

BANG-YEN CHEN

[1970a] "On the total absolute curvature of manifolds immersed in riemannian manifold II", *Kōdai Math. Sem. Rep.* **22** (1970), 89–97.

[1970b] "On the total absolute curvature of manifolds immersed in Riemannian manifolds III", *Kōdai Math. Sem. Rep.* **22** (1970), 385–400.

CHANG SHING CHEN

[1972] "On tight isometric immersion of codimension two", *Amer. J. Math.* **94** (1972), 974–990.

[1973] "More on tight isometric immersions of codimension two", *Proc. Amer. Math. Soc.* **40** (1973), 545–553.

[1979] "Tight embedding and projective transformation", *Amer. J. Math.* **101** (1979), 1083–1102.

CHANG SHING CHEN AND WILLIAM F. POHL

[1974] "On classification of tight surfaces in a Euclidean 4-space", preprint, Univ. Minnesota, 1974.

SHIING-SHEN CHERN

[1991] "An introduction to Dupin submanifolds", pp. 95–102 in *Differential geometry: A symposium in honour of Manfredo do Carmo* (Rio de Janeiro, 1988), edited by H. B. Lawson and K. Tenenblat, Pitman Monographs Surveys Pure Appl. Math. **52**, Longman Sci. Tech., Harlow, 1991.

SHIING-SHEN CHERN AND RICHARD K. LASHOF

[1957] "On the total curvature of immersed manifolds", *Amer. J. Math.* **79** (1957), 306–318.

[1958] "On the total curvature of immersed manifolds II", *Michigan Math. J.* **5** (1958), 5–12.

QUO-SHIN CHI, GARY R. JENSEN, AND RUI JIA LIAO

[1995] "Isoparametric functions and flat minimal tori in $\mathbb{CP}^2$", *Proc. Amer. Math. Soc.* **123**:9 (1995), 2849–2854.

LESLIE COGHLAN

[1987] "Tight stable surfaces I", *Proc. Roy. Soc. Edinburgh* Sect. A **107**:3-4 (1987), 213–232.

[1989a] "Some general constructions of tight mappings", *Proc. Roy. Soc. Edinburgh* Sect. A **111**:3-4 (1989), 231–247.

[1989b] "Tight stable surfaces. II", *Proc. Roy. Soc. Edinburgh* Sect. A **111** (1989), 213–229.

[1991] "Some remarks on tight hypersurfaces", *Proc. Roy. Soc. Edinburgh* Sect. A **119**:3-4 (1991), 279–285.

EUGENE CURTIN

[1991] "Manifolds with intermediate tautness", *Geom. Dedicata* **38**:3 (1991), 245–255.

[1994] "Tautness for manifolds with boundary", *Houston J. Math.* **20**:3 (1994), 409–424.

R. VAN DAMME AND L. ALBOUL

[1995] "Tight triangulations", pp. 517–526 in *Mathematical methods for curves and surfaces* (Ulvik, Norway, 1994), edited by M. Daehlen et al., Vanderbilt Univ. Press, Nashville, TN, 1995.

[1996] "Polyhedral metrics in surface reconstruction", pp. xiv+569 in *The mathematics of surfaces VI* (Uxbridge, 1994), edited by G. Mullineux, Inst. Math. Appl. Conf. Ser. New Ser. **58**, Oxford U. Press, New York, 1996.

JOSEPH E. D'ATRI

[1979] "Certain isoparametric families of hypersurfaces in symmetric spaces", *J. Differential Geom.* **14**:1 (1979), 21–40.

FILIP DEFEVER, RYSZARD DESZCZ, AND LEOPOLD VERSTRAELEN

[1993] "The compact cyclides of Dupin and a conjecture of B.-Y. Chen", *J. Geom.* 46:1-2 (1993), 33–38.

[1994] "The Chen-type of the noncompact cyclides of Dupin", *Glasgow Math. J.* 36:1 (1994), 71–75.

WENDELIN DEGEN

[1982] "Surfaces with a conjugate net of conics in projective space", *Tensor* (N.S.) 39 (1982), 167–172.

[1986] "Die zweifachen Blutelschen Kegelschnittflächen", *Manuscripta Math.* 55:1 (1986), 9–38.

RYSZARD DESZCZ AND SAHNUR YAPRAK

[1994] "Curvature properties of Cartan hypersurfaces", *Colloq. Math.* 67:1 (1994), 91–98.

JOSEF DORFMEISTER AND ERHARD NEHER

[1983a] "An algebraic approach to isoparametric hypersurfaces I, II", *Tôhoku Math. J.* 35 (1983), 187–224, 225–247.

[1983b] "Isoparametric triple systems of algebra type", *Osaka J. Math.* 20:1 (1983), 145–175.

[1983c] "Isoparametric triple systems of FKM-type I", *Abh. Math. Sem. Univ. Hamburg* 53 (1983), 191–216.

[1983d] "Isoparametric triple systems of FKM-type II", *Manuscripta Math.* 43:1 (1983), 13–44.

[1985] "Isoparametric hypersurfaces, case $g = 6$, $m = 1$", *Comm. Algebra* 13:11 (1985), 2299–2368.

[1990] "Isoparametric triple systems with special Z-structure", *Algebras Groups Geom.* 7:1 (1990), 21–94.

JOHANNES J. DUISTERMAAT

[1983] "Convexity and tightness for restrictions of Hamiltonian functions to fixed point sets of an antisymplectic involution", *Trans. Amer. Math. Soc.* 275 (1983), 417–429.

PIERRE CHARLES FRANÇOIS DUPIN

[1822] *Applications de géométrie et de méchanique: à la marine, aux ponts et chaussées, etc.*, Bachelier, Paris, 1822.

JAMES EELLS, JR. AND NICOLAAS H. KUIPER

[1962] "Manifolds which are like projective planes", *Publ. Math. Inst. Hautes Études Sci.* 14 (1962), 181–222.

KAZUYUKI ENOMOTO

[1978] "Total absolute curvature of immersed submanifolds of spheres", *Kōdai Math. J.* 1 (1978), 258–263.

JOST-HINRICH ESCHENBURG AND CARLOS OLMOS

[1994] "Rank and symmetry of Riemannian manifolds", *Comment. Math. Helv.* **69**:3
(1994), 483–499.

JOST-HINRICH ESCHENBURG AND VIKTOR SCHROEDER

[1991] "Tits distance of Hadamard manifolds and isoparametric hypersurfaces",
Geom. Dedicata **40**:1 (1991), 97–101.

HEIKO EWERT

[1996] "A splitting theorem for equifocal submanifolds in simply connected compact
symmetric spaces", preprint, Universität zu Köln, 1996.

[1997] *Equifocal submanifolds in Riemannian symmetric spaces*, Ph.D. thesis, Universität zu Köln, 1997.

FUQUAN FANG

[1995a] "On the topology of isoparametric hypersurfaces with four distinct principal
curvatures", preprint, Nankai Institute of Mathematics, 1995.

[1995b] "Topology of Dupin hypersurfaces with six principal curvatures", preprint,
University of Bielefeld, 1995.

[1996] "Multiplicities of principal curvatures of isoparametric hypersurfaces", preprint,
Max-Planck–Institut für Mathematik, Bonn, 1996.

ISTVÁN FÁRY

[1949] "Sur la courbure totale d'une courbe gauche faisant un nœud", *Bull. Soc. Math.
France* **77** (1949), 128–138.

YVES FÉLIX, STEPHEN HALPERIN, AND JEAN-CLAUDE THOMAS

[1993] "Elliptic spaces II", *Enseign. Math.* (2) **39**:1-2 (1993), 25–32.

WERNER FENCHEL

[1929] "Über die Krümmung und Windung geschlossener Raumkurven", *Math. Ann.*
101 (1929), 238–252.

[1930] "Geschlossene Raumkurven mit vorgeschriebenem Tangentenbild", *Jahresber.
der Deutschen Math.-Verein.* **39** (1930), 183–185.

[1940] "On total curvatures of Riemannian manifolds I", *J. London Math. Soc.* **15**
(1940), 15–22.

[1951] "On the differential geometry of closed space curves", *Bull. Amer. Math. Soc.*
57 (1951), 44–54.

E. V. FERAPONTOV

[1995a] "Dupin hypersurfaces and integrable Hamiltonian systems of hydrodynamic
type, which do not possess Riemann invariants", *Differential Geom. Appl.* **5**:2 (1995),
121–152.

[1995b] "Isoparametric hypersurfaces in spheres, integrable nondiagonalizable systems
of hydrodynamic type, and N-wave systems", *Differential Geom. Appl.* **5**:4 (1995),
335–369.

DIRK FERUS

[1967] "Über die absolute Totalkrümmung höher-dimensionaler Knoten", *Math. Ann.* **171** (1967), 81–86.

[1968] *Totale Absolutkrümmung in Differentialgeometrie und -topologie*, Lecture Notes in Math. **66**, Springer, Berlin, 1968.

[1980] "Symmetric submanifolds of Euclidean space", *Math. Ann.* **247** (1980), 81–93.

[1982] "The tightness of extrinsic symmetric submanifolds", *Math. Z.* **181** (1982), 563–565.

DIRK FERUS, HERMANN KARCHER, AND HANS FRIEDRICH MÜNZNER

[1981] "Cliffordalgebren und neue isoparametrische Hyperflächen", *Math. Z.* **177** (1981), 479–502.

SUSANA FORNARI

[1981] "A bound for total absolute curvature of surfaces in $\mathbb{R}^3$", *An. Acad. Brasil. Ciências* **53**:2 (1981), 255–256.

RALPH H. FOX

[1950] "On the total curvature of some tame knots", *Ann. of Math.* (2) **52** (1950), 258–260.

HANS FREUDENTHAL

[1985] "Oktaven, Ausnahmegruppen und Oktavengeometrie", *Geom. Dedicata* **19**:1 (1985), 7–63.

THOMAS FRIEDRICH

[1974a] "Berandete Mannigfaltigkeiten in euklidischen Räumen", *Demonstratio Math.* **7** (1974), 327–336.

[1974b] "A characterization of the disc D^{nn}", *Colloq. Math.* **32** (1974), 77–81.

[1974c] "Über die n-te totale Absolutkrümmung der Ordnung β einer Immersion", *Colloq. Math.* **31** (1974), 83–86.

[1975] "m-Funktionen und ihre Anwendung auf die totale Absolutkrümmung", *Math. Nachr.* **67** (1975), 281–301.

CHIH CHY FWU

[1980] "Total absolute curvature of submanifolds in compact symmetric spaces of rank one", *Math. Z.* **172**:3 (1980), 245–254.

MARTIN VAN GEMMEREN

[1996] "Total absolute curvature and tightness of noncompact manifolds", *Trans. Amer. Math. Soc.* **348**:6 (1996), 2413–2426.

[noye] "Tightness of manifolds with H-spherical ends". to appear in *Composition Math.*

PETER GRITZMANN

[1981] "Tight polyhedral realisations of closed 2-dimensional manifolds in $\mathbb{R}^3$", *J. Geom.* **17** (1981), 69–76.

NATHANIEL GROSSMAN

[1972] "Relative Chern-Lashof theorems", *J. Differential Geometry* **7** (1972), 607–614.

KARSTEN GROVE

[1993] "Critical point theory for distance functions", pp. 357–385 in *Differential geometry: Riemannian geometry* (Los Angeles, 1990), Proc. Sympos. Pure Math. **54** (part 3), Amer. Math. Soc., Providence, RI, 1993.

KARSTEN GROVE AND STEPHEN HALPERIN

[1987] "Dupin hypersurfaces, group actions and the double mapping cylinder", *J. Differential Geom.* **26**:3 (1987), 429–459.

[1991] "Elliptic isometries, condition C and proper maps", *Arch. Math. (Basel)* **56**:3 (1991), 288–299.

FRANÇOIS HAAB

[1992] "Immersions tendues de surfaces dans $\mathbf{E}^3$", *Comment. Math. Helv.* **67**:2 (1992), 182–202.

[1993] "Surfaces tendues dans E^3 et nombre d'ensembles Morse-singuliers", preprint, 1993/94.

[1994] "Surfaces tendues dans E^4", preprint, Depto. de Matemática, Universidade Federal de Minas Gerais, Belo Horizonte, MG, Brazil, 1994/95.

FRANÇOIS HAAB AND NICOLAAS H. KUIPER

[1992] "On the normal Gauss map of a tight smooth surface in $\mathbb{R}^3$", pp. 161–175 in *Chern—a great geometer of the twentieth century*, edited by S.-T. Yau, Internat. Press, Hong Kong, 1992.

JÖRG HAHN

[1984] "Isoparametric hypersurfaces in the pseudo-Riemannian space forms", *Math. Z.* **187**:2 (1984), 195–208.

[1988] "Isotropy representations of semisimple symmetric spaces and homogeneous hypersurfaces", *J. Math. Soc. Japan* **40**:2 (1988), 271–288.

HAZARIAN HAIG

[1982] "On a class of immersions in the euclidean space", *Boll. Un. Mat. Ital.* (6) **1**:1 (1982), 393–402.

CARLOS EDGARD HARLE

[1982] "Isoparametric families of submanifolds", *Bol. Soc. Brasil. Mat.* **13**:2 (1982), 35–48.

JAMES J. HEBDA

[1981] "Manifolds admitting taut hyperspheres", *Pacific J. Math.* **97** (1981), 119–124.

[1984] "Some new tight embeddings which cannot be made taut", *Geom. Dedicata* **17**:1 (1984), 49–60.

[1988] "The possible cohomology of certain types of taut submanifolds", *Nagoya Math. J.* **111** (1988), 85–97.

ERNST HEINTZE AND XIAOBO LIU

[1997] "A splitting theorem for isoparametric submanifolds in Hilbert space", *J. Differential Geom.* **45**:2 (1997), 319–335.

ERNST HEINTZE AND CARLOS OLMOS

[1992] "Normal holonomy groups and s-representations", *Indiana Univ. Math. J.* **41**:3 (1992), 869–874.

ERNST HEINTZE, CARLOS OLMOS, AND GUDLAUGUR THORBERGSSON

[1991] "Submanifolds with constant principal curvatures and normal holonomy groups", *Internat. J. Math.* **2**:2 (1991), 167–175.

ERNST HEINTZE, RICHARD S. PALAIS, CHUU-LIAN TERNG, AND GUDLAUGUR THORBERGSSON

[1994] "Hyperpolar actions and k-flat homogeneous spaces", *J. Reine Angew. Math.* **454** (1994), 163–179.

[1995] "Hyperpolar actions on symmetric spaces", pp. 214–245 in *Geometry, topology, and physics for Raoul Bott*, edited by S.-T. Yau, Conf. Proc. Lecture Notes Geom. Topology **4**, Internat. Press, Cambridge, MA, 1995.

BRADLEY W. HEMPSTEAD

[1970] *Tight immersions in higher dimensions*, Ph.D. thesis, University of Minnesota, 1970.

HEINZ HOPF

[1925] "Über die curvatura integra geschlossener Hyperflächen", *Math. Ann.* **95** (1925), 340–367.

WU-YI HSIANG AND H. BLAINE LAWSON, JR.

[1971] "Minimal submanifolds of low cohomogeneity", *J. Differential Geometry* **5** (1971), 1–38.

WU-YI HSIANG, RICHARD S. PALAIS, AND CHUU-LIAN TERNG

[1985] "Geometry and topology of isoparametric submanifolds in Euclidean spaces", *Proc. Nat. Acad. Sci. U.S.A.* **82**:15 (1985), 4863–4865.

[1988] "The topology of isoparametric submanifolds", *J. Differential Geom.* **27**:3 (1988), 423–460.

TOSHIHIKO IKAWA

[1984] "Morse functions and submanifolds of hyperbolic space", *Rocky Mountain J. Math.* **14**:2 (1984), 301–303.

TÔRU ISHIHARA

[1982] "Total curvature of manifolds in self-immersed manifolds", *Manuscripta Math.* **39** (1982), 201–218.

[1986] "Total curvatures of Kaehler manifolds in complex projective spaces", *Geom. Dedicata* **20**:3 (1986), 307–318.

THOMAS IVEY

[1995] "Surfaces with orthogonal families of circles", *Proc. Amer. Math. Soc.* **123**:3 (1995), 865–872.

ANDRZEJ JANKOWSKI AND RYSZARD RUBINSZTEIN

[1972] "Functions with non-degenerate critical points on manifolds with boundary", *Comment. Math. Prace Mat.* **16** (1972), 99–112.

LUQUESIO P. JORGE AND FRANCESCO MERCURI

[1984] "Minimal immersions into space forms with two principal curvatures", *Math. Z.* **187**:3 (1984), 325–333.

JYOICHI KANEKO

[1986] "Wave equation and Dupin hypersurface", *Mem. Fac. Sci. Kyushu Univ. Ser. A* **40**:1 (1986), 51–55.

S. M. B. KASHANI

[1992] "On quadratic isoparametric submanifolds", *Bull. Iranian Math. Soc.* **18**:2 (1992), 31–39.

[1993a] "Isoparametric functions and submanifolds", *Glasgow Math. J.* **35**:2 (1993), 145–152.

[1993b] "Quadratic isoparametric systems in $\mathbb{R}_p^{n+m}$", *Glasgow Math. J.* **35**:2 (1993), 135–143.

EDMUND F. KELLY

[1971] "0-tight equivariant imbeddings of symmetric spaces", *Bull. Amer. Math. Soc.* **77** (1971), 580–583.

[1972] "Tight equivariant imbeddings of symmetric spaces", *J. Differential Geometry* **7** (1972), 535–548.

U-HANG KI AND HISAO NAKAGAWA

[1987] "A characterization of the Cartan hypersurface in a sphere", *Tôhoku Math. J. (2)* **39**:1 (1987), 27–40.

[1991] "Hypersurfaces with harmonic Weyl tensor of a real space form", *J. Korean Math. Soc.* **28**:2 (1991), 229–244.

U-HANG KI AND YOUNG JIN SUH

[1990] "On real hypersurfaces of a complex space form", *Math. J. Okayama Univ.* **32** (1990), 207–221.

[1994] "On a characterization of real hypersurfaces of type A in a complex space form", *Canad. Math. Bull.* **37**:2 (1994), 238–244.

U-HANG KI AND RYOICHI TAKAGI

[1992] "Real hypersurfaces in $P_n(\mathbb{C})$ with constant principal curvatures", *Math. J. Okayama Univ.* **34** (1992), 233–240.

MAKOTO KIMURA

[1986a] "Real hypersurfaces and complex submanifolds in complex projective space", *Trans. Amer. Math. Soc.* **296**:1 (1986), 137–149.

[1986b] "Real hypersurfaces of a complex projective space", *Bull. Austral. Math. Soc.* **33**:3 (1986), 383–387.

[1987a] "Sectional curvatures of holomorphic planes on a real hypersurface in $P^n(\mathbb{C})$", *Math. Ann.* **276**:3 (1987), 487–497.

[1987b] "Some real hypersurfaces of a complex projective space", *Saitama Math. J.* **5** (1987), 1–5. Errata in **10** (1992), 33–34.

[1993] "$O(p) \times O(q)$-invariant minimal hypersurfaces in hyperbolic space", *Nihonkai Math. J.* **4**:2 (1993), 233–238.

MAKOTO KIMURA AND SADAHIRO MAEDA

[1989] "On real hypersurfaces of a complex projective space", *Math. Z.* **202**:3 (1989), 299–311.

[1991] "On real hypersurfaces of a complex projective space II", *Tsukuba J. Math.* **15**:2 (1991), 547–561.

[1993] "On real hypersurfaces of a complex projective space III", *Hokkaido Math. J.* **22**:1 (1993), 63–78.

NORBERT KNARR AND LINUS KRAMER

[1995] "Projective planes and isoparametric hypersurfaces", *Geom. Dedicata* **58**:2 (1995), 193–202.

SHOSHICHI KOBAYASHI

[1967] "Imbeddings of homogeneous spaces with minimum total curvature", *Tôhoku Math. J.* (2) **19** (1967), 63–70.

MASAHIRO KON

[1979] "Pseudo-Einstein real hypersurfaces in complex space forms", *J. Diff. Geom.* **14** (1979), 339–354.

[1980] "Real minimal hypersurfaces in a complex projective space", *Proc. Amer. Math. Soc.* **79** (1980), 285–288.

MAREK KOSSOWSKI

[1989] "The S^2-valued Gauss maps and split total curvature of a space-like codimension-2 surface in Minkowski space", *J. London Math. Soc.* (2) **40**:1 (1989), 179–192.

WOLFGANG KÜHNEL

[1977] "Total curvature of manifolds with boundary in E^n", *J. London Math. Soc.* (2) **15**:1 (1977), 173–182.

[1978] "$(n - 2)$-tightness and curvature of submanifolds with boundary", *Intern. J. Math. Math. Sci.* **1** (1978), 421–431.

[1979a] "A lower bound for the i-th total absolute curvature of an immersion", *Colloq. Math.* **41** (1979), 253–255.

[1979b] "Total absolute curvature of polyhedral manifolds with boundary in $\mathbb{E}^n$", *Geom. Dedicata* **8** (1979), 1–12.

[1980] "Tight and 0-tight polyhedral embeddings of surfaces", *Invent. Math.* **58** (1980), 161–177.

[1981] "0-tight surfaces with boundary and the total curvature of curves in surfaces", *Colloq. Math.* **45** (1981), 251–256.

[1992] "Tightness, torsion, and tubes", *Ann. Global Anal. Geom.* **10**:3 (1992), 227–236.

[1994a] "Manifolds in the skeletons of convex polytopes, tightness, and generalized Heawood inequalities", pp. 241–247 in *Polytopes: abstract, convex and computational* (Scarborough, ON, 1993), edited by T. Bisztriczky et al., NATO Adv. Sci. Inst. Ser. C Math. Phys. Sci. **440**, Kluwer, Dordrecht, 1994.

[1994b] "Tensor products of spheres", pp. 106–109 in *Geometry and topology of submanifolds, VI* (Leuven/Brussels, 1993), edited by F. Dillen et al., World Scientific, River Edge, NJ, 1994.

[1995] *Tight polyhedral submanifolds and tight triangulations*, Lecture Notes in Math. **1612**, Springer, Berlin, 1995.

[1996] "Centrally-symmetric tight surfaces and graph embeddings", *Beiträge zur Algebra und Geometrie* **37** (1996), 347–354.

WOLFGANG KÜHNEL AND THOMAS F. BANCHOFF

[1983] "The 9-vertex complex projective plane", *Math. Intelligencer* **5**:3 (1983), 11–22.

WOLFGANG KÜHNEL AND ULRICH PINKALL

[1985] "Tight smoothing of some polyhedral surfaces", pp. 227–239 in *Global differential geometry and global analysis* (Berlin, 1984), edited by D. Ferus et al., Lecture Notes in Math. **1156**, Springer, Berlin, 1985.

[1986] "On total mean curvatures", *Quart. J. Math. Oxford Ser.* (2) **37**:148 (1986), 437–447.

NICOLAAS H. KUIPER

[1959a] "Immersions with minimal total absolute curvature", pp. 75–88 in *Colloque Géom. Diff. Globale* (Bruxelles, 1958), Centre Belge Rech. Math., Louvain, 1959.

[1959b] "Sur les immersions à courbure totale minimale", pp. 1–5 in *Séminaire de topologie et de géométrie différentielle dirigé par Charles Ehresmann*, 1ʳᵉ année, 1957/58, Institut Henri Poincaré, Paris, 1959.

[1960a] "La courbure d'indice k et les applications convexes", pp. 1–15 in *Séminaire de topologie et de géométrie différentielle dirigé par Charles Ehresmann*, vol. 2, Institut Henri Poincaré, Paris, 1960.

[1960b] "On surfaces in euclidean three-space", *Bull. Soc. Math. Belg.* **12** (1960), 5–22.

[1961a] "A continuous function with two critical points", *Bull. Amer. Math. Soc.* **67** (1961), 281–285.

[1961b] "Convex immersions of closed surfaces in E^3. Nonorientable closed surfaces in E^3 with minimal total absolute Gauss-curvature", *Comment. Math. Helv.* **35** (1961), 85–92.

[1962] "On convex maps", *Nieuw Arch. Wisk.* (3) **10** (1962), 147–164.

[1966] "C^r-functions near non-degenerate critical points", mimeoraphed notes, Warwick Univ., Coventry, 1966.

[1967] "Der Satz von Gauss–Bonnet für Abbildungen in E^N und damit verwandte Probleme", *Jber. Deutsch. Math.-Verein.* **69**:2, Abt. 1 (1967), 77–88.

[1968] "Non-degenerate piecewise linear functions", *Rev. Roumaine Math. Pures Appl.* **13** (1968), 993–1000.

[1970] "Minimal total absolute curvature for immersions", *Invent. Math.* **10** (1970), 209–238.

[1971a] "Morse relations for curvature and tightness", pp. 77–89 in *Proceedings of Liverpool Singularities Symposium* (Liverpool, 1969/1970), vol. 2, edited by C. T. C. Wall, Lecture Notes in Math. **209**, Springer, Berlin, 1971.

[1971b] "Tight topological embeddings of the Moebius band", *J. Differential Geometry* **6** (1971/72), 271–283.

[1975] "Stable surfaces in euclidean three space", *Math. Scand.* **36** (1975), 83–96.

[1976a] "Curvature measures for surfaces in E^{n}", pp. 195–199 in *Lobaschevski Colloquium* (Kazan, USSR), 1976.

[1976b] "La courbure totale absolue minimale des surfaces dans E^3", *Bull. Soc. Math. France, Suppl. Mém.* **46** (1976), 11.

[1980] "Tight embeddings and maps: Submanifolds of geometrical class three in E^{n}", pp. 97–145 in *The Chern Symposium* (Berkeley, 1979), edited by W.-Y. Hsiang et al., Springer, New York, 1980.

[1983a] "Polynomial equations for tight surfaces", *Geom. Dedicata* **15**:2 (1983), 107–113.

[1983b] "There is no tight continuous immersion of the Klein bottle into $\mathbb{R}^3$", preprint, Inst. Hautes Études Sci., 1983.

[1984a] "Geometry in total absolute curvature theory", pp. 377–392 in *Perspectives in mathematics* (Oberwolfach, 1984), edited by W. Jäger et al., Birkhäuser, Basel, 1984.

[1984b] "Taut sets in three space are very special", *Topology* **23**:3 (1984), 323–336.

[1985] "Minimal total curvature. A look at old and new results", *Uspekhi Mat. Nauk* **40**:4 (1985), 49–54. Lecture at the International conference on current problems in algebra and analysis (Moscow-Leningrad, 1984). Translated from the English by M. P. Dolbinin and A. M. Volkhonskiĭ; English version appears in *Russ. Math. Surv.* **40**:4 (1985), 49–55.

[1993] "Tight and taut immersions", pp. 169–172 in *Encyclopaedia of Mathematics*, vol. 9, edited by M. Hazewinkel, Kluwer, Dordrecht, 1993.

[1997] "Geometry in curvature theory", pp. 1–50 in *Tight and Taut Submanifolds*, edited by T. E. Cecil and S.-s. Chern, Cambridge U. Press, 1997.

NICOLAAS H. KUIPER AND WILLIAM H. MEEKS, III

[1983] "Sur la courbure des surfaces nouées dans R^3", pp. 215–217 in *Third Schnepfenried geometry conference* (Schnepfenried, 1982), vol. 1, Astérisque **107-108**, Soc. Math. France, Paris, 1983.

[1984] "Total curvature for knotted surfaces", *Invent. Math.* **77**:1 (1984), 25–69.

[1987] "The total curvature of a knotted torus", *J. Differential Geom.* **26**:3 (1987), 371–384.

NICOLAAS H. KUIPER AND WILLIAM F. POHL

[1977] "Tight topological embeddings of the real projective plane in E^5", *Invent. Math.* **42** (1977), 177–199.

RÉMI LANGEVIN

[1979] "Feuilletages tendues", *Bull. Soc. Math. France* **107** (1979), 271–281.

[1981] "Tight foliations", pp. 181–186 in *Global Differential Geometry and Global Analysis* (Berlin, 1979), edited by D. Ferus et al., Lecture Notes in Math. **838**, Springer, Berlin, 1981.

RÉMI LANGEVIN AND CLAUDIO POSSANI

[1989] "Courbure totale de feuilletages et enveloppes", *C. R. Acad. Sci. Paris Sér. I Math.* **309**:13 (1989), 821–824.

RÉMI LANGEVIN AND HAROLD ROSENBERG

[1976] "On curvature integrals and knots", *Topology* **15**:4 (1976), 405–416.

[1996] "Fenchel type theorems for submanifolds of S^n", *Comment. Math. Helv.* **71**:4 (1996), 594–616.

WILLIAM STANLEY LASTUFKA

[1981] "Tight topological immersions of surfaces in Euclidean space", *J. Diff. Geom.* **16** (1981), 373–400.

H. BLAINE LAWSON, JR.

[1970] "Rigidity theorems in rank-1 symmetric spaces", *J. Differential Geometry* **4** (1970), 349–357.

TULLIO LEVI-CIVITA

[1937] "Famiglie di superficie isoparametrische nell'ordinario spacio euclideo", *Atti Accad. Naz. Lincei Rend. Cl. Sci. Fis. Mat. Natur.* **26** (1937), 355–362.

ZHEN QI LI

[1988] "Isoparametric hypersurfaces in $\mathbb{CP}^n$ with constant principal curvatures", *Chinese Ann. Math. Ser. B* **9**:4 (1988), 485–493.

HEINRICH LIEBMANN

[1929] "Elementarer Beweis des Fenchelschen Satzes über die Krümmung geschlossener Raumkurven", *Sitzungsberichte Preuss. Akad. Wiss.* (1929), 392–393.

JOHN A. LITTLE AND WILLIAM F. POHL

[1971] "On tight immersions of maximal codimension", *Invent. Math.* **13** (1971), 179–204.

SADAHIRO MAEDA

[1983] "Real hypersurfaces of complex projective spaces", *Math. Ann.* **263**:4 (1983), 473–478.

[1984] "Real hypersurfaces of a complex projective space II", *Bull. Austral. Math. Soc.* **30**:1 (1984), 123–127.

[1994] "Ricci tensors of real hypersurfaces in a complex projective space", *Proc. Amer. Math. Soc.* **122**:4 (1994), 1229–1235.

YOSHIAKI MAEDA

[1976] "On real hypersurfaces of a complex projective space", *J. Math. Soc. Japan* **28**:3 (1976), 529–540.

MARTIN A. MAGID

[1985] "Lorentzian isoparametric hypersurfaces", *Pacific J. Math.* **118**:1 (1985), 165–197.

JAMES CLERK MAXWELL

[1867] "On the cyclide", *Quart. J. of Pure and Appl. Math.* **34** (1867). Reprinted as pp. 144–159 in *Scientific papers*, vol. 2, Cambridge U. Press, 1890.

ANDREW MCLENNAN

[1990] "Round immersions", preprint, Department of Economics, Univ. of Minnesota, 1990.

JOHN W. MILNOR

[1950a] *On the relationship between the Betti numbers of a hypersurface and an integral of its Gaussian curvature*, Junior thesis, Princeton University, 1950. Reprinted as pp. 15–26 in *Collected papers*, Vol. I, Publish or Perish, Houston, TX, 1994.

[1950b] "On the total curvature of knots", *Ann. of Math.* (2) **52** (1950), 248–257.

[1953] "On the total curvature of closed space curves", *Math. Scand.* **1** (1953), 289–296.

[1963] *Morse Theory*, Ann. Math. Stud. **51**, Princeton U. Press, 1963.

VICENTE MIQUEL

[1994] "Compact Hopf hypersurfaces of constant mean curvature in complex space forms", *Ann. Global Anal. Geom.* **12**:3 (1994), 211–218.

REIKO MIYAOKA

[1982] "Complete hypersurfaces in the space form with three principal curvatures", *Math. Z.* **179**:2 (1982), 345–354. Correction in *Bol. Soc. Brasil. Mat.* **18** (1987), 83–94.

[1984a] "Compact Dupin hypersurfaces with three principal curvatures", *Math. Z.* **187**:4 (1984), 433–452.

[1984b] "Taut embeddings and Dupin hypersurfaces", pp. 15–23 in *Differential geometry of submanifolds* (Kyoto, 1984), edited by K. Kenmotsu, Lecture Notes in Math. **1090**, Springer, Berlin, 1984.

[1987]   "Dupin hypersurfaces with four principal curvatures", preprint, Tokyo Institute of Technology, 1987.

[1989a] "Dupin hypersurfaces and a Lie invariant", *Kodai Math. J.* **12**:2 (1989), 228–256.

[1989b] "Dupin hypersurfaces with six principal curvatures", *Kodai Math. J.* **12**:3 (1989), 308–315.

[1991a] "Lie contact structures and conformal structures", *Kodai Math. J.* **14**:1 (1991), 42–71.

[1991b] "Lie contact structures and normal Cartan connections", *Kodai Math. J.* **14**:1 (1991), 13–41.

[1993a] "The linear isotropy group of $G_2/SO(4)$, the Hopf fibering and isoparametric hypersurfaces", *Osaka J. Math.* **30**:2 (1993), 179–202.

[1993b] "A note on Lie contact manifolds", pp. 169–187 in *Progress in differential geometry*, Adv. Stud. Pure Math., 22, Math. Soc. Japan, Tokyo, 1993.

REIKO MIYAOKA AND TETSUYA OZAWA

[1989]   "Construction of taut embeddings and Cecil–Ryan conjecture", pp. 181–189 in *Geometry of manifolds* (Matsumoto, 1988), edited by K. Shiohama, Perspect. Math. **8**, Academic Press, Boston, MA, 1989.

SEBASTIÁN MONTIEL

[1985]   "Real hypersurfaces of a complex hyperbolic space", *J. Math. Soc. Japan* **37**:3 (1985), 515–535.

SEBASTIÁN MONTIEL AND ALFONSO ROMERO

[1986]   "On some real hypersurfaces of a complex hyperbolic space", *Geom. Dedicata* **20**:2 (1986), 245–261.

MARSTON MORSE

[1959]   "Topologically non-degenerate functions on a compact n-manifold M", *J. Analyse Math.* **7** (1959), 189–208.

[1960]   "The existence of polar non-degenerate functions on differentiable manifolds", *Ann. of Math.* (2) **71** (1960), 352–383.

[1973]   "F-deformations and F-tractions", *Proc. Nat. Acad. Sci. U.S.A.* **70** (1973), 1634–1635.

[1975]   "Topologically nondegenerate functions", *Fund. Math.* **88**:1 (1975), 17–52.

MARSTON MORSE AND STEWART S. CAIRNS

[1969]   *Critical point theory in global analysis and differential topology: An introduction*, Pure and Applied Mathematics **33**, Academic Press, New York, 1969.

MARSTON MORSE AND GEORGE B. VAN SCHAACK

[1934]   "The critical point theory under general boundary conditions", *Ann. of Math.* (2) **35** (1934), 545–571.

HUGH R. MORTON

[1979] "A criterion for an embedded surface in $\mathbb{R}^n$ to be unknotted", pp. 93–98 in *Topology of low-dimensional manifolds* (Chelwood Gate, Sussex, 1977), edited by R. Fenn, Lecture Notes in Math. **722**, Springer, Berlin, 1979.

SIMON MULLEN

[1994] "Isoparametric systems on symmetric spaces", pp. 152–154 in *Geometry and topology of submanifolds, VI* (Leuven/Brussels, 1993), edited by F. Dillen et al., World Sci. Publishing, River Edge, NJ, 1994.

HANS FRIEDRICH MÜNZNER

[1980] "Isoparametrische Hyperflächen in Sphären", *Math. Ann.* **251** (1980), 57–71.

[1981] "Isoparametrische Hyperflächen in Sphären II: Über die Zerlegung der Sphäre in Ballbündel", *Math. Ann.* **256** (1981), 215–232.

HIROO NAITOH AND MASARU TAKEUCHI

[1984] "Symmetric submanifolds of symmetric spaces", *Sûgaku* **36**:2 (1984), 137–156. Translation in *Sugaku Expositions* bf 2:2 (1989), 157–188.

ROSS NIEBERGALL

[1991] "Dupin hypersurfaces in R^5 I", *Geom. Dedicata* **40**:1 (1991), 1–22.

[1992] "Dupin hypersurfaces in $\mathbb{R}^5$ II", *Geom. Dedicata* **41**:1 (1992), 5–38.

[1994] "Tight analytic immersions of highly connected manifolds", *Proc. Amer. Math. Soc.* **120**:3 (1994), 907–916.

ROSS NIEBERGALL AND PATRICK J. RYAN

[1993] "Isoparametric hypersurfaces—the affine case", pp. 201–214 in *Geometry and topology of submanifolds, V* (Leuven/Brussels, 1992), edited by F. Dillen et al., World Sci. Publishing, River Edge, NJ, 1993.

[1994a] "Affine isoparametric hypersurfaces", *Math. Z.* **217**:3 (1994), 479–485.

[1994b] "Focal sets in affine geometry", pp. 155–164 in *Geometry and topology of submanifolds, VI* (Leuven/Brussels, 1993), edited by I. V. d. W. Franki Dillen and L. Verstraelen, World Sci. Publishing, River Edge, NJ, 1994.

[1996a] "Affine Dupin surfaces", *Trans. Amer. Math. Soc.* **348**:3 (1996), 1093–1115.

[1996b] "Semi-parallel and semi-symmetric real hypersurfaces in complex space forms", preprint, McMaster University, 1996.

ROSS NIEBERGALL AND GUDLAUGUR THORBERGSSON

[1996] "Tight immersions and local differential geometry", preprint, University of Northern British Columbia and Universität Köln, 1996.

LOUIS NIRENBERG

[1963] "Rigidity of a class of closed surfaces", pp. 177–193 in *Nonlinear Problems* (Madison, WI, 1962), edited by R. E. Langer, U. of Wisconsin Press, Madison, 1963.

KATSUMI NOMIZU

[1973] "Some results in E. Cartan's theory of isoparametric families of hypersurfaces",
        Bull. Amer. Math. Soc. **79** (1973), 1184–1188.

[1975] "Élie Cartan's work on isoparametric families of hypersurfaces", pp. 191–200
        in *Differential geometry* (Stanford, 1973), vol. 1, edited by S.-s. Chern and R.
        Osserman, Proc. Sympos. Pure Math. **27**, Amer. Math. Soc., Providence, 1975.

[1981] "On isoparametric hypersurfaces in Lorentzian space forms", *Jap. J. Math.*
        (N.S.) **7** (1981), 217–226.

KATSUMI NOMIZU AND LUCIO L. RODRÍGUEZ

[1972] "Umbilical submanifolds and Morse functions", *Nagoya Math. J.* **48** (1972),
        197–201.

YONG-GEUN OH

[1991] "Tight Lagrangian submanifolds in $\mathbb{CP}^n$", *Math. Z.* **207**:3 (1991), 409–416.

MASAFUMI OKUMURA

[1975] "On some real hypersurfaces of a complex projective space", *Trans. Amer.
        Math. Soc.* **212** (1975), 355–364.

[1978] "Compact real hypersurface with constant mean curvature of a complex
        projective space", *J. Diff. Geom.* **13** (1978), 43–50.

CARLOS OLMOS

[1990] "The normal holonomy group", *Proc. Amer. Math. Soc.* **110**:3 (1990), 813–818.

[1992] "On the holonomy group of the normal connection", pp. 587–595 in *Differential
        geometry and its applications* (Eger, 1989), edited by J. Szenthe and L. Tamássy,
        Colloq. Math. Soc. János Bolyai **56**, North-Holland, Amsterdam, 1992.

[1993] "Isoparametric submanifolds and their homogeneous structures", *J. Differential
        Geom.* **38**:2 (1993), 225–234.

[1994] "Homogeneous submanifolds of higher rank and parallel mean curvature", *J.
        Differential Geom.* **39**:3 (1994), 605–627.

[1995] "Orbits of rank one and parallel mean curvature", *Trans. Amer. Math. Soc.*
        347:8 (1995), 2927–2939.

CARLOS OLMOS AND CRISTIÁN SÁNCHEZ

[1991] "A geometric characterization of the orbits of *s*-representations", *J. Reine
        Angew. Math.* **420** (1991), 195–202.

ROBERT OSSERMAN

[1981] "The total curvature of algebraic surfaces", pp. 249–257 in *Contributions to
        analysis and geometry* (Baltimore, 1980), edited by D. N. Clark et al., Johns Hopkins
        Univ. Press, Baltimore, 1981.

TOMINOSUKE ÔTSUKI

[1966] "On the total curvature of surfaces in Euclidean spaces", *Japan. J. Math.* **35**
        (1966), 61–71.

TETSUYA OZAWA

[1982a] "Curvatures of PL-complexes in R^N and tightness", preprint, 1982.

[1982b] "On tight PL-manifolds", preprint, 1982.

[1982c] "Relations between tightness and k-tightness for PL-manifolds in Euclidean spaces", preprint, 1982.

[1983] "Products of tight continuous functions", *Geom. Dedicata* **14**:3 (1983), 209–213.

[1986] "On critical sets of distance functions to a taut submanifold", *Math. Ann.* **276**:1 (1986), 91–96.

HIDEKI OZEKI AND MASARU TAKEUCHI

[1975] "On some types of isoparametric hypersurfaces in spheres I", *Tôhoku Math. J.* (2) **27**:4 (1975), 515–559.

[1976] "On some types of isoparametric hypersurfaces in spheres II", *Tôhoku Math. J.* (2) **28**:1 (1976), 7–55.

RICHARD S. PALAIS AND CHUU-LIAN TERNG

[1986] "Reduction of variables for minimal submanifolds", *Proc. Amer. Math. Soc.* **98**:3 (1986), 480–484.

[1987a] "A general theory of canonical forms", *Trans. Amer. Math. Soc.* **300**:2 (1987), 771–789.

[1987b] "Geometry of canonical forms", pp. 133–151 in *The legacy of Sonya Kovalevskaya* (Cambridge and Amherst, MA, 1985), edited by L. Keen, Contemp. Math. **64**, Amer. Math. Soc., Providence, 1987.

[1988] *Critical point theory and submanifold geometry*, Lecture Notes in Math. **1353**, Springer, Berlin, 1988.

KWANG SUNG PARK

[1989] "Isoparametric families on projective spaces", *Math. Ann.* **284**:3 (1989), 503–513.

CHIA KUEI PENG AND ZI XIN HOU

[1989] "A remark on the isoparametric polynomials of degree 6", pp. 222–224 in *Differential geometry and topology* (Tianjin, 1986–87), edited by B. Jiang et al., Lecture Notes in Math. **1369**, Springer, Berlin, 1989.

CHIA KUEI PENG AND ZI ZHOU TANG

[1997] "On representing homotopy classes of spheres by harmonic maps", *Topology* **36**:4 (1997), 867–879.

[noye] "The Brouwer degrees of the gradient maps of isoparametric functions". to appear in *Science in China*.

W. D. PEPE

[1969] "On the total curvature of C^1 hypersurfaces in E^{n+1}", *Amer. J. Math.* **91** (1969), 984–1002.

ULRICH PINKALL

[1981] *Dupinsche Hyperflächen*, Ph.D. thesis, Univ. Freiburg, 1981.

[1985a] "Dupin hypersurfaces", *Math. Ann.* **270**:3 (1985), 427–440.

[1985b] "Dupinsche Hyperflächen in E^4", *Manuscripta Math.* **51**:1-3 (1985), 89–119.

[1985c] *Totale Absolutkrümmung immersierter Flächen*, Habilitationsschrift, Friedrich-Wilhelm-Universität, Bonn, 1985.

[1986a] "Curvature properties of taut submanifolds", *Geom. Dedicata* **20**:1 (1986), 79–83.

[1986b] "Cyclides of Dupin", pp. 28–30 in *Mathematical models from the collections of universities and museums*, vol. 2, edited by G. Fischer, Vieweg, Braunschweig, 1986.

[1986c] "Tight surfaces and regular homotopy", *Topology* **25**:4 (1986), 475–481.

ULRICH PINKALL AND GUDLAUGUR THORBERGSSON

[1989a] "Deformations of Dupin hypersurfaces", *Proc. Amer. Math. Soc.* **107**:4 (1989), 1037–1043.

[1989b] "Taut 3-manifolds", *Topology* **28**:4 (1989), 389–401.

[1990] "Examples of infinite-dimensional isoparametric submanifolds", *Math. Z.* **205**:2 (1990), 279–286.

HELMUT RECKZIEGEL

[1976] "Krümmungsflächen von isometrischen Immersionen in Räume konstanter Krümmung", *Math. Ann.* **223**:2 (1976), 169–181.

[1979a] "Completeness of curvature surfaces of an isometric immersion", *J. Diff. Geom.* **14** (1979), 7–20.

[1979b] "On the eigenvalues of the shape operator of an isometric immersion into a space of constant curvature", *Math. Ann.* **243** (1979), 71–82.

LUCIO L. RODRÍGUEZ

[1973] *The two-piece property and relative tightness for surfaces with boundary*, Ph.D. thesis, Brown University, 1973.

[1976] "The two-piece-property and convexity for surfaces with boundary", *J. Differential Geometry* **11**:2 (1976), 235–250.

[1977] "Convexity and tightness of manifolds with boundary", pp. 510–541 in *Geometry and topology* (Rio de Janeiro, 1976), edited by J. Palis and M. do Carmo, Lecture Notes in Math. **597**, Springer, Berlin, 1977.

ARNOUD C. M. VAN ROOIJ

[1965] "The total curvature of curves", *Duke Math. J.* **32** (1965), 313–324.

PATRICK J. RYAN

[1969] "Homogeneity and some curvature conditions for hypersurfaces", *Tôhoku Math. J.* (2) **21** (1969), 363–388.

[1971] "Hypersurfaces with parallel Ricci tensor", *Osaka J. Math.* **8** (1971), 251–259.

[1972] "A class of complex hypersurfaces", *Colloq. Math.* **26** (1972), 175–182, 385.

B. A. SALEEMI

[1987] "Morse covers and tight immersions", *Punjab Univ. J. Math. (Lahore)* **20** (1987), 1–6.

CRISTIÁN U. SÁNCHEZ

[1985] "k-symmetric submanifolds of $\mathbb{R}^N$", *Math. Ann.* **270**:1 (1985), 297–316. Erratum in **279** (1987), 169–172.

[1988] "Extrinsic k-symmetric submanifolds are tight", *Rev. Un. Mat. Argentina* **34** (1988/1990), 122–131. Conference in Honor of Mischa Cotlar (Buenos Aires, 1988).

[1990] "The tightness of certain almost complex submanifolds", *Proc. Amer. Math. Soc.* **110**:3 (1990), 807–811.

SHIGEO SASAKI

[1959] "On the total curvature of a closed curve", *Japan. J. Math.* **29** (1959), 118–125.

BENIAMINO SEGRE

[1938] "Famiglie di ipersuperficie isoparametrische negli spazi euclidei ad un qualunque numero di dimensioni", *Atti Accad. Naz. Lincei Rend. Cl. Sci. Fis. Mat. Natur.* **27** (1938), 203–207.

VLADIMIR V. SHARKO

[1987] "On the equivalence of exact Morse functions I", pp. 137–146 in *Fourteen papers translated from the Russian*, Amer. Math. Soc. Transl. (2) **134**, 1987.

RICHARD W. SHARPE

[1988] "Total absolute curvature and embedded Morse numbers", *J. Differential Geom.* **28**:1 (1988), 59–92.

[1989] "A proof of the Chern-Lashof conjecture in dimensions greater than five", *Comment. Math. Helv.* **64**:2 (1989), 221–235.

BRUCE SOLOMON

[1990a] "The harmonic analysis of cubic isoparametric minimal hypersurfaces, I: Dimensions 3 and 6", *Amer. J. Math.* **112**:2 (1990), 157–203.

[1990b] "The harmonic analysis of cubic isoparametric minimal hypersurfaces, II: Dimensions 12 and 24", *Amer. J. Math.* **112**:2 (1990), 205–241.

[1992] "Quartic isoparametric hypersurfaces and quadratic forms", *Math. Ann.* **293**:3 (1992), 387–398.

E. SPARLA

[1997a] "A new lower bound theorem for combinatorial $2k$-manifolds", 1997. To appear in *Graphs and Comb.*

[1997b] "An upper and a lower bound theorem for combinatorial 4-manifolds", 1997. To appear in *Disc. Comput. Geom.*

WOLF STRÜBING

[1979] "Symmetric submanifolds of Riemannian manifolds", *Math. Ann.* **245** (1979), 37–44.

[1986] "Isoparametric submanifolds", *Geom. Dedicata* **20**:3 (1986), 367–387.

YOUNG JIN SUH AND RYOICHI TAKAGI

[1991]   "A rigidity for real hypersurfaces in a complex projective space", *Tôhoku Math. J.* (2) **43**:4 (1991), 501–507.

DAN SUNDAY

[1976]   "The total curvature of knotted spheres", *Bull. Amer. Math. Soc.* **82**:1 (1976), 140–142.

JÁNOS SZENTHE

[1968]   "On the total curvature of closed curves in Riemannian manifolds", *Publ. Math. Debrecen* **15** (1968), 99–105.

SHIN-SHENG TAI

[1968]   "Minimum imbeddings of compact symmetric spaces of rank one", *J. Diff. Geom.* **2** (1968), 55–66.

HITOSHI TAKAGI

[1985]   "A condition for isoparametric hypersurfaces of S^n to be homogeneous", *Tôhoku Math. J.* (2) **37**:2 (1985), 241–250.

RYOICHI TAKAGI

[1973]   "On homogeneous real hypersurfaces in a complex projective space", *Osaka J. Math.* **10** (1973), 495–506.

[1975a] "Real hypersurfaces in a complex projective space with constant principal curvatures", *J. Math. Soc. Japan* **27** (1975), 43–53.

[1975b] "Real hypersurfaces in a complex projective space with constant principal curvatures II", *J. Math. Soc. Japan* **27**:4 (1975), 507–516.

[1976]   "A class of hypersurfaces with constant principal curvatures in a sphere", *J. Differential Geometry* **11**:2 (1976), 225–233.

[1990]   "Real hypersurfaces in a complex projective space", *Proc. Topology and Geometry Research Center (Taegu)* **1** (1990), 1–17.

RYOICHI TAKAGI AND TSUNERO TAKAHASHI

[1972]   "On the principal curvatures of homogeneous hypersurfaces in a sphere", pp. 469–481 in *Differential geometry (in honor of Kentaro Yano)*, edited by S. Kobayashi et al., Kinokuniya, Tokyo, 1972.

MASARU TAKEUCHI

[1981]   "Parallel submanifolds of space forms", pp. 429–447 in *Manifolds and Lie groups* (Notre Dame, IN, 1980), edited by J. Hano et al., Progr. Math. **14**, Birkhäuser, Boston, 1981.

[1988]   "Basic transformations of symmetric R-spaces", *Osaka J. Math.* **25**:2 (1988), 259–297.

[1991]   "Proper Dupin hypersurfaces generated by symmetric submanifolds", *Osaka J. Math.* **28**:1 (1991), 153–161.

MASARU TAKEUCHI AND SHOSHICHI KOBAYASHI

[1968] "Minimal imbeddings of R-spaces", *J. Differential Geometry* **2** (1968), 203–215.

ZI ZHOU TANG

[1991] "Isoparametric hypersurfaces with four distinct principal curvatures", *Chinese Sci. Bull.* **36**:15 (1991), 1237–1240.

[1996] "Harmonic polynomial morphisms between Euclidean spaces", preprint, Academia Sinica, Beijing, 1996.

CHUU-LIAN TERNG

[1985] "Isoparametric submanifolds and their Coxeter groups", *J. Differential Geom.* **21**:1 (1985), 79–107.

[1986] "Convexity theorem for isoparametric submanifolds", *Invent. Math.* **85**:3 (1986), 487–492.

[1987] "Submanifolds with flat normal bundle", *Math. Ann.* **277**:1 (1987), 95–111.

[1989] "Proper Fredholm submanifolds of Hilbert space", *J. Differential Geom.* **29**:1 (1989), 9–47.

[1991] "Variational completeness and infinite-dimensional geometry", pp. 279–293 in *Geometry and topology of submanifolds, III* (Leeds, 1990), edited by L. Verstraelen and A. West, World Sci. Publishing, River Edge, NJ, 1991.

[1993] "Recent progress in submanifold geometry", pp. 439–484 in *Differential geometry: partial differential equations on manifolds* (Los Angeles, 1990), edited by R. Greene and S. T. Yau, Proc. Sympos. Pure Math. **54** (part 1), Amer. Math. Soc., Providence, RI, 1993.

CHUU-LIAN TERNG AND GUDLAUGUR THORBERGSSON

[1995] "Submanifold geometry in symmetric spacs", *J. Differential Geom.* **42**:3 (1995), 665–718.

[1997] "Taut immersions into complete Riemannian manifolds", pp. 181–228 in *Tight and Taut Submanifolds*, edited by T. E. Cecil and S.-s. Chern, Cambridge U. Press, 1997.

EBERHARD TEUFEL

[1980a] "Anwendungen der differentialtopologischen Berechnung der totalen Krümmung und totalen Absolutkrümmung in der sphärischen Differentialgeometrie", *Manuscripta Math.* **32** (1980), 239–262.

[1980b] "Eine differentialtopologische Berechnung der totalen Krümmung und totalen Absolutkrümmung in der sphärischen Differentialgeometrie", *Manuscripta Math.* **31** (1980), 119–147.

[1982] "Differential topology and the computation of total absolute curvature", *Math. Ann.* **258** (1982), 471–480.

[1986] "On the total absolute curvature of closed curves in spheres", *Manuscripta Math.* **57**:1 (1986), 101–108.

[1988] "On the total absolute curvature of immersions into hyperbolic spaces", pp. 1201–1209 in *Topics in differential geometry* (Debrecen, 1984), vol. 2, Colloq. Math. Soc. János Bolyai **46**, North-Holland, Amsterdam, 1988.

[1992] "The isoperimetric inequality and the total absolute curvature of closed curves in spheres", *Manuscripta Math.* **75**:1 (1992), 43–48.

GUDLAUGUR THORBERGSSON

[1983a] "Dupin hypersurfaces", *Bull. London Math. Soc.* (2) **15**:5 (1983), 493–498.

[1983b] "Highly connected taut submanifolds", *Math. Ann.* **265**:3 (1983), 399–405.

[1985a] "Straffe Untermannigfaltigkeiten in konvexen Hyperflächen", *Manuscripta Math.* **54**:1-2 (1985), 1–15.

[1985b] "Tight and taut immersions of highly connected manifolds", pp. 251–254 in *Proceedings of the nineteenth Nordic congress of mathematicians* (Reykjavík, 1984), Vísindafél. Ísl. **44**, Icel. Math. Soc., Reykjavík, 1985.

[1986a] *Geometrie hochzusammenhängender Untermannigfaltigkeiten*, Bonner Mathematische Schriften **170**, Universität Bonn Math. Institut, Bonn, 1986. Habilitationsschrift, Friedrich-Wilhelms-Universität, Bonn, 1985.

[1986b] "Tight immersions of highly connected manifolds", *Comment. Math. Helv.* **61**:1 (1986), 102–121.

[1988] "Homogeneous spaces without taut embeddings", *Duke Math. J.* **57**:1 (1988), 347–355.

[1989] "Isoparametric submanifolds", *Note Mat.* **9**:suppl. (1989), 33–38. Conference on Differential Geometry and Topology (Italian) (Lecce, 1989).

[1991a] "Isoparametric foliations and their buildings", *Ann. of Math.* (2) **133**:2 (1991), 429–446.

[1991b] "Tight analytic surfaces", *Topology* **30**:3 (1991), 423–428.

CHOU TIAN

[1982] "On the total curvature of submanifolds", pp. 1451–1464 in *Beijing Symposium on Differential Geometry and Differential Equations*, vol. 3, edited by S.-S. Chern and W. T. Wu, Science Press, Beijing, 1982.

FRANCO TRICERRI AND LIEVEN VANHECKE

[1990] "Cartan hypersurfaces and reflections", *Nihonkai Math. J.* **1**:2 (1990), 203–208.

YÔTARÔ TSUKAMOTO

[1974] "On the total absolute curvature of closed curves in manifolds of negative curvature", *Math. Ann.* **210** (1974), 313–319.

L. VERHÓCZKI

[1992] "Isoparametric submanifolds of general Riemannian manifolds", pp. 691–705 in *Differential geometry and its applications* (Eger, 1989), edited by J. Szenthe and L. Tamássy, Colloq. Math. Soc. János Bolyai **56**, North-Holland, Amsterdam, 1992.

[1995] "Special isoparametric orbits in Riemannian symmetric spaces", *Geom. Dedicata* **55**:3 (1995), 305–317.

MICHEAL H. VERNON

[1989]   "Some families of isoparametric hypersurfaces and rigidity in a complex hyperbolic space", *Trans. Amer. Math. Soc.* **312**:1 (1989), 237–256.

[1991]   "Semi-symmetric hypersurfaces of anti-de Sitter spacetime that are S^1-invariant", *Tensor* (N.S.) **50**:2 (1991), 99–105.

K. VOSS

[1955]   "Eine Bemerkung über die Totalkrümmung geschlossener Raumkurven", *Arch. Math.* **6** (1955), 259–263.

[1981]   "Eine Verallgemeinerung der Dupinschen Zykliden", Tagungsbericht 41/1981, geometrie, Mathematisches Forschungsinstitut Oberwolfach, 1981.

M. VYAL'YAS

[1986]   "Dupin hypersurfaces with a holonomic net of curvature lines in E_4", *Tartu Riikl. Ül. Toimetised* **734** (1986), 20–30.

CHANG PING WANG

[1992]   "Surfaces in Möbius geometry", *Nagoya Math. J.* **125** (1992), 53–72.

[1995]   "Möbius geometry for hypersurfaces in S^4", *Nagoya Math. J.* **139** (1995), 1–20.

QI MING WANG

[1982]   "Isoparametric hypersurfaces in complex projective spaces", pp. 1509–1523 in *Symposium on Differential Geometry and Differential Equations* (Beijing, 1980), vol. 3, Science Press, Beijing, 1982.

[1983]   "Real hypersurfaces with constant principal curvatures in complex projective spaces I", *Sci. Sinica Ser. A* **26**:10 (1983), 1017–1024.

[1986]   "Isoparametric maps of Riemannian manifolds and their applications", pp. 79–103 in *Advances in science of China. Mathematics*, vol. 2, edited by C. H. Gu and Y. Wang, Wiley, New York, 1986.

[1987]   "Isoparametric functions on Riemannian manifolds I", *Math. Ann.* **277**:4 (1987), 639–646.

[1988]   "On the topology of Clifford isoparametric hypersurfaces", *J. Differential Geom.* **27**:1 (1988), 55–66.

XIN MIN WANG

[1993]   "Dupin hypersurfaces with constant mean curvatures", pp. 247–253 in *Differential geometry* (Shanghai, 1991), edited by C. H. G. et al., World Sci. Publishing, River Edge, NJ, 1993.

BERND WEGNER

[1973]   "Existence of four concurrent normals to a smooth closed hypersurface of E^n", *Amer. Math. Monthly* **80** (1973), 782–785.

[1984]   "Morse theory for distance functions to affine subspaces of Euclidean spaces", pp. 165–168 in *Proceedings of the conference on differential geometry and its applications. Part 1, Differential Geometry* (Nové Město na Moravě, 1983), edited by O. Kowalski, Charles Univ., Prague, 1984.

[1989] "Über das kritische Verhalten zylindrischer Abstandsfunktionen auf Unterman-
        nigfaltigkeiten euklidischer Räume", *Serdica* **15**:1 (1989), 19–27.

JOEL L. WEINER

[1974] "Total curvature and total absolute curvature of immersed submanifolds of
        spheres", *J. Differential Geometry* **9** (1974), 391–400.

ALAN WEST

[1989] "Isoparametric systems", pp. 222–230 in *Geometry and topology of subman-
        ifolds* (Marseille, 1987), edited by J.-M. Morvan and L. Verstraelen, World Sci.
        Publishing, Teaneck, NJ, 1989.

[1992] *An introduction to isoparametric submanifolds and other topics*, World Scien-
        tific, London, 1992.

[1993] "Isoparametric systems on symmetric spaces", pp. 281–287 in *Geometry and
        topology of submanifolds, V* (Leuven/Brussels, 1992), edited by F. Dillen et al.,
        World Sci. Publishing, River Edge, NJ, 1993.

BRIAN WHITE

[1987] "Complete surfaces of finite total curvature", *J. Differential Geom.* **26**:2 (1987),
        315–326. Corrections in **28**:2 (1988), 359–360.

JAMES H. WHITE

[1974] "Minimal total absolute curvature for orientable surfaces with boundary", *Bull.
        Amer. Math. Soc.* **80** (1974), 361–362.

THOMAS J. WILLMORE

[1971] "Tight immersions and total absolute curvature", *Bull. London Math. Soc.* (2)
        3 (1971), 129–151.

[1982] *Total curvature in Riemannian geometry*, Ellis Horwood, West Sussex, Eng-
        land, 1982.

THOMAS J. WILLMORE AND B. A. SALEEMI

[1966] "The total absolute curvature of immersed manifolds", *J. London Math. Soc.*
        41 (1966), 153–160.

J. P. WILSON

[1965] "The total absolute curvature of an immersed manifold", *J. London Math. Soc.*
        40 (1965), 362–366.

[1969] "Some minimal imbeddings of homogenous spaces", *J. London Math. Soc.* (2)
        1 (1969), 335–340.

PETER WINTGEN

[1977] *Zur Integralkrümmung verknoteter Sphären*, Habilitationsschrift, Humboldt-
        Universität Berlin, 1977.

[1978a] "On the total curvature of knotted spheres", *Bull. Acad. Polon. Sci., Ser. Sci.
        math. astr. phys.* **26** (1978), 249–253.

[1978b] "On the total curvature of surfaces in E^4", *Colloq. Math.* **39** (1978), 284–296.

[1980a] "Totale Absolutkrümmung von Hyperflächen", *Beiträge zur Algebra und Geometrie* **10** (1980), 87–96.

[1980b] "Über die totale Absolutkrümmung verknoteter Sphären", *Beiträge zur Algebra und Geometrie* **9** (1980), 131–147.

[1981] "Courbure totale des surfaces ayant des singularités", *C. R. Acad. Sci. Paris, Sér. I* **292** (1981), 515–517.

[1984] "On total absolute curvature of nonclosed submanifolds", *Ann. Global Anal. Geom.* **2**:1 (1984), 55–87.

BING-LE WU

[1992] "Isoparametric submanifolds of hyperbolic spaces", *Trans. Amer. Math. Soc.* **331**:2 (1992), 609–626.

[1994a] "A finiteness theorem for isoparametric hypersurfaces", *Geom. Dedicata* **50**:3 (1994), 247–250.

[1994b] "Hyper-isoparametric submanifolds in Riemannian symmetric spaces", preprint, Univ. of Pennsylvania, 1994.

[1995] "Equifocal focal hypersurfaces of rank one symmetric spaces", preprint, Northeastern Univ., 1995.

LIANG XIAO

[1995] "Isoparametric submanifolds of complex projective spaces", *Acta Math. Sinica* **38**:6 (1995), 845–856.

QIANG ZHAO

[1993] "Isoparametric submanifolds of hyperbolic space", *Chinese J. Contemp. Math.* **14**:4 (1993), 339–346.

Index

List of Publications

NICOLAAS H. KUIPER

[1941] NHK, "Lijnen in R_4" ("Lines in R_4"), *Nieuw Archief voor Wiskunde* (2) **21** (1941), 124–143.

[1946] NHK, *Onderzoekingen over lijnenmeetkunde (Investigation on line geometry)*, Dissertatie (thesis), Universiteit Leiden, 1946.

[1947] NHK, "Een stelling uit de vlakke meetkunde" ("A theorem from plane geometry"), in *Handelingen van het 30. Nederlandse Natuur- en Geneeskundig Congres*, 1947.

[1948a] NHK, "On differentiable linesystems of one dual variable I", *Indag. Math.* **10** (1948), 361–369.

[1948b] NHK, "On differentiable linesystems of one dual variable II", *Indag. Math.* **10** (1948), 388–394.

[1949a] NHK, "Een sluitingstheorema" ("A closure theorem"), *Simon Stevin* **27** (1949), 6–15.

[1949b] NHK, "On conformally-flat spaces in the large", *Ann. of Math.* (2) **50** (1949), 916–924.

[1950a] NHK, "Compact spaces with a local structure determined by the group of similarity transformations in E^n", *Indag. Math.* **12** (1950), 411–418.

[1950b] NHK, "Distribution modulo 1 of some continuous functions", *Indag. Math.* **12**:5 (1950), 460–466.

[1950c] NHK, "Einstein spaces and connections I", *Indag. Math.* **12** (1950), 505–512.

[1950d] NHK, "Einstein spaces and connections II", *Indag. Math.* **12** (1950), 513–521.

[1950e] NHK, "On compact conformally Euclidean spaces of dimension > 2", *Ann. of Math.* (2) **52** (1950), 478–490.

[1950f] NHK, "On linear families of involutions", *Amer. J. Math.* **72** (1950), 425–441.

[1950g] NHK, "Symmetrie en waarschijnlijkheid" ("Symmetry and probability"), 1950. Inaugural lecture, Wageningen.

[1951a] NHK, "On the holonomic groups of the vector displacement in Riemannian spaces", *Indag. Math.* **13** (1951), 445–451. Erratum in **14** (1952), 191.

[1951b] NHK, "Over grafische voorstelling" ("On graphical representation"), *Ned. Vereniging voor statistiek. Medisch Biologische* (3) **6** (July 1951), 57–61.

[1951c] NHK, "Sur les propriétés conformes des espaces d'Einstein" ("On conformal properties of Einstein spaces"), pp. 165–166 in *Colloque de géométrie différentielle* (Louvain, 1951), Georges Thone, Liège, 1951.

[1952a] Shiing-shen Chern and NHK, "Some theorems on the isometric imbedding of compact Riemann manifolds in euclidean space", *Ann. of Math.* (2) **56** (1952), 422–430.

[1952b] NHK, "Variantie-analyse" ("Analysis of variance"), *Statistica Neerlandica* **6** (1952), 149–194.

[1953a] NHK, "Analyse van vectoren" ("Vector analysis"), *Biometrisch Contact* **5** (1953), 22–35.

[1953b] NHK, "On 4^3-factorial design with confounding", *Netherlands Journ. of Agr. Sc.* **1** (1953), 11–14.

[1953c] NHK, "Sur les surfaces localement affines" ("On locally affine surfaces"), pp. 79–87 in *Géométrie différentielle* (Strasbourg, 1953), Centre National de la Recherche Scientifique, Paris, 1953.

[1953d] NHK, "Sur l'immersion isométrique" ("On isometric immersion"), pp. 3 in *Colloque de topologie et géométrie différentielle* (Strasbourg, 1952), La Bibliothèque Nationale et Universitaire de Strasbourg, 1953.

[1953e] NHK and L. C. A. Corsten, "Open plaatsen in variantie-schema's" ("Open places in variational schemes"), *Mededelingen van de Landbouwhogeschool te Wageningen* **53**:2 (1953), 25–30.

[1954a] NHK, "Algebraic geometry", lecture notes, University of Michigan, Ann Arbor, 1954.

[1954b] NHK, "Een opmerking over het aanpassen van functies met een groot aantal waarnemingsuitkomsten" ("Note on the fitting of a function to a large number of observations"), *Statistica Neerl.* **8**:1 (1954), 1–6.

[1954c] NHK, "Een vlakke meetkunde" ("A plane geometry"), *Simon Stevin* **30**:2 (1954), 94–105.

[1954d] NHK, "Locally projective spaces of dimension one", *Michigan Math. J.* **2** (1954), 95–97.

[1954e] NHK, "On convex locally-projective spaces", pp. 200–213 in *Convegno Internazionale di Geometria Differenziale* (Italia, 1953), Edizioni Cremonese, Roma, 1954.

[1955a] NHK, "On C^1-isometric imbeddings I", *Indag. Math.* **17**:4 (1955), 545–556.

[1955b] NHK, "On C^1-isometric imbeddings II", *Indag. Math.* **17**:5 (1955), 683–689.

[1955c] NHK and Kentaro Yano, "On geometric objects and Lie groups of transformations", *Indag. Math.* **17** (1955), 411–420.

[1956a] NHK, *Differentiaal- en integraalrekening* (*Differential and integral calculus*), Veeman en Zonen, Wageningen, 1956. 2nd edition, 1960.

[1956b] NHK, "Eine charakteristische Eigenschaft der Kurven zweiter Ordnung" ("A characteristic property of curves of second order"), *Math.-Phys. Semesterber.* **5** (1956), 138–140.

[1956c] NHK, "Groups of motions of order $\frac{1}{2}n(n - 1) + 1$ in Riemannian n-spaces", *Indag. Math.* **18** (1956), 313–318.

[1956d] NHK and Kentaro Yano, "Two algebraic theorems with applications", *Indag. Math.* **18** (1956), 319–328.

[1957a] NHK, "On convex sets and lines in the plane", *Indag. Math.* **19** (1957), 272–283.

[1957b] NHK, "A real analytic non-desarguesian plane", *Nieuw Arch. Wisk.* (3) **5** (1957), 19–24.

[1958a] NHK, "Ausgewählte Kapitel der Riemannschen Geometrie" ("Selected chapters in Riemannian geometry"), notes, Math. Institut Bonn, 1958.

[1958b] NHK, "Capita selecta uit de differentiaalmeetkunde" ("Selected chapters in differential geometry"), notes, Math. Centrum, Amsterdam, 1958.

[1958c] NHK, "Differentiaal-meetkunde" ("Differential geometry"), *Euclides* (1958), 289–300.

[1958d] NHK, "On some algebraic isometric imbeddings", *Simon Stevin* **32** (1958), 23–28.

[1959a] NHK, "Alternative proof of a theorem of Birnbaum and Pyke", *Ann. Math. Statist.* **30** (1959), 251–252.

[1959b] NHK, *Analytische meetkunde (verklaard met lineaire algebra)* (*Analytic geometry (explained by means of linear algebra)*), N. V. Noord-Hollandsche Uitgevers-maatschappij, Amsterdam, 1959.

[1959c] NHK, "De barycentrische Calcul en het ontstaan van vectoren" ("Barycentric calculus and the origin of vectors"), *Euclides* (Groningen) **35** (1959/60), 113–126.

[1959d] NHK, *Dictaat wiskundige verwerking van waarnemingsuitkomsten* (*Lecture notes on the mathematical treatment of observational results*), Uitg. Stud. Inlichtingen Dienst, Wageningen, 1959.

[1959e] NHK, "Een symbool voor isomoor" ("A symbol for isomorphism"), *Statistica Neerlandica* **13** (1959), 295–317.

[1959f] NHK, "Immersions with minimal total absolute curvature", pp. 75–88 in *Colloque Géom. Diff. Globale* (Bruxelles, 1958), Centre Belge Rech. Math., Louvain, 1959.

[1959g] NHK, "In memoriam Prof. Dr. M. J. van Uven, 1878–1959", *Statistica Neerlandica* **13:3** (1959), 255–258.

[1959h] NHK, "Isometric and short imbeddings", *Indag. Math.* **21** (1959), 11–25.

[1959i] NHK, "Sur les immersions à courbure totale minimale", pp. 1–5 in *Séminaire de topologie et de géométrie différentielle dirigé par Charles Ehresmann*, 1re année, 1957/58, Institut Henri Poincaré, Paris, 1959.

[1960a] NHK, "La courbure d'indice k et les applications convexes", pp. 1–15 in *Séminaire de topologie et de géométrie différentielle dirigé par Charles Ehresmann*, vol. 2, Institut Henri Poincaré, Paris, 1960.

[1960b] NHK, "On surfaces in euclidean three-space", *Bull. Soc. Math. Belg.* **12** (1960), 5–22.

[1960c] NHK, "On the random cumulative frequency function", *Indag. Math.* **22** (1960), 32–37.

[1960d] NHK, "Random variables and random vectors", *Bulletin de l'Institute agronomique et des stations de recherches, Gembloux* (1960), 344–355.

[1960e] NHK, "Tests concerning random points on a circle", *Indag. Math.* **22** (1960), 38–47.

[1961a] James Eells, Jr. and NHK, "Closed manifolds which admit nondegenerate functions with three critical points", *Indag. Math* **23** (1961), 411–417.

[1961b] NHK, "A continuous function with two critical points", *Bull. Amer. Math. Soc.* **67** (1961), 281–285.

[1961c] NHK, "Convex immersions of closed surfaces in E^3. Nonorientable closed surfaces in E^3 with minimal total absolute Gauss-curvature", *Comment. Math. Helv.* **35** (1961), 85–92.

[1961d] NHK, "Welke gevolgen voor het V.H.M.O. brengt de moderne ontwikkeling der wiskundige wetenschappen met zich mede" ("What consequences does the development of modern mathematical sciences have for the V.H.M.O.?"), *Euclides* **37** (1961), 257–284.

[1962a] James Eells, Jr. and NHK, "An invariant for certain smooth manifolds", *Ann. Mat. Pura Appl.* (4) **60** (1962), 93–110.

[1962b] James Eells, Jr. and NHK, "Manifolds which are like projective planes", *Publ. Math. Inst. Hautes Études Sci.* **14** (1962), 181–222.

[1962c] NHK, "Efficiency en abstractie in de wiskunde" ("Efficiency and abstraction in science"), 1962. Inaugural lecture, Noord-Hollandsche Uitg. Mif., Amsterdam.

[1962d] NHK, *Linear algebra and geometry*, North-Holland, Amsterdam, 1962. Translation of [1959b] by A. van der Sluis. 2nd revised edition, 1965.

[1962e] NHK, "On a metric in the space of random variables", *Statistica Neerlandica* **16** (1962), 231–235.

[1962f] NHK, "On convex maps", *Nieuw Arch. Wisk.* (3) **10** (1962), 147–164.

[1962g] NHK, "Random variables rather than distribution functions", *Mededelingen van de Landbouwhogeschool* **61**:12 (1962), 1–18.

[1962h] NHK and S. H. Justesen, "Eine statistische Bemerkung über das Gesetz des Minimums" ("A statistical observation on the Law of the Minimum"), *Biometrische Zeitschrift* **4**:2 (1962), 97–99.

[1963a] NHK, "Lofzang op de meetkunde" ("Song of praise for geometry"), *Euclides* **39** (1963/64), 33–47. Voordracht Vakantiecursus Mathematisch Centrum, Amsterdam 1962.

[1963b] NHK, "Twintig jaren studiekring voor statistische techniek" ("Twenty years of the Study Group for Statistical Techniques"), *Statistica Neerlandica* **17** (1963), 175–185.

[1964a] NHK, "Double normals of convex bodies", *Israel J. Math.* **2** (1964), 71–80.

[1964b] NHK, "On the general linear group in Hilbert space", preprint ZW-011, Math. Centrum Amsterdam, Afd. Zuivere Wisk., 1964. In Dutch.

[1965a] NHK, "The homotopy type of the unitary group of Hilbert space", *Topology* **3** (1965), 19–30.

[1965b] NHK, "On the smoothings of trangulated and combinatorial manifolds", pp. 3–22 in *Differential and combinatorial topology (a symposium in honor of Marston Morse)*, edited by S. S. Cairns, Princeton Univ. Press, Princeton, NJ, 1965.

[1966a] NHK, "C^r-functions near non-degenerate critical points", mimeoraphed notes, Warwick Univ., Coventry, 1966.

[1966b] NHK, *Lineaire algebra en meetkunde (Linear algebra and geometry)*, 2nd revised ed., N. V. Noord-Hollandsche Uitgeversmaatschappij, Amsterdam, 1966. See [1959b] and [1962d].

[1966c] NHK and R. K. Lashof, "Microbundles and bundles, I: Elementary theory", *Invent. Math.* **1** (1966), 1–17.

[1966d] NHK and R. K. Lashof, "Microbundles and bundles, II: Semisimplical theory", *Invent. Math.* **1** (1966), 243–259.

[1967a] NHK, "Algebraic equations for nonsmoothable 8-manifolds", *Publ. Math. Inst. Hautes Études Sci.* **33** (1967), 139–155.

[1967b] NHK, "Der Satz von Gauss–Bonnet für Abbildungen in E^N und damit verwandte Probleme" ("The Gauss–Bonnet theorem for maps in E^N and related applied problems"), *Jber. Deutsch. Math.-Verein.* **69**:2, Abt. 1 (1967), 77–88.

[1968a] NHK, "Non-degenerate piecewise linear functions", *Rev. Roumaine Math. Pures Appl.* **13** (1968), 993–1000.

[1968b] NHK, "Over equivalentie van functies" ("On equivalence of functions"), pp. 112–116 in *Conf. Kon. Nederl. Akad. Wetensch.*, Verslag gewone vergadering Afd. Natuurk., Amsterdam, 1968/69.

[1969a] Dan Burghelea and NHK, "Hilbert manifolds", *Ann. of Math.* (2) **90** (1969), 379–417.

[1969b] James Eells, Jr. and NHK, "Homotopy negligible subsets", *Compositio Math.* **21** (1969), 155–161.

[1969c] NHK, "Smoothings of $S^n \times S^1$", preprint, 1969.

[1970a] NHK, "Curvature and critical points", mimeographed lecture notes, Univ. Liverpool, 1970.

[1970b] NHK, "Minimal total absolute curvature for immersions", *Invent. Math.* **10** (1970), 209–238.

[1970c] NHK and Besseline Terpstra-Keppler, "Differentiable closed embeddings of Banach manifolds", pp. 118–125 in *Essays on topology and related topics (Mémoires dédiés à Georges de Rham)*, Springer, New York, 1970.

[1971a] NHK, "Connection and convexity", *Nederl. Akad. Wetensch. Verslag Afd. Natuurk.* **80** (1971), 39–40. In Dutch.

[1971b] NHK, "The differential topology of separable Banach manifolds", pp. 85–90 in *Actes du Congrès International des Mathématiciens* (Nice, 1970), vol. 2, Gauthier-Villars, Paris, 1971.

[1971c] NHK (editor), *Manifolds—Amsterdam 1970: Proceedings of the Nuffic Summer School on Manifolds*, Lecture Notes in Math. **197**, Springer, Berlin, 1971.

[1971d] NHK, "Morse relations for curvature and tightness", pp. 77–89 in *Proceedings of Liverpool Singularities Symposium* (Liverpool, 1969/1970), vol. 2, edited by C. T. C. Wall, Lecture Notes in Math. **209**, Springer, Berlin, 1971.

[1971e] NHK, "Tight topological embeddings of the Moebius band", *J. Differential Geometry* **6** (1971/72), 271–283.

[1971f] NHK, *Varietés hilbertiennes: aspects géométriques (Hilbert manifolds: geometric aspects)*, Séminaire de Mathématiques Supérieures **38**, Les Presses de l'Université de Montréal, Montreal, Que., 1971. Suivi de deux textes de Dan Burghelea.

[1972] NHK, "C^1-equivalence of functions near isolated critical points", pp. 199–218 in *Symposium on infinite-dimensional topology* (Baton Rouge, LA, 1967), Ann. of Math. Studies **69**, Princeton Univ. Press, Princeton, NJ, 1972.

[1973a] NHK, "Sur les variétés riemanniennes très pincées" ("On strongly pinched Riemannian manifolds"), pp. 201–218 in *Séminaire Bourbaki 1971/1972*, exposé n° 410, Lecture Notes in Math. **317**, Springer, Berlin, 1973.

[1973b] NHK and J. W. Robbin, "Topological classification of linear endomorphisms", *Ann. Inst. Fourier Invent. Math.* **23**:2 (1973), 93–95.

[1973c] NHK and J. W. Robbin, "Topological classification of linear endomorphisms", *Invent. Math.* **19** (1973), 83–106.

[1974a] NHK, "A generalisation of convexity; open problems", pp. 37–42 in *International Conference on Prospects in Mathematics* (Kyoto, 1973), Res. Inst. Math. Sci., Kyoto Univ., Kyoto, 1974.

[1974b] NHK, "The quotient space of $\mathbb{CP}(2)$ by complex conjugation is the 4-sphere", *Math. Ann.* **208** (1974), 175–177.

[1975a] NHK, "The topology of the solutions of a linear differential equation on $\mathbb{R}^n$", pp. 195–203 in *Manifolds* (Tokyo, 1973), Univ. Tokyo Press, Tokyo, 1975.

[1975b] NHK, "Physics International", *Physics Bulletin* **26** (July 1975). Description of the work at the I.H.E.S.

[1975c] NHK, "Stable surfaces in euclidean three space", *Math. Scand.* **36** (1975), 83–96.

[1975d] NHK, "Topological conjugacy of real projective transformations", pp. 57–59 in *Dynamical systems* (Warwick, 1974), edited by A. Manning, Lecture Notes in Math. **468**, Springer, Berlin, 1975.

[1975e] NHK, "The topology of the solutions of a linear homogeneous differential equation on $\mathbb{R}^n$", pp. 417–418 in *Differential geometry* (Stanford, 1973), edited by S. S. Chern and R. Osserman, Proceedings of Symposia in Pure Mathematics **27** (part 1), American Mathematical Society, Providence, RI, 1975.

[1976a] César Camacho, NHK, and Jacob Palis, "La topologie du feuilletage d'un champ de vecteurs holomorphes près d'une singularité" ("The topology of the foliation of a holomorphic vector field near a singularity"), *C. R. Acad. Sci. Paris Sér. A-B* **282**:17 (1976), Ai, A959–A961.

[1976b] NHK, "Curvature measures for surfaces in E^n", pp. 195–199 in *Lobaschevski Colloquium* (Kazan, USSR), 1976.

[1976c] NHK, "La courbure totale absolue minimale des surfaces dans E^3" ("The total absolute minimal curvature of surfaces in E^3"), *Bull. Soc. Math. France, Suppl. Mém.* **46** (1976), 11.

[1976d] NHK, "Topological conjugacy of real projective transformations", *Topology* **15**:1 (1976), 13–22.

[1977] NHK and William F. Pohl, "Tight topological embeddings of the real projective plane in E^5", *Invent. Math.* **42** (1977), 177–199.

[1978a] César Camacho, NHK, and Jacob Palis, "The topology of holomorphic flows with singularity", *Publ. Math. Inst. Hautes Études Sci.* **48** (1978), 5–38.

[1978b] NHK, "Commentary on L. E. J. Brouwer (on triangulation)", pp. 14–18 in *Two decades of Mathematics in the Netherlands, 1920-1940*, edited by E. M. J. Bertin et al., Mathematical Centre, Amsterdam, 1978.

[1978c] NHK, "La topologie des singularités hyperboliques des actions de $\mathbb{R}^2$" ("The topology of hyperboliques singularities of the actions of $\mathbb{R}^2$"), pp. 131–150 in *Journées singulières de Dijon* (Dijon, 1978), Astérisque **59-60**, Soc. Math. France, Paris, 1978.

[1979a] NHK, "Connelly's flexing sphere", in *Mathematical calendar*, Springer, Berlin, 1979.

[1979b] NHK, "A short history of triangulation and related matters", pp. 61–79 in *Bicentennial Congress Wiskundig Genootschap* (Amsterdam, 1978), vol. 1, Math. Centre Tracts **100**, Math. Centre, Amsterdam, 1979.

[1979c] NHK, "Sphères polyédriques flexibles dans E^3, d'après Robert Connelly" ("Flexible polyhedral spheres in E^3, after Robert Connelly"), pp. 147–168 in *Séminaire Bourbaki* 1977/78, exposé n° 514, Lecture Notes in Math. **710**, Springer, Berlin, 1979.

[1980a] NHK, "On the topology of $\mathbb{R}^2$-actions near a singular point", pp. 554–562 in *Atas do 12° Colóquio Brasileiro de Mathemática* (Poços de Caldas, Brazil, 1979), vol. 2, Inst. Mat. Pura e Aplicada, Rio de Janeiro, 1980.

[1980b] NHK, "Tight embeddings and maps: Submanifolds of geometrical class three in E^n", pp. 97–145 in *The Chern Symposium* (Berkeley, 1979), edited by W.-Y. Hsiang et al., Springer, New York, 1980.

[1981] Thomas F. Banchoff and NHK, "Geometrical class and degree for surfaces in three-space", *J. Diff. Geom.* **16** (1981), 559–576.

[1983a] NHK, "Polynomial equations for tight surfaces", *Geom. Dedicata* **15**:2 (1983), 107–113.

[1983b] NHK, "There is no tight continuous immersion of the Klein bottle into $\mathbb{R}^3$", preprint, Inst. Hautes Études Sci., 1983.

[1983c] NHK, "The topology of linear C^m-flows on $\mathbb{C}^n$", pp. 448–462 in *Geometric dynamics* (Rio de Janeiro, 1981), Lecture Notes in Math. **1007**, Springer, Berlin, 1983.

[1983d] NHK and William H. Meeks, III, "Sur la courbure des surfaces nouées dans R^3" ("On the curvature of embedded surfaces in R^3"), pp. 215–217 in *Third Schnepfenried geometry conference* (Schnepfenried, 1982), vol. 1, Astérisque **107-108**, Soc. Math. France, Paris, 1983.

[1984a] NHK, "Geometry in total absolute curvature theory", pp. 377–392 in *Perspectives in mathematics* (Oberwolfach, 1984), edited by W. Jäger et al., Birkhäuser, Basel, 1984.

[1984b] NHK, "Taut sets in three space are very special", *Topology* **23**:3 (1984), 323–336.

[1984c] NHK and William H. Meeks, III, "Total curvature for knotted surfaces", *Invent. Math.* **77**:1 (1984), 25–69.

[1985] NHK, "Minimal total curvature. A look at old and new results", *Uspekhi Mat. Nauk* **40**:4 (1985), 49–54. Lecture at the International conference on current problems in algebra and analysis (Moscow-Leningrad, 1984). Translated from the English by M. P. Dolbinin and A. M. Volkhonskiĭ; English version appears in *Russ. Math. Surv.* **40**:4 (1985), 49–55.

[1987a] NHK, "A new knot invariant", *Math. Ann.* **278** (1987), 193–209.

[1987b] NHK and William H. Meeks, III, "The total curvature of a knotted torus", *J. Differential Geom.* **26**:3 (1987), 371–384.

[1988] NHK, "Hyperbolic 4-manifolds and tesselations", *Inst. Hautes Études Sci. Publ. Math.* **68** (1988), 47–76.

[1990] NHK, "Fairly symmetric hyperbolic manifolds", pp. 165–204 in *Geometry and topology of submanifolds, II* (Avignon, 1988), edited by M. Boyom et al., World Sci. Publishing, Teaneck, NJ, 1990.

[1991] NHK, "On gradient curves of an analytic function near a critical point", preprint, Inst. Hautes Études Sci., 1991.

[1992] François Haab and NHK, "On the normal Gauss map of a tight smooth surface in $\mathbb{R}^3$", pp. 161–175 in *Chern—a great geometer of the twentieth century*, edited by S.-T. Yau, Internat. Press, Hong Kong, 1992.

[1993] NHK, "Tight and taut immersions", pp. 169–172 in *Encyclopaedia of Mathematics*, vol. 9, edited by M. Hazewinkel, Kluwer, Dordrecht, 1993.

[1997] NHK, "Geometry in curvature theory", pp. 1–50 in *Tight and Taut Submanifolds*, edited by T. E. Cecil and S.-s. Chern, Cambridge U. Press, 1997.

Tight and Taut Submanifolds
MSRI Publications
Volume 32, 1997

Dissertations
of N. H. Kuiper's
Doctoral Students

L. C. A. CORSTEN: "Vectors, a tool in statistical regression theory", Veenman, Wageningen, 1957.

FRANKLIN E. ESSED: "Estimation of standing timber", Landbouwhogeschool te Wageningen, 1957. Published by Veenman, Wageningen (Mededelingen van de Landbouwhogeschool Wageningen 57:5).

M. T. G. MEULENBERG: "Vraaganalyse voor landbouwprodukten uit tijdreeksen" ("Time-series analysis of agricultural products"), Veenman en Zonen, Wageningen, 1962.

FLORIS TAKENS: "The minimal number of critical points of a function on a compact manifold and the Lusternik–Schnirelman category", Amsterdam, 1969. Published in *Inventiones Math.* **6** (1968), 197–244.

NICOLE MOULIS: "Approximation de fonctions différentiables sur certains espaces de Banach: Sur les variétés hilbertiennes et les fonctions non-dégénérées", Paris, 1970.

ROALD RAMER: "Integration on infinite dimensional manifolds", Amsterdam, 1974.

EDUARD J. N. LOOIJENGA: "Structural stability of smooth families of C^∞-functions", Amsterdam, 1974.

DIRK SIERSMA: "Classification and deformation of singularities", University of Amsterdam, 1974.